Birkhäuser

Trends in Mathematics

Trends in Mathematics is a series devoted to the publication of volumes arising from conferences and lecture series focusing on a particular topic from any area of mathematics. Its aim is to make current developments available to the community as rapidly as possible without compromise to quality and to archive these for reference.

Proposals for volumes can be submitted using the Online Book Project Submission Form at our website www.birkhauser-science.com.

Material submitted for publication must be screened and prepared as follows:

All contributions should undergo a reviewing process similar to that carried out by journals and be checked for correct use of language which, as a rule, is English. Articles without proofs, or which do not contain any significantly new results, should be rejected. High quality survey papers, however, are welcome.

We expect the organizers to deliver manuscripts in a form that is essentially ready for direct reproduction. Any version of TeX is acceptable, but the entire collection of files must be in one particular dialect of TeX and unified according to simple instructions available from Birkhäuser.

Furthermore, in order to guarantee the timely appearance of the proceedings it is essential that the final version of the entire material be submitted no later than one year after the conference.

More information about this series at http://www.springer.com/series/4961

Modern Trends in Hypercomplex Analysis

Swanhild Bernstein • Uwe Kähler • Irene Sabadini
Franciscus Sommen
Editors

Birkhäuser

Editors
Swanhild Bernstein
Institute of Applied Analysis
TU Bergakademie Freiberg
Freiberg, Germany

Uwe Kähler
Departamento de Matemática
Universidade de Aveiro
Aveiro, Portugal

Irene Sabadini
Dipartimento di Matematica
Politecnico di Milano
Milano, Italy

Franciscus Sommen
Department of Mathematical Analysis
Ghent University
Gent, Belgium

This work is published under the auspices of the International Society of Analysis, its Applications and Computation (ISAAC)

ISSN 2297-0215 ISSN 2297-024X (electronic)
Trends in Mathematics
ISBN 978-3-319-42528-3 ISBN 978-3-319-42529-0 (eBook)
DOI 10.1007/978-3-319-42529-0

Library of Congress Control Number: 2016959577

Mathematics Subject Classification (2010): 30G35, 30G25, 22E46, 32A50, 68U10

Printed on acid-free paper

This book is published under the trade name Birkhäuser, www.birkhauser-science.com
The registered company is Springer International Publishing AG
The registered company address is: Gewerbestrasse 11, 6330 Cham, Switzerland

Contents

Preface

The 10th International ISAAC Congress (International Society for Analysis, its Applications, and Computations), was held at the University of Macau, China from August 3 to August 8, 2015. It has been a tradition that at these conferences there is a session "Clifford and Quaternionic Analysis", starting from the first International ISAAC Congress held at the University of Delaware in 1997. Its tradition of mixing speakers from the area and from related fields as well as using the opportunity of inviting local speakers has not only made it one of the largest sessions, but also contributed to show how active and interesting the field is to other mathematical communities. For a branch of mathematics which started as an active research field in form of a generalization of complex analysis only in the 1970s, these sessions are crucial in promoting and advertising the area, in particular, showing new and interesting directions. It is obvious that only a small part of the talks could find its way into a single volume. Therefore, the papers in this volume present a careful selection of the contributions presented during the session. The editors hope that the present choice of several different aspects and direction of hypercomplex analysis will give the interested reader many new ideas and promising new directions. Among the new directions, the editors would like to point out the overviews on quaternionic spectral theory, Clifford differential forms, and script geometry. That also classic topics in Clifford analysis are well alive and running can be seen in the papers by de Ridder, Eelbode, Eriksson, Ferreira, Kheyfits, Ren and their co-authors. Moreover, also applications are an important part of the field as the papers by Baratchart, Cerejeiras, Guerlebeck, Grigoriev and their co-authors demonstrate.

The editors express their gratitude to the contributors to this volume and to the work of the anonymous referees without which this volume would never have seen the light. They also thank warmly professor Tao Qian as chairman of the local organizing committee of the Conference, and the University of Macau for the organization of such a wonderful event. Furthermore, they would like to thank professor Luigi Rodino, president of the ISAAC society, for his work and dedication to make ISAAC the foremost international organization in the area of Mathematical Analysis.

May 2016, The Editors

Modern Trends in Hypercomplex Analysis
Trends in Mathematics, 1–17

Cauchy–Pompeiu Formula for Multi-meta-weighted-monogenic Functions of first class

Eusebio Ariza García and Antonio Di Teodoro

Abstract. In this paper we give a Cauchy–Pompeiu type integral formula for a class of functions called **multi-meta-weighted-monogenic** using a distance calculated via the quadratic form associated with an elliptic operator. This is used for the construction of the kernel over the domain $\mathbb{R}^{m+1}$, constructed by fixing the real part for all products of

$$\mathbb{R}^{m+1} = \mathbb{R}^{m_1} \times \mathbb{R}^{m_2} \times \cdots \times \mathbb{R}^{m_n}.$$

Also, we present a section where we discuss the inhomogeneous meta-n-weighted-monogenic equation and a distributional solution for this equation is obtained. In some special cases, the distributional solution becomes a classical solution.

Mathematics Subject Classification (2010). 30A05; 15A66; 30G35.

Keywords. Monogenic functions, meta-monogenic functions, multi-meta-weighted-monogenic functions, meta-n-weighted-monogenic functions, Clifford type algebras.

1. Actual state of theory of multi-monogenic functions

The theory of multi-monogenic functions generalizes the theory of holomorphic functions in several complex variables to the case of monogenic functions. This theory (Cauchy's integral formula, Hartog's extension theorem, Cousin problem, and so on) can be found in [11] as an extension of the works [10, 21] to the case of holomorphic functions.

In addition to the construction of this theory, Tutschke and Hung Son [23] discuss a theory of multi-monogenic functions in the case that the dimension 2^m of the corresponding algebra of Clifford type is defined by

$$m + 1 = \sum_{j=1}^{n} m_j.$$

With the help of Clifford algebras depending on parameters in [22], the authors discuss the case when all of the factor spaces $\mathbb{R}^{m_1+1}$ have the same real part. On the other hand, in [1] the multi-meta-monogenic theory worked out when for each partial space $\mathbb{R}^{m_j+1}$ have its own real part and m_j imaginary units.

Another approach to the theory of monogenic functions in several vector variables, in the special case $q_{ij} = 1$, was developed by Sommen and collaborators. The main difference between the works of Sommen et al. and Tutschke and Hung Son is the motivation. While for Tutschke and Hung Son the problem is how to define the dimension of the Clifford algebra in such a way that the dimensions m_j of the n given Euclidean spaces R^{m_j} have an equivalent influence on the final choice of the dimension m. That is, no space R^{m_j} is preferred in comparison with the other spaces. On the other hand, for Sommen and collaborators the idea is to define axial algebras to extend the Clifford structure, see [3, 4, 6, 13, 18, 19, 20].

Another difference between these works is the use of the first-order differential operator. Whereas Tutschke and Hung Son use the Cauchy–Riemann operator and its consequences in lower dimension [22, 23], Sommen et al. use the Dirac operator and the simple factorization of the Laplace operator, and as consequence, the applications in physics [6, 3, 13]. Despite the use of the Dirac operator in physics problems, the modification of the Clifford structure for a more general structure like Clifford depending on parameter algebra, allows us to use the Cauchy–Riemann operator with the identification of the real part with the time parameter, in order to obtain operators as D'Alembert, Heat, among others. See [9, 16, 26].

Finally, the extension of the theory of multi-monogenic functions in several variables in the Sommen–Soucek approach has allowed to obtain many properties in the direction of axial algebras, as series expansion, harmonic spherical, polynomial representation and separately monogenic functions. See [6, 7, 13, 18, 19, 20].

In the ideas of Tutschke and Hung Son many things are yet to be covered.

Recently, we contributed to the development of this theory constructing integral representation formulas using an algebraic structure of Clifford type, where the Cauchy–Riemann operator is constructed placing for each partial space $\mathbb{R}^{m_j+1}$, its own real part and m_j imaginary units. To do that, the multi-meta-φ-monogenic of second class operator was introduced. See [2].

When we discuss the so-called Dirac operator, with constant parameters, one can, physically, model Dirac fermions realized in a homogeneous material. Consider the description of such Dirac fermions by means of a space-dependent velocity that allows the extension of the analysis to heterogeneous material, i.e., situations in which the sample is composed of two or more materials attached to each other. In these circumstances, the boundary or matching conditions at the interfaces are an important ingredient in determining the physical properties of the sample. Mathematically speaking, this corresponds to determining a boundary value problem and the conditions for the existence and uniqueness of solutions to the corresponding Dirac equation. See [17].

A natural mathematical extension of these previous situations is to consider a case where the parameters (or weights) of the modified Dirac operator are Clifford or real-valued functions. In this case, it is interesting to discuss the relevant mathematical and physical implications for the associated boundary value problem. See [8].

That is the reason for considering weights in the structure of the Dirac operator. As a first step, we propose integral representation formulas for constant parameter (or weights). This method combined with the ideas of multi-monogenic functions and separately monogenic functions allows us to construct integral representation of the non-constant weighted Dirac operator.

In this paper we give a Cauchy-type integral formula for a new class of functions called **multi-meta-weighted-monogenic** over a bounded domain $\Omega \subset \mathbb{R}^{m+1}$ constructed by fixing the real part for all factors of

$$\mathbb{R}^{m+1} = \mathbb{R}^{m_1} \times \mathbb{R}^{m_2} \times \cdots \times \mathbb{R}^{m_n}.$$

In order to do this, we first introduce a more general class of functions, called meta-n-weighted-monogenic functions. We obtain a Cauchy–Pompeiu integral formula for this class of functions. Second, motivated by the fact that multi-meta-weighted-monogenic functions are a particular class of meta-n-weighted-monogenic functions, we have that the Cauchy–Pompeiu integral formula obtained for the **multi-meta-weighted-monogenic functions** is also valid for the **meta-n-weighted-monogenic functions**.

Additionally, in this work we present a section where we discuss the inhomogeneous meta-n-weighted-monogenic equation and obtain a distributional solution for such an equation.

Although the integral formula is developed only for an elliptic operator, we believe that this contribution represents a first step in the extension of the theory.

2. Preliminaries and notations

2.1. Clifford algebras depending on parameters

Let $\mathcal{A}_l$, for $l \geq 1$, be the classical Clifford algebra with basis $\{e_A\}$, $A = (\alpha_1, \dots, \alpha_k)$ with $1 \leq \alpha_1 < \cdots < \alpha_k \leq l$, and the multiplication rules given by

$$\begin{cases} e_0^2 = e_0, & \\ e_i^2 = -e_0, & \text{for} \quad i = 1, \dots, l, \\ e_i e_k + e_k e_i = 0, & \text{for} \quad i, k = 1, \dots, n \quad \text{and} \quad i \neq k, \end{cases}$$

where e_0 is the identity element. Thus, $\mathcal{A}_l$ is 2^l-dimensional as a real space and non-commutative for $l \geq 2$. See [5].

A generalization of the Clifford algebra is given by fixing a set of real-valued functions $\alpha_i(p)$ and $\gamma_{ik}(p)$, $i, k = 1, \dots, l$, $i \neq k$, possibly depending on a parameter

p where p can be a real number or a real-valued function, and considering the more general multiplication rules for the elements of the basis $\{e_A\}$, (see [24, 25]),

$$\begin{cases} e_0^2 = 1, & \\ e_i^2 = -\alpha_i, & \text{for} \quad i = 1, \dots, l, \\ e_i e_k + e_k e_i = 2\gamma_{ik}, & \text{for} \quad i, k = 1, \dots, l \quad \text{and} \quad i \neq k. \end{cases}$$

The obtained algebra is called a Clifford algebra depending on parameters, and it is denoted by $\mathcal{A}_l(2, \alpha_i, \gamma_{ik})$. If the α_i's and γ_{ik}'s are constant, this algebra is usually denoted by $\mathcal{A}_{l,2}^*$.

The uniquely defined Cauchy–Riemann

$$D = \partial_0 + \sum_{i=1}^{l} e_i \partial_i,$$

where $\partial_i = \partial_{x_i}$ is the partial differentiation with respect to x_i for $i = 0, 1, \dots, l$. A continuously differentiable $\mathcal{A}_{l,2}^*$-valued function is said to be *monogenic* if it satisfies the equation $Du = 0$.

If the operator considered in $\mathcal{A}_{l,2}^*$ is the weighted-monogenic operator given by

$$\mathfrak{D}_q = q_0 \partial_0 + \sum_{i=1}^{l} q_i e_i \partial_i,$$

where $q = (q_0, q_1, \dots, q_l)$ is a constant vector in $\mathbb{R}^{l+1}$, a continuously differentiable $\mathcal{A}_{l,2}^*$-valued function is said to be *weighted-monogenic* if it satisfies the equation $\mathfrak{D}_q u = 0$.

Let Ω be a bounded domain in $\mathbb{R}^{l+1}$ with sufficiently smooth boundary. If

$$u : \mathbb{R}^{l+1} \to \mathcal{A}_{l,2}^*$$

is a weighted-monogenic function in Ω, then we get the homogeneous second-order differential equation

$$\overline{\mathfrak{D}_q} \mathfrak{D}_q u = q_0^2 \partial_0^2 u + \sum_{i=1}^{l} \alpha_i q_i^2 \partial_i^2 u - 2 \sum_{i<k} \gamma_{ik} q_i q_k \partial_i \partial_k u = 0 \tag{2.1}$$

where

$$\overline{\mathfrak{D}_q} = q_0 \partial_0 - \sum_{i=1}^{l} q_i e_i \partial_i.$$

Since the coefficients α_k and γ_{ik} are real, the differential equation (2.1) is uncoupled, that is, each real-valued component u_A of u satisfies this differential equation. If the α_k are supposed to be positive and the absolute values of the γ_{ik} are not too large, then (2.1) is elliptic.

Without lost of generality, we can assume that $q_0 = 1$, thus the coefficient matrix of the differential equation (2.1) is

$$\mathbf{B} = \begin{pmatrix} 1 & 0 & 0 & \cdots & 0 \\ 0 & \alpha_1 q_1^2 & -\gamma_{12} q_1 q_2 & \cdots & -\gamma_{1l} q_1 q_l \\ 0 & -\gamma_{21} q_1 q_2 & \alpha_2 q_2^2 & & -\gamma_{2l} q_2 q_l \\ \vdots & \vdots & \vdots & \ddots & \vdots \\ 0 & -\gamma_{l1} q_1 q_l & -\gamma_{l2} q_2 q_l & \cdots & \alpha_l q_l^2 \end{pmatrix}. \tag{2.2}$$

We will assume that the determinant of (2.2) is different from zero. This is satisfied, for instance, if it is the elliptic case. Then (2.2) has an inverse matrix having the form

$$\mathbf{B}^{-1} = \begin{pmatrix} 1 & 0 & \dots & 0 \\ 0 & A_{11} & \dots & A_{1l} \\ \vdots & \vdots & \ddots & \vdots \\ 0 & A_{l1} & \dots & A_{ll} \end{pmatrix},$$

where $A_{ik} = A_{ki}$. We must assume that $q_k \neq 0$ for each $k \in \mathbb{N}$. Using these coefficients, we define for two points $\xi = (\xi_0, \xi_1, \dots, \xi_n)$ and $x = (x_0, x_1, \dots, x_m)$ of $\mathbb{R}^{l+1}$ a non-Euclidean distance ϱ by

$$\varrho^2 = (x_0 - \xi_0)^2 + \sum_{i,k=1}^{l} A_{ik}(x_i - \xi_i)(x_k - \xi_k). \tag{2.3}$$

Using this distance, the function $\mathcal{E}(x, \xi)$ is constructed, for a bounded domain Ω in $\mathbb{R}^{l+1}$ with sufficiently smooth boundary $\partial\Omega$, by

$$\mathcal{E}(x, \xi) = \frac{1}{\omega_{l+1}} \frac{1}{\varrho^{l+1}} \left((x_0 - \xi_0) - \sum_{i,k=1}^{l} e_i q_i A_{ik}(x_k - \xi_k) \right), \tag{2.4}$$

for $x, \xi \in \mathbb{R}^{l+1}$.

Following the idea given in [25], it is easy to see that $E(x, \xi)$ is a (left and right) weighted-monogenic function.

Denoting

$$\wp(x, \xi) = \sum_{s=0}^{l} \frac{\lambda_s}{q_s} (x_s - \xi_s), \tag{2.5}$$

we define the function $\mathcal{E}_\lambda(x, \xi)$ by

$$\mathcal{E}_\lambda(x, \xi) = \exp\left(\wp(x, \xi)\right) \mathcal{E}(x, \xi). \tag{2.6}$$

Note that the function $\mathcal{E}_\lambda(x, \xi)$ satisfies the equation

$$\mathfrak{D}_q \mathcal{E}_\lambda = \lambda \mathcal{E}_\lambda,$$

where $\lambda = \sum_{i=0}^{l} \lambda_i e_i$. In fact,

$$\begin{aligned}\mathfrak{D}_q \mathcal{E}_\lambda(x,\xi) &= \mathfrak{D}_q \left(\exp\left(\wp(x,\xi)\right)\right) \cdot \mathcal{E}(x,\xi) + \exp\left(\wp(x,\xi)\right) \cdot \mathfrak{D}_q \mathcal{E}(x,\xi) \\ &= \sum_{i=0}^{l} \frac{\lambda_i}{q_i} q_i e_i \cdot \exp\left(\wp(x,\xi)\right) \cdot E(x,\xi) \\ &= \lambda \mathcal{E}_\lambda(x,\xi).\end{aligned}$$

In view of this, we have that

$$\mathfrak{D}_q \mathcal{E}_\lambda(x,\xi) - \lambda \mathcal{E}_\lambda(x,\xi) = 0,$$

that is,

$$\mathfrak{D}_{q,\lambda} \mathcal{E}_\lambda(x,\xi) = 0,$$

where $\mathfrak{D}_{q,\lambda} = \mathfrak{D}_q - \lambda$ is the, so-called, meta-weighted-monogenic operator. This mean that $\mathcal{E}_\lambda(x,\xi)$ is a (left) meta-weighted-monogenic function.

Remark 2.1. $\mathcal{E}_\lambda(x,\xi)$ is also a right solution of $\mathfrak{D}_{q,\lambda}$.

3. Clifford-algebra-valued functions in several variables

Consider the function u defined in the (real) Euclidean space

$$\mathbb{R}^{k+1} \times \mathbb{R}^{m_2+1} \times \cdots \times \mathbb{R}^{m_n+1}$$

with dimension equal to

$$k + \sum_{j=2}^{n} m_j + n.$$

Function u depends on n variables

$$x^{(1)} = (x_0^{(1)}, \ldots, x_k^{(1)}) \text{ and } x^{(j)} = (x_0^{(j)}, \ldots, x_{m_j}^{(j)}), \quad j = 2, \ldots, n.$$

Suppose we take the same real part for all of the factor spaces, in such a way that

$$x_0^{(1)} = x_0^{(2)} = \cdots = x_0^{(n)} = x_0$$

is the common real part. In this case we put $k = m_1 - 1$ and the functions can be considered as defined in the space

$$\mathbb{R}^{m_1} \times \mathbb{R}^{m_2} \times \cdots \times \mathbb{R}^{m_n}.$$

Remark 3.1. A second possibility is to use different real parts. This case is discussed in [2].

We define $\varphi^{(1)}$- and $\varphi^{(j)}$-Cauchy–Riemann operators, for $j = 2, \ldots, n$, respectively, by

$$\mathfrak{D}_{\varphi^{(1)}} = \varphi_0 \partial_0 + \sum_{i=1}^{m_1-1} \varphi_i e_i \partial_i$$

and

$$\mathfrak{D}_{\varphi^{(j)}} = \varphi_0 \partial_0 + \sum_{i=m_1+\cdots+m_{j-1}}^{m_1+\cdots+m_j-1} \varphi_i e_i \partial_i,$$

where ∂_0 corresponds to the derivative with respect to the common real part x_0, ∂_i corresponds to the derivative with respect to x_i,

$$\varphi^{(1)} = \varphi_0 + \sum_{i=1}^{m_1-1} \varphi_i e_i,$$

$$\varphi^{(j)} = \varphi_0 + \sum_{i=m_1+\cdots+m_{j-1}}^{m_1+\cdots+m_j-1} \varphi_i e_i,$$

$m+1 = m_1 + \cdots + m_n$ and φ_i can be real-valued functions defined in $\mathbb{R}^{m+1}$.

Remark 3.2. For each $j = 2, \ldots, n$, operator $\mathfrak{D}_{\varphi^{(j)}}$ can be rewritten in the form

$$\mathfrak{D}_{\varphi^{(j)}} = e_0 \varphi_0(x) \partial_0 + \sum_{i=0}^{m_j-1} e_{i+a_j} \varphi_{i+a_j}(x) \partial_{i+a_j} \tag{3.1}$$

and

$$\varphi^{(j)} = \varphi_0 + \sum_{i=0}^{m_j-1} \varphi_{i+a_j} e_{i+a_j},$$

where $a_j = m_1 + \cdots + m_{j-1}$.

4. Clifford type algebra and the associated multi-meta-weighted-monogenic operator

Define

$$m+1 = \sum_{j=1}^{n} m_j, \tag{4.1}$$

consider $\mathbb{R}^{m+1}$ and the corresponding algebra of Clifford type, which dimension is 2^m, defined by following algebraic structure. Denote the basis vectors of

$$\mathbb{R}^{m+1} = \mathbb{R}^{m_1} \times \mathbb{R}^{m_2} \times \cdots \times \mathbb{R}^{m_n}.$$

by

$$\{e_0 = 1, \ldots, e_{m_1-1}; e_{m_1}, \ldots, e_{m_1+m_2-1}; \ldots e_{m-m_n}, \ldots, e_m\}.$$

Let s be one of the indices:

$$s = m_1, m_1 + m_2, \ldots, m_1 + \cdots + m_{n-1}, \tag{4.2}$$

whereas k and l are indices (between 1 and m) which are different from these $n-1$ indices s. Then the algebra $\mathcal{A}_m(\sigma_1)$ is defined by the structure relations

$$\begin{cases} e_s^2 = \alpha_s, \quad e_k^2 = -\alpha_k, \\ e_k e_l + e_l e_k = 2\gamma_{lk}, \end{cases} \tag{4.3}$$

where α_k and $\gamma_{lk} = \gamma_{kl}$ are real numbers, s is as (4.2) and $l, k = 1, 2, \ldots, m$.

The $\mathfrak{D}_{\varphi^{(j)},\lambda^{(j)}}$ operator is given by

$$\mathfrak{D}_{\varphi^{(j)},\lambda^{(j)}} = \mathfrak{D}_{\varphi^{(j)}} - \lambda^{(j)}, \tag{4.4}$$

for $j = 1, \ldots, n$, where

$$\lambda^{(1)} = \lambda_0 + \sum_{i=1}^{m_1-1} \lambda_i e_i,$$

$$\lambda^{(j)} = \lambda_0 + \sum_{i=0}^{m_j-1} \lambda_{i+a_j} e_{i+a_j}, \quad \text{for} \quad j = 2, \ldots, n,$$

λ_i are real numbers and $\mathfrak{D}_{\varphi^{(j)}}$ is given by (3.1) .

Definition 4.1. A function $u \in C^1(\Omega, \mathcal{A}_m(\sigma_1))$ satisfying the system

$$\mathfrak{D}_{\varphi^{(j)},\lambda^{(j)}} u = 0,$$

for each $j = 1, \ldots, n$, is said to be **multi-meta-weighted-monogenic of first class** or multi-meta-weighted-monogenic (for short).

Remark 4.2. In order to clarify the purpose of the paper. We work with a multi-algebraic structure based in fixing the real part for all products of

$$\mathbb{R}^{m+1} = \mathbb{R}^{m_1} \times \mathbb{R}^{m_2} \times \cdots \times \mathbb{R}^{m_n}.$$

this induces the treatment of the theory to find integral representation via elliptic operators, based on a specific kind of operator according to this algebra. On the other hand, other possibility to construct a multi-algebraic structure is putting, for each subdomain, his own real parts,

$$\mathcal{R} = \mathbb{R}^{m_1+1} \times \mathbb{R}^{m_2+1} \times \cdots \times \mathbb{R}^{m_n+1}.$$

This case is discussed in [2].

In the next section we will construct our kernel using a distance calculated via the quadratic form associated to an elliptic second-order operator.

5. n-weighted-monogenic functions

Let Ω be a bounded domain in $\mathbb{R}^{m+1} = \mathbb{R}^{m_1} \times \cdots \times \mathbb{R}^{m_n}$. Consider the Clifford type algebra $\mathcal{A}_m(\sigma_1)$ defined by (4.3) , $\lambda^{(j)}$ and $\varphi^{(j)}$ as in Subsection 4.

Definition 5.1. Let $\mathfrak{D}_\varphi$ be the operator given by

$$\mathfrak{D}_\varphi = n\varphi_0\partial_0 + \sum_{i=1}^{m} \varphi_i e_i \partial_i. \tag{5.1}$$

A continuously differentiable $\mathcal{A}_m(\sigma_1)$-valued function u defined in Ω and satisfying $\mathfrak{D}_\varphi u = 0$ is said to be an ***n*-weighted-monogenic function**.

Remark 5.2. Note that

$$\mathfrak{D}_\varphi = \sum_{j=1}^{n} \mathfrak{D}_{\varphi^{(j)}}.$$

Therefore, if u is a multi-weighted-monogenic function, then u is n-weighted-monogenic as well.

5.1. Green's integral formula for $\mathfrak{D}_\varphi$

Let $\Omega^{(j)}$ be a bounded domain in $\mathbb{R}^{m_j}$, $u = \sum_A u_A e_A$ and $v = \sum_B v_B e_B$ $\mathcal{A}_m(\sigma_1)$-valued functions defined in $\Omega = \Omega^{(1)} \times \Omega^{(2)} \times \cdots \times \Omega^{(n)} \subset \mathbb{R}^{m+1}$ and continuously differentiable in $\overline{\Omega}$. Using a similar argument to that given in [25], it can be proved the following Green type integral formula

$$\int_\Omega (v\mathfrak{D}_\varphi \cdot u + v \cdot \mathfrak{D}_\varphi u)\, dx = |N_\varphi| \int_{\partial\Omega} v \cdot d\tau \cdot u, \tag{5.2}$$

where

$$N_\varphi = (\varphi_0 N_0, \varphi_1 N_1, \dots, \varphi_m N_m),$$

$d\tau = \widehat{N}_\varphi d\mu$ and $\widehat{N}_\varphi = \dfrac{N_\varphi}{|N_\varphi|}$.

Let us write the left-hand side of (5.2) as

$$\begin{aligned}&\int_\Omega (v\mathfrak{D}_\varphi \cdot u + v \cdot (\mathfrak{D}_\varphi u - \lambda u + \lambda u))\, dx \\ &= \int_\Omega (v(\mathfrak{D}_\varphi + \lambda) \cdot u + v \cdot (\mathfrak{D}_\varphi - \lambda)u)\, dx.\end{aligned}$$

where

$$\lambda = n\lambda_0 e_0 + \sum_{i=1}^{m} \lambda_i e_i, \tag{5.3}$$

$\lambda_i \in \mathbb{R}$ for $i = 0, 1, \dots, m$. Doing $\mathfrak{D}_{\varphi,\lambda} = \mathfrak{D}_\varphi - \lambda$ and $\mathfrak{D}_{\varphi,-\lambda} = \mathfrak{D}_\varphi + \lambda$, we obtain the formula

$$\int_\Omega (v\mathfrak{D}_{\varphi,-\lambda} \cdot u + v \cdot \mathfrak{D}_{\varphi,\lambda} u)\, dx = \int_{\partial\Omega} v \cdot d\tau \cdot u. \tag{5.4}$$

We will call this formula the *Green type formula for operator* $\mathfrak{D}_{\varphi,\lambda}$.

Remark 5.3. A solution of $\mathfrak{D}_{\varphi,\lambda} u = 0$ is called a **meta-*n*-weighted-monogenic function**.

Remark 5.4. Note that if u is a multi-meta-weighted-monogenic function, then u is meta-n-weighted-monogenic. In fact,

$$\mathfrak{D}_{\varphi,\lambda} = \sum_{j=1}^{n} \mathfrak{D}_{\varphi^{(j)},\lambda^{(j)}}.$$

5.2. Solution for $\mathfrak{D}_{\varphi,\lambda}$

Consider the operator $\mathfrak{D}_\varphi$ given by (5.1). Let Ω be a bounded domain in $\mathbb{R}^{m+1}$ with sufficiently smooth boundary. If $u : \mathbb{R}^{m+1} \to \mathcal{A}_m(\sigma_1)$ is a n-weighted-monogenic function in Ω, then we get the homogeneous second-order differential equation

$$\overline{\mathfrak{D}_\varphi}\mathfrak{D}_\varphi u = n^2\varphi_0^2\partial_0^2 u + \sum_{i=1}^{m} \alpha_i\varphi_i^2\partial_i^2 u - 2\sum_{j=2}^{n} \alpha_{a_j}\varphi_{a_j}^2\partial_{a_j}^2 u - 2\sum_{i<k} \gamma_{ik}\varphi_i\varphi_k\partial_i\partial_k u = 0, \tag{5.5}$$

where

$$\overline{\mathfrak{D}_\varphi} = n\varphi_0\partial_0 - \sum_{i=1}^{m} \varphi_i e_i \partial_i.$$

Since the coefficients α_k and γ_{ik} are real, the differential equation (5.5) is uncoupled; that is, each real-valued component u_A of u satisfies this differential equation. If the α_k are supposed to be positive and the absolute values of the γ_{ik} are not too large, then (5.5) is elliptic.

We can assume, without lost of generality, that $\varphi_0 = 1$, thus the coefficient matrix of the differential equation (5.5) is

$$\mathbf{B} = \begin{pmatrix} n^2 & 0 & 0 & \cdots & 0 \\ 0 & \alpha_1\varphi_1^2 & -\gamma_{12}\varphi_1\varphi_2 & \cdots & -\gamma_{1m}\varphi_1\varphi_m \\ 0 & -\gamma_{21}\varphi_1\varphi_2 & \alpha_2\varphi_2^2 & & -\gamma_{2m}\varphi_2\varphi_m \\ \vdots & \vdots & \vdots & \ddots & \vdots \\ 0 & -\gamma_{m1}\varphi_1\varphi_m & -\gamma_{m2}\varphi_2\varphi_m & \cdots & \alpha_m\varphi_m^2 \end{pmatrix}, \tag{5.6}$$

where the $(s+1)(s+1)$-entrie is of the form $b_{ss} = -\alpha_s\varphi_s^2$, being s given by (4.2). Assuming that the determinant of $\mathbf{B}$ is different from zero (this is the case for elliptic operators), $\mathbf{B}$ has an inverse matrix having the form

$$\mathbf{B}^{-1} = \begin{pmatrix} \frac{1}{n^2} & 0 & \cdots & 0 \\ 0 & A_{11} & \cdots & A_{1m} \\ \vdots & \vdots & \ddots & \vdots \\ 0 & A_{m1} & \cdots & A_{mm} \end{pmatrix},$$

where $A_{ik} = A_{ki}$ and we must assume that $\varphi_k \neq 0$ for each

$$k \in \{1, 2, \ldots, m\}.$$

Using the coefficients of $\mathbf{B}^{-1}$ we define, for two points $\xi = (\xi_0, \xi_1, \ldots, \xi_m)$ and $x = (x_0, x_1, \ldots, x_m)$ of $\mathbb{R}^{m+1}$, the non-Euclidean distance ϱ by

$$\varrho^2 = \frac{1}{n^2}(x_0 - \xi_0)^2 + \sum_{i,k=1}^{m} A_{ik}(x_i - \xi_i)(x_k - \xi_k). \tag{5.7}$$

With this distance, we define the function $\mathfrak{E}(x, \xi)$ by

$$\mathfrak{E}(x,\xi) = \frac{1}{\omega_{m+1}} \frac{1}{\varrho^{m+1}} \left(\frac{1}{n}(x_0 - \xi_0) - \sum_{i,k=1}^{m} e_i \varphi_i A_{ik}(x_k - \xi_k) \right). \tag{5.8}$$

It is easy to see that

$$\omega_{m+1}\mathfrak{D}_\varphi \mathfrak{E}(x,\xi) = \mathfrak{D}_\varphi \left(\frac{1}{\varrho^{m+1}} \right) \cdot K(x,\xi) + \frac{1}{\varrho^{m+1}} \cdot \mathfrak{D}_\varphi K(x,\xi),$$

where $K(x,\xi) = \frac{1}{n}(x_0 - \xi_0) - \sum_{i,k=1}^{m} e_i \varphi_i A_{ik}(x_k - \xi_k)$. A straightforward calculation gives

$$\begin{aligned} \omega_{m+1}\mathfrak{D}_\varphi \mathfrak{E}(x,\xi) &= -\frac{m+1}{\varrho^{m+3}} \overline{K(x,\xi)} \cdot K(x,\xi) + \frac{1}{\varrho^{m+1}}(1+m) \\ &= -\frac{m+1}{\varrho^{m+3}} \varrho^2 + \frac{1}{\varrho^{m+1}}(1+m) \\ &= 0, \end{aligned}$$

i.e., $\mathfrak{E}(x,\xi)$ is a (left) n-weighted-monogenic function.

Remark 5.5. $\mathfrak{E}(x,\xi)$ is a right n-weighted-monogenic function also.

Denoting

$$\wp(x,\xi) = \lambda_0(x_0 - \xi_0) + \sum_{j=1}^{m} \frac{\lambda_j}{\varphi_j}(x_j - \xi_j), \tag{5.9}$$

we define the function $\mathfrak{E}_\lambda(x,\xi)$ by

$$\mathfrak{E}_\lambda(x,\xi) = \exp\left(\wp(x,\xi)\right) \mathfrak{E}(x,\xi). \tag{5.10}$$

Note that the function $\mathfrak{E}_\lambda(x,\xi)$ satisfies the equation

$$\mathfrak{D}_\varphi \mathfrak{E}_\lambda = \lambda \mathfrak{E}_\lambda,$$

where λ is given by (5.3), i.e., $\lambda = n\lambda_0 + \sum_{i=1}^{m} \lambda_i e_i$.

In fact, since $\exp(\wp(x,\xi))$ is a real-valued function, we have

$$\begin{aligned}\mathfrak{D}_\varphi \mathfrak{E}_\lambda(x,\xi) &= \mathfrak{D}_\varphi\left(\exp(\wp(x,\xi))\right)\cdot \mathfrak{E}(x,\xi) + \exp(\wp(x,\xi))\cdot \mathfrak{D}_\varphi \mathfrak{E}(x,\xi)\\ &= \left[n\lambda_0 + \sum_{j=1}^{m}\frac{\lambda_j}{\varphi_j}\varphi_j e_j\right]\cdot \exp(\wp(x,\xi))\,\mathfrak{E}(x,\xi)\\ &= \lambda \mathfrak{E}_\lambda(x,\xi).\end{aligned}$$

In view of this, we have that

$$\mathfrak{D}_\varphi \mathfrak{E}_\lambda(x,\xi) - \lambda\mathfrak{E}_\lambda(x,\xi) = 0,$$

that is,

$$\mathfrak{D}_{\varphi,\lambda}\mathfrak{E}_\lambda(x,\xi) = 0.$$

This means that $\mathfrak{E}_\lambda(x,\xi)$ is a left meta-n-weighted-monogenic function.

Remark 5.6. $\mathfrak{E}_\lambda(x,\xi)$ is also a right solution of $\mathfrak{D}_{\varphi,\lambda}$. A similar calculation shows that

$$\mathfrak{E}_{-\lambda} := \exp(-\wp(x,\xi))\,\mathfrak{E}(x,\xi)$$

is such that $\mathfrak{D}_{\varphi,-\lambda}\mathfrak{E}_{-\lambda}(x,\xi) = \mathfrak{E}_{-\lambda}(x,\xi)\mathfrak{D}_{\varphi,-\lambda} = 0$.

5.3. Cauchy–Pompeiu formula for $\mathfrak{D}_{\varphi,\lambda}$

Now we are going to show that $\mathfrak{E}_\lambda(x,\xi)$ turns out to be a fundamental solution for $\mathfrak{D}_{\varphi,\lambda}$ with the singularity at $x=\xi$ (see [15] for more details of fundamental solutions).

First, since the differential equation (5.5) is supposed to be elliptic, the non-Euclidean distance ϱ defined by (5.7) can be estimated by $\varrho \geq cr$, where c is a constant and r is the Euclidean distance of x and ξ. Thus we have

$$\begin{aligned}|\mathfrak{E}_\lambda(x,\xi)| &\leq \frac{1}{\omega_{m+1}}\frac{1}{\varrho^{m+1}}\left(\frac{|x_0-\xi_0|}{n} + \sum_{i,j=1}^{m}|\varphi_i||A_{ij}||x_j-\xi_j|\right)e^{\wp(x,\xi)}\\ &\leq \frac{\text{const}}{r^m}.\end{aligned}$$

Therefore $\mathfrak{E}_\lambda(x,\xi)$ has a weak singularity at $x=\xi$. The same is true for $\mathfrak{E}_{-\lambda}(x,\xi)$.

Now, let $U_\varepsilon(\xi)$ be an ε-neighbourhood of ξ. Applying the formula (5.4) on $\Omega_\varepsilon = \Omega - \overline{U_\varepsilon(\xi)}$ and $v = \mathfrak{E}_{-\lambda}(x,\xi)$ we obtain the expression

$$\int_{\Omega_\varepsilon}\mathfrak{E}_{-\lambda}(x,\xi)\cdot\mathfrak{D}_{\varphi,\lambda}u\,dx = \int_{\partial\Omega}\mathfrak{E}_{-\lambda}(x,\xi)\cdot d\tau\cdot u + \int_{|x-\xi|=\varepsilon}\mathfrak{E}_{-\lambda}(x,\xi)\cdot d\tau\cdot u. \quad (5.11)$$

Consider the points in $|x-\xi|=\varepsilon$ represented in the form $x-\xi=\varepsilon y$, where y is a point of the unit sphere. Let ϱ_0 be the non-Euclidean distance between the points y and $(0,0,\ldots,0)$. Then (5.7) implies that $\varrho = \varepsilon\varrho_0$ and

$$\mathfrak{E}_{-\lambda}(y,0) = e^{-\varepsilon\wp(y,0)}\frac{1}{\omega_{m+1}}\cdot\frac{1}{\varrho_0^{m+1}\varepsilon^m}\left(\frac{1}{n}y_0 - \sum_{i,j=1}^{m}\varphi_i e_i A_{ij} y_j\right).$$

Moreover, we have $d\mu = \varepsilon^m d\mu_1$, where $d\mu_1$ is the measure element of the unit sphere. And so the integral

$$-\int\limits_{|x-\xi|=\varepsilon} \mathfrak{E}_{-\lambda}(x,\xi)\cdot d\tau$$

is independent on ε. It depends (continuously) only on the values of the constants α_j, γ_{ij} in the structure relations. We denote this value by $c(\alpha_j,\gamma_{ij})$. On the other hand, since u is continuous then

$$\begin{aligned}\lim_{\varepsilon\to 0}\int\limits_{|x-\xi|=\varepsilon} \mathfrak{E}_{-\lambda}(x,\xi)\cdot d\tau\cdot u(x) &= \lim_{\varepsilon\to 0}\int\limits_{|x-\xi|=\varepsilon} \mathfrak{E}_{-\lambda}(x,\xi)\cdot d\tau\cdot (u(x)-u(\xi)+u(\xi))\\ &= -c(\alpha_j,\gamma_{ij})\cdot u(\xi).\end{aligned}$$

Hence, we have established the next result.

Theorem 5.7. *In interior points ξ of Ω, each function u twice continuously differentiable with values in the Clifford algebra $\mathcal{A}_n(2,\alpha_j,\gamma_{ij})$ can be represented by*

$$c(\alpha_j,\gamma_{ij})\cdot u(\xi) = \int_{\partial\Omega} \mathfrak{E}_{-\lambda}(x,\xi)\cdot d\tau\cdot u(x) - \int\limits_{\Omega} \mathfrak{E}_{-\lambda}(x,\xi)\cdot \mathfrak{D}_{\varphi,\lambda}u(x)\,dx \tag{5.12}$$

Remark 5.8. As a corollary of this formula we have **the Cauchy type formula**

$$c(\alpha_j,\gamma_{ij})\cdot u(\xi) = \int_{\partial\Omega} \mathfrak{E}_{-\lambda}(x,\xi)\cdot d\tau\cdot u(x).$$

In a complete analogous way, we can obtain the Cauchy–Pompeiu formula

Theorem 5.9. *In interior points ξ of Ω each function ν twice continuously differentiable with values in the Clifford algebra $\mathcal{A}(\sigma_1)$ can be represented by*

$$\nu(\xi)\cdot c(\alpha_j,\gamma_{ij}) = \int_{\partial\Omega} \nu(x)\cdot d\tau\cdot \mathfrak{E}_{\lambda}(x,\xi) - \int_{\Omega} \nu(x)\mathfrak{D}_{\varphi,-\lambda}\cdot \mathfrak{E}_{\lambda}(x,\xi)dx.$$

Remark 5.10. Remark 5.4 implies that these integral representations are also valid for multi-meta-weighted-monogenic functions.

5.4. Example

Consider the case $n=2$, $m_1=2$, $m_2=1$. Then $m=2$ and functions u are to be defined in $\mathbb{R}^{m+1}=\mathbb{R}^2\times\mathbb{R}^1$. We have

$$u = u(x_0,x_1,x_2) = u_0(x_0,x_1,x_2) + u_1(x_0,x_1,x_2)e_1 + u_2(x_0,x_1,x_2)e_2,$$

where 1, e_1, e_2 is the standard basis of $\mathbb{R}^3$. If $\alpha_1=\alpha_2=1$ and $\gamma_{12}=0$, the Clifford type algebra $\mathcal{A}_2$ is generated by the structure relations

$$\begin{cases} e_1^2=-1, \quad e_2^2=1,\\ e_1e_2+e_2e_1=0.\end{cases}$$

Note that

$$\mathfrak{D}_\varphi = 2\partial_0 + \varphi_1 e_1\partial_1 + \varphi_2 e_2\partial_2$$

and

$$\mathfrak{D}_{\varphi}^{(1)} = \partial_0 + \varphi_1 e_1 \partial_1, \quad \mathfrak{D}_{\varphi}^{(2)} = \partial_0 + \varphi_2 e_2 \partial_2,$$

where φ_1 and φ_2 are real numbers. Thus, if $u = u_0 + u_1 e_1 + u_2 e_2$, we have that $\mathfrak{D}_{\varphi}^{(1)} u = 0$ and $\mathfrak{D}_{\varphi}^{(2)} u = 0$ are equivalent, respectively, to the systems

$$\begin{cases} \partial_0 u_0 - \varphi_1 \partial_1 u_1 = 0, & \partial_0 u_1 - \varphi_1 \partial_1 u_0 = 0, \\ \partial_0 u_2 = 0, & \varphi_1 \partial_1 u_2 = 0, \end{cases}$$

and

$$\begin{cases} \partial_0 u_0 + \varphi_2 \partial_2 u_2 = 0, & \partial_0 u_1 = 0, \\ \partial_0 u_2 + \varphi_2 \partial_2 u_0 = 0, & \varphi_1 \partial_2 u_1 = 0. \end{cases}$$

This implies that a multi-weighted-monogenic function must satisfy

$$u(x_0, x_1, x_2) = c_0 x_0 + c_1 x_1 e_1 + c_2 x_2 e_2,$$

with $c_0, c_1, c_2 \in \mathbb{R}$. However, a 2-weighted-monogenic function is not necessarily of this form. In fact,

$$u(x_0, x_1, x_2) = 2(x_1 + x_2) + (x_2 - x_0)\varphi_1 e_1 + (x_1 - x_0)\varphi_2 e_2$$

and

$$\mathfrak{E}(x_0, x_1, x_2) = \frac{1}{\omega_3} \frac{1}{\varrho^3} \left(\frac{1}{2} x_0 - \frac{1}{\varphi_1} x_1 e_1 + \frac{1}{\varphi_2} x_2 e_2 \right)$$

are two 2-weighted-monogenic functions but not multi-weighted-monogenic, where ϱ is the distance defined by

$$\varrho^2 = \frac{1}{4} x_0^2 + \frac{1}{\varphi_1^2} x_1^2 - \frac{1}{\varphi_2^2} x_1^2$$

and φ_2 is small enough.

6. Distributional solution for the inhomogeneous meta-n-weighted-monogenic equation

The theory developed above is useful to solve some types of differential equations. In particular, the equation $u\mathfrak{D}_{\varphi,-\lambda} = h$ with h a continuous function in Ω.

In fact, we have the result

Theorem 6.1. *Let h be a $\mathcal{A}(\sigma_1)$-valued continuous function in Ω. Then*

$$u(x) = -\int_{\Omega} h(\xi) \cdot c^{-1}(\alpha_j, \gamma_{ij}) \cdot \mathfrak{E}_{-\lambda}(x, \xi)\, d\xi \tag{6.1}$$

is a solution of the equation $u\mathfrak{D}_{\varphi,-\lambda} = h$ in Ω, where $\mathfrak{E}_{-\lambda}$ defined as in (5.10) and $c(\alpha_j, \gamma_{ij})$ is supposed to be invertible.

Proof. Let ϕ be a test function k times continuously differentiable. Replacing u with ϕ in the Cauchy–Pompeiu formula (5.12), we obtain

$$c(\alpha_j, \gamma_{ij}) \cdot \phi(\xi) = \int_\Omega \mathfrak{E}_{-\lambda}(x, \xi) \cdot \mathfrak{D}_{\varphi,\lambda}\phi(x)\, dx.$$

Taking into account Fubini's theorem for weakly singular integrands, we have

$$\int_{\Omega_x} u(x) \cdot \mathfrak{D}_{\varphi,\lambda}\phi(x)\, dx = -\int_{\Omega_x}\int_{\Omega_\xi} h(\xi) \cdot c^{-1}(\alpha_j, \gamma_{ij}) \cdot \mathfrak{E}_{-\lambda}(x, \xi) \cdot \mathfrak{D}_{\varphi,\lambda}\phi(x)\, d\xi\, dx$$

$$= -\int_{\Omega_\xi} h(\xi) \cdot c^{-1}(\alpha_j, \gamma_{ij}) \cdot \left[\int_{\Omega_x} \mathfrak{E}_{-\lambda}(x, \xi) \cdot \mathfrak{D}_{\varphi,\lambda}\phi(x)\, dx\right] d\xi = -\int_{\Omega_\xi} h(\xi)\, \phi(\xi)\, d\xi.$$

Thus

$$\int_{\Omega_x} [u(x) \cdot \mathfrak{D}_{\varphi,\lambda}\phi(x) + h(x) \cdot \phi(x)]\, dx = 0. \tag{6.2}$$

Using (5.4),

$$\int_{\Omega_x} [u(x)\, \mathfrak{D}_{\varphi,-\lambda} \cdot \phi(x) + u(x) \cdot \mathfrak{D}_{\varphi,\lambda}\phi(x)]\, dx = 0$$

is also true. The last equation and (6.2) give us, for any test function ϕ, that

$$\int_{\Omega_x} [-u(x)\, \mathfrak{D}_{\varphi,-\lambda} \cdot \phi(x) + h(x) \cdot \phi(x)]\, dx$$

$$= \int_{\Omega_x} [-u(x)\, \mathfrak{D}_{-\lambda} + h(x)] \cdot \phi(x)\, dx = 0.$$

Therefore, from the Fundamental Lemma of variational calculus, it follows that $-u(x)\, \mathfrak{D}_{-\lambda} + h(x) = 0$. □

Remark 6.2. If h is not continuous but it is integrable in Ω, then u defined by (6.1) is a distributional solution of $u\mathfrak{D}_{\varphi,-\lambda} = h$ in Ω.

7. Concluding remarks

- An alternative to the approach presented in this work is to iterate, in each $\Omega^{(j)}$, a Cauchy integral formula given for weighted-monogenic functions. This approach is discussed in [2, 6] and can be adapted to this case.
- A disadvantage of this method, as can be seen in section 5.2, is that the method works for elliptic operators because we construct our kernel through the use of a distance calculated via the quadratic form associated to an elliptic second-order operator.
- A wide quantity of problems can be solved using the integral representations given in this work. For example, initial value problems and boundary value problems involving the weighted monogenic differential operators, the study of properties of the different integral operators derived from our integral representations and its applications, representation in power series of the different weighted-monogenic functions defined here and its applications, etc.

- We believe that this theory of multi-monogenic functions, in connection with the theory of wavelets, allows us to solve problems related with solar energy, phototherapy, climate change. See [12, 14].

Acknowledgment

The authors would like to express their sincere gratitude to Professor Wolfgang Tutschke for his suggestions in this investigation. Also, The authors are very thankful to the anonymous referee for his/her valuable remarks and suggestions that have improved this article. The referee has also pointed out additional relevant references.

References

[1] Ariza, E. and Di Teodoro, A., *Multi M q-monogenic Function in Different Dimension.* In Hypercomplex Analysis: New Perspectives and Applications, pp. 61–73. (2014). Springer International Publishing.

[2] Ariza, E. and Di Teodoro, A., *An integral representation formula for Multi Meta-φ-monogenic functions of second class.* Submitted. (2016).

[3] Bernardes, G., *Monogenic functions of higher spin in two vector variables: Cauchy–Kowalewski theorem and polynomial solutions for some simple examples.* Advances in Applied Clifford Algebras. Vol. 11(S2), pp. 9–20. (2001).

[4] Bernardes, G. and Sommen, F., *Monogenic functions of higher spin by Cauchy–Kovalevska extension of real-analytic functions.* Complex Variables Theory Appl. Vol. 39, pp. 305–325. (1999).

[5] Brackx, F. Delanghe, R. and Sommen, F., *Clifford Analysis.* (1982). Pitman Research Notes.

[6] Delanghe, R., Sommen, F. and Soucek, V., *Clifford algebra and spinor-valued functions: a function theory for the Dirac operator.* Vol. 53. (2012). Springer Science & Business Media.

[7] Colombo, F., Sabadini, I. and Struppa, D., *Noncommutative Functional Calculus.* Progress in Mathematics, (2011). Birkhäuser Mathematics, Birkhäuser.

[8] Di Teodoro, A., Franquiz, R. and López A.,*A Modified Dirac Operator in Parameter-Dependent Clifford Algebra: A Physical Realization.* Advances in Applied Clifford Algebras. Vol. 25, Issue 2, pp. 303–320. (2015).

[9] Folland, G., *Introduction to Partial Differential Equations,* (1995). Princeton University Press, Second edition.

[10] Hörmander, L., *An Introduction to Complex Analysis in Several Variables.* North-Holland Mathematical Library, 3rd ed., Vol. 7. (1990). North-Holland Publishing Co., Amsterdam.

[11] Hung Son, L., *Recent Trends in Theory of Regular Functions Taking Value in Clifford Algebra.* Proceedings of the International Conference of Applied Mathematics (ICAM), Hanoi, pp. 113–122. (2004). SAS International Publications, Delhi.

[12] Iqbal, M., *An Introduction to Solar Radiation.* (1983). Elsevier.

[13] Jank, G. and Sommen, F., *Clifford analysis, biaxial symmetry and pseudoanalytic functions.* Complex Variables Theory Appl., 3-4, 13, 195–212. (1990).

[14] Mitrea M., *Clifford wavelets, singular integrals, and Hardy spaces.* (1994). Springer-Verlag Berlin.

[15] Miranda, C., *Partial differential equations of elliptic type.* (1970). Springer.

[16] Obolashvili, E., *Partial Differential Equations in Clifford Analysis*, (1999). First edition, Chapman and Hall/CRC.

[17] Peres N., *Scattering in one-dimensional heterostructure described by the Dirac equation.* J. Phys.: Condens. Matter 21, 095501 (2009).

[18] Sommen, F., *Clifford Analysis in Two and Several Clifford Variables.* Appl. Anal. 73. 1-2, 225–253. (1999).

[19] Sommen, F., *Monogenic Functions of Higher Spin Z.* Anal. Anwendungen. 15. 2, 279–282. (1996).

[20] Sommen, F. and Bernardes, G., *Multivariable Monogenic Functions of Higher Spin.* Journal of Natural Geometry. 18. 1-2, 101–114. (2000).

[21] Shabat, D.V., *Introduction to Complex Analysis, Part II, Functions of several variables.* (Translated from the third (1985) Russian edition by J.S. Joel), Translations of Mathematical Monographs. Vol. 110. (1992). American Mathematical Society. Providence, RI.

[22] Tutschke, W. and Hung Son, L., *A New Concept of Separately Holomorphic and Separately Monogenic Functions. Algebraic structures in partial differential equations related to complex and Clifford analysis.* 67–78, (2010). Ho Chi Minh City Univ. Educ. Press, Ho Chi Minh City.

[23] Tutschke, W. and Hung Son, L., *Multi-monogenic functions in different dimensions.* Complex Variables and Elliptic Equations: An International Journal. Vol. 58, 2, 293–298. (2013).

[24] Tutschke, W. and Vanegas, C. J., *Clifford algebras depending on parameters and their applications to partial differential equations.* Contained in *Some topics on value distribution and differentiability in complex and p-adic analysis.* (2008). Science Press, Beijing, 430–449.

[25] Tutschke, W. and Vanegas, C. J., *Métodos del análisis complejo en dimensiones superiores.* (2008). Ediciones IVIC, Caracas.

[26] Vladimirov. V., *Methods of the theory of functions of many complex variables* (transl. from the Russian), (1966). The M.I.T. Press, Cambridge, Mass.-London.

Eusebio Ariza García and Antonio Di Teodoro
School of Information Science and
Technology Department of Mathematics
Yachay Tech
Yachay City of Knowledge
San Miguel de Urcuquí
Ecuador-100119
e-mail: `eariza@yachaytech.edu.ec`
`aditeodoro@yachaytech.edu.ec`

Modern Trends in Hypercomplex Analysis
Trends in Mathematics, 19–33

Greedy Algorithms and Rational Approximation in One and Several Variables

Laurent Baratchart, Wei-Xiong Mai and Tao Qian

Abstract. We will review the recent development of rational approximation in one and several real and complex variables. The concept rational approximation is closely related to greedy algorithms, based on a dictionary consisting of Szegő kernels in the present context.

Mathematics Subject Classification (2010). 45P05, 30H.

Keywords. Rational orthogonal system, Takenaka–Malmquist system, Szegő kernel, Hardy spaces.

1. Introduction

Traditionally, rational approximation is mostly restricted to approximation to functions of one complex variable in various domains. Recently a number of adaptive or sparse representation methods were developed, including in principle the greedy algorithms and those based on learning theory, including the SVM method ([25, 26]), that all fit into the concept of rational approximation. This article mainly concerns the greedy algorithm method in relation to a dictionary in the context at hand. Under the greedy algorithm method, rational approximation is generalized to include approximation in several complex and real variables. This idea, in particular, is applicable to function spaces with a Cauchy structure, as well as to reproducing kernel Hilbert spaces. Below we discuss this concept in a number of individual contexts.

2. Preliminaries on Greedy algorithm in Hilbert spaces

Let $\mathbb{H}$ be a Hilbert space with a dictionary, where, by a dictionary, we mean a set $\mathbb{D} \subset \mathbb{H}$ satisfying (i) $e \in \mathbb{D}$ implies $\|e\| = 1$; and (ii) $\overline{\text{span}\{\mathbb{D}\}} = \mathbb{H}$.

Supported in part by MYRG116(Y1-L3)-FST13- QT; and FDCT 098/2012/A3.

The most basic greedy algorithm would be the following. Let $\rho \in (0,1]$ be fixed. Let $f \in \mathbb{H}$ and $g_1 = f$. Choose $e_1 \in \mathbb{H}$ such that

$$|\langle g_1, e_1\rangle| \geq \rho \sup\{|\langle g_1, e\rangle| \mid e \in \mathbb{D}\}.$$

When $\rho \in (0,1)$, a desired e_1 always exists. While for $\rho = 1$, a desired e_1 satisfying the above requirement may not exist. In the context of the present paper, where a Cauchy structure prevails, the desired e_1 for $\rho = 1$ always exists. Our discussion, however, is for the general case $\rho \in (0,1]$. Subsequently, we have the decomposition

$$f = \langle g_1, e_1\rangle e_1 + g_2,$$

where g_2 is the *standard remainder*:

$$g_2 = f - \langle g_1, e_1\rangle e_1.$$

Obviously, g_2 is orthogonal with $\langle g_1, e_1\rangle e_1$, and hence,

$$\|f\|^2 = \|\langle g_1, e_1\rangle e_1\|^2 + \|g_2\|^2 = |\langle g_1, e_1\rangle|^2 + \|g_2\|^2.$$

Therefore, to minimize $\|g_2\|^2$ is to maximize $|\langle g_1, e_1\rangle|^2$. If we apply the same reduction to g_2, we obtain the standard remainder g_3, where

$$f = \langle g_1, e_1\rangle e_1 + \langle g_2, e_2\rangle e_2 + g_3,$$

and

$$\|f\|^2 = |\langle g_1, e_1\rangle|^2 + |\langle g_2, e_2\rangle|^2 + \|g_3\|^2,$$

where e_2 is chosen so that

$$|\langle g_2, e_2\rangle| \geq \rho \sup\{|\langle g_2, e\rangle| \mid e \in \mathbb{D}\}.$$

Repeating the same procedure on g_3 we get g_4, and so on. Inductively we obtain

$$f = \sum_{k=1}^{n} \langle g_k, e_k\rangle e_k + g_{k+1}$$

and

$$\|f\|^2 = \sum_{k=1}^{n} |\langle g_k, e_k\rangle|^2 + \|g_{k+1}\|^2,$$

where e_k is chosen to make

$$|\langle g_k, e_k\rangle| \geq \rho \sup\{|\langle g_k, e\rangle| \mid e \in \mathbb{D}\},$$

while $e_1, \ldots, e_{k-1}$ were previously chosen.

The theory of greedy algorithms ([18], [40]) asserts that

$$f = \sum_{k=1}^{\infty} \langle g_k, e_k\rangle e_k \quad \text{and} \quad \|f\|^2 = \sum_{k=1}^{\infty} |\langle g_k, e_k\rangle|^2.$$

The above-described algorithm is the so-called *General Greedy Algorithm*. A refinement of the general greedy algorithm is the so-called *Orthogonal Greedy Algorithm*.

The difference is only that the g_k are replaced by the *orthogonal remainders* $\tilde{g}_k$, defined through the relation

$$f = \sum_{k=1}^{n} \langle \tilde{g}_k, B_k \rangle B_k + \tilde{g}_{k+1}, \tag{2.1}$$

where $\{B_1, \ldots, B_{k-1}, B_k\}$ is the G-S orthogonalization of $\{B_1, \ldots, B_{k-1}, e_k\}$, and e_k is chosen so that

$$|\langle \tilde{g}_k, e_k \rangle| \geq \rho \sup\{|\langle \tilde{g}_k, e \rangle| \mid e \in \mathbb{D}\}.$$

Among the well-known greedy algorithms, the most effective one would be the just described Orthogonal Greedy Algorithm. We note that neither the General Greedy Algorithm, nor the Orthogonal Greedy Algorithm can repeatedly select dictionary elements: Repeated selections or even selecting one in the linear span of the already selected will give nil contribution to the energy approximation.

When the space is a reproducing kernel Hilbert space, consisting of functions defined, say, in a set $\mathbf{D}$, and if the dictionary consists of the normalized reproducing kernels e_a, where a ranges over a set $\mathbf{D} \subset \mathbf{C}^{\mathbf{N}}$ for some N, then for any $f \in \mathbb{H}$,

$$\langle f, e_a \rangle = r(a) f(a),$$

where $r(a) > 0$ is the normalizing constant that makes $\|e_a\| = 1$.

In such a case, at each recursive step of the General Greedy and Orthogonal Greedy Algorithms, one seeks a suitable e_a such that

$$r(a)|f(a)| \geq \rho \sup\{r(b)|f(b)| \mid b \in \mathbf{D}\}.$$

Numerically this is easy to achieve through computation based on the information on the known function f.

We will show that under certain assumption there exists a variation of the Orthogonal Greedy Algorithm that allows repeated selection of the variable a. Repeating the selection of the variable a corresponds to selecting directional derivatives, of order one and higher, of the dictionary elements e_a.

This new greedy algorithm proposed in [37], called *Pre-Orthogonal Greedy Algorithm* (P-OGA), is formulated as follows.

Let $\{e_{a_1}, \ldots, e_{a_{n-1}}\}$ be the $(n-1)$-tuple of the previously selected dictionary elements, and $\{B_1, \ldots, B_{n-1}\}$ its G-S orthogonalization. Sometimes $B_k = B_{\{a_1,\ldots,a_k\}}$ is more precisely written as $B^{a_k}_{\{a_1,\ldots,a_{k-1}\}}$. The selection criterion for a_n is

$$|\langle f_n, B^{a_n}_{\{a_1,\ldots,a_{n-1}\}} \rangle| \geq \rho \sup\{|\langle f_n, B^{a}_{\{a_1,\ldots,a_{n-1}\}} \rangle| \mid a \in \mathbf{D}\}, \tag{2.2}$$

where f_n is the standard remainder with respect to the orthonormal system $\{B_1, \ldots, B_{n-1}\}$. We note that under such a selection criterion f_k is different from $\tilde{g}_k$ defined through (2.1).

Now we add an assumption under which the machinery that we design will work. Namely, we assume for any $f \in \mathbb{H}$ and $a_1, \ldots, a_{n-1} \in \mathbf{D}$ that

$$\lim_{a \to \partial \mathbf{D}} |\langle f, B^{a}_{\{a_1,\ldots,a_{n-1}\}} \rangle| = 0, \tag{2.3}$$

where $\partial\mathbf{D}$ indicates the boundary of $\mathbf{D}$ in $C^N \cup \infty$. We show that under this assumption, if the dictionary is suitably extended, then the threshold $\rho = 1$ can be reached, and the following equality holds:

$$|\langle f_n, B^{a_n}_{\{a_1,\ldots,a_{n-1}\}}\rangle| = \max\{|\langle f_n, B^{a}_{\{a_1,\ldots,a_{n-1}\}}\rangle| \mid a \in \mathbf{D}\}. \tag{2.4}$$

The extension of the dictionary consists in adjoining the directional derivatives of the kernels e_a with respect to a, in all directions and for all $a \in \mathbf{D}$. This of course only makes sense if $\mathbf{D}$ is open and e_a has some smoothness with respect to a. Hereafter we assume that $a \to e_a$ is smooth as a function $\mathbf{D} \to \mathbb{H}$, in particular the derivatives of e_a again lie in $\mathbb{H}$ as limits of divided differences in $\mathbb{H}$. In all examples that we shall deal with, e_a is even analytic with respect to a, and this warrants the discussion below.

In fact, the Cauchy–Schwarz inequality gives

$$|\langle f_n, B^{a}_{\{a_1,\ldots,a_{n-1}\}}\rangle| \le \|f_n\|.$$

Thus, there exists a sequence of points, $a^{(l)}$, converging to an interior or a boundary point of $\mathbf{D}$, $\lim_{l\to\infty} a^{(l)} = a_n$, such that

$$\lim_{l\to\infty} |\langle f_n, B^{a^{(l)}}_{\{a_1,\ldots,a_{n-1}\}}\rangle| = \sup\{|\langle f_n, B^{a}_{\{a_1,\ldots,a_{n-1}\}}\rangle| \mid a \in \mathbf{D}\}. \tag{2.5}$$

With the assumption (2.3) the limiting point a_n of $a^{(l)}$ must be an interior point of $\mathbf{D}$ unless $f_n \equiv 0$. In the latter case, our contention trivially holds. Otherwise, the relation (2.5) becomes

$$\lim_{l\to\infty} |\langle f_n, B^{a^{(l)}}_{\{a_1,\ldots,a_{n-1}\}}\rangle| = \max\{|\langle f_n, B^{a}_{\{a_1,\ldots,a_{n-1}\}}\rangle| \mid a \in \mathbf{D}.\} \tag{2.6}$$

Next, we compute

$$\lim_{l\to\infty} B^{a^{(l)}}_{\{a_1,\ldots,a_{n-1}\}}.$$

Denote by $P_{\{a_1,\ldots,a_{n-1}\}}$ the projection operator from $\mathbb{H}$ to $\mathrm{span}\{B_1,\ldots,B_{n-1}\}$. Now, there are two cases.

1. The limiting point e_{a_n} is not in span $\{B_1,\ldots,B_{n-1}\}$. In such a case, a_n, in particular, does not coincide with any of $a_1,\ldots,a_{n-1}$, and $B^{a_n}_{\{a_1,\ldots,a_{n-1}\}}$ is just given by

$$B^{a_n}_{\{a_1,\ldots,a_{n-1}\}} = \frac{e_{a_n} - P_{\{a_1,\ldots,a_{n-1}\}}e_{a_n}}{\|e_{a_n} - P_{a_1,\ldots,a_{n-1}}e_{a_n}\|}, \tag{2.7}$$

and

$$\{B_1,\ldots,B_{n-1}, B^{a_n}_{\{a_1,\ldots,a_{n-1}\}}\}$$

is the G-S orthogonalization of

$$\{B_1,\ldots,B_{n-1}, e_{a_n}\}.$$

2. The limiting point e_{a_n} lies in span$\{B_1,\ldots,B_{n-1}\}$. In particular, if a_n coincides with one of $a_1,\ldots,a_{n-1}$, we are in this case. We note that none of the $e_{a^{(l)}}$ is in span$\{B_1,\ldots,B_{n-1}\}$, for, otherwise, $B^{a^{(l)}}_{\{a_1,\ldots,a_{n-1}\}}=0$, and $|\langle f_n, B^{a^{(l)}}_{\{a_1,\ldots,a_{n-1}\}}\rangle|$ will have no contribution towards the maximum. We consequently have

$$\begin{aligned}
B^{a^{(l)}}_{\{a_1,\ldots,a_{n-1}\}} &= \frac{e_{a^{(l)}} - P_{\{a_1,\ldots,a_{n-1}\}}e_{a^{(l)}}}{\|e_{a^{(l)}} - P_{a_1,\ldots,a_{n-1}}e_{a^{(l)}}\|}\\
&= \frac{\left(e_{a^{(l)}} - P_{\{a_1,\ldots,a_{n-1}\}}e_{a^{(l)}}\right) - \left(e_{a_n} - P_{\{a_1,\ldots,a_{n-1}\}}e_{a_n}\right)}{\|\left(e_{a^{(l)}} - P_{a_1,\ldots,a_{n-1}}e_{a^{(l)}}\right) - \left(e_{a_n} - P_{\{a_1,\ldots,a_{n-1}\}}e_{a_n}\right)\|}\\
&= \frac{\left(e_{a^{(l)}} - e_{a_n}\right) - P_{\{a_1,\ldots,a_{n-1}\}}\left(e_{a^{(l)}} - e_{a_n}\right)}{\|\left(e_{a^{(l)}} - e_{a_n}\right) - P_{\{a_1,\ldots,a_{n-1}\}}\left(e_{a^{(l)}} - e_{a_n}\right)\|}\\
&= \frac{\left(\frac{e_{a^{(l)}}-e_{a_n}}{\|e_{a^{(l)}}-e_{a_n}\|}\right) - P_{\{a_1,\ldots,a_{n-1}\}}\left(\frac{e_{a^{(l)}}-e_{a_n}}{\|e_{a^{(l)}}-e_{a_n}\|}\right)}{\|\left(\frac{e_{a^{(l)}}-e_{a_n}}{\|e_{a^{(l)}}-e_{a_n}\|}\right) - P_{\{a_1,\ldots,a_{n-1}\}}\left(\frac{e_{a^{(l)}}-e_{a_n}}{\|e_{a^{(l)}}-e_{a_n}\|}\right)\|}.
\end{aligned}$$

Extracting a subsequence if necessary, we may suppose that $(e_{a^{(l)}} - e_{a_n})/\|e_{a^{(l)}} - e_{a_n}\|$ converges to some unit vector $\nu \in \mathbb{H}$. If ν is not in span$\{B_1,\ldots,B_{n-1}\}$, then we can take the limit in the above expression as $a^{(l)} \to a_n$, to obtain that $B^{a^{(l)}}_{\{a_1,\ldots,a_{n-1}\}}$ converges to $B^{\partial_v a_n}_{\{a_1,\ldots,a_{n-1}\}}$, where the notation ∂_v indicates that we computed the directional derivative of e_a in the direction v.

When $\nu \in$ span$\{B_1,\ldots,B_{n-1}\}$, then the above expression has indeterminate limiting form 0/0, and higher-order derivatives must be computed that we will not discuss further. Observe there may be several directions ν, accounting for the fact that a_n needs not be unique. In particular, the directions along the real or purely imaginary axes induce partial derivatives of the reproducing kernels ([37]).

The just described theory mainly grows out from the study of the Hardy space H^2 on the open unit disc. It is then applied, at least in part or adaptively, to other contexts in one or several complex variables, as well as in the Clifford algebra setting. That helps to form a concept of rational approximation in various contexts. In the following sections we will briefly review the particulars of each individual context.

3. The Hardy $H^2(\mathbf{D})$ case

The so-called AFD and Pre-Orthogonal Greedy Algorithm were originated from this context. In this section **D** represents the open unit disc. Among the equivalent definitions of the Hardy $H^2(\mathbf{D})$ space we will cite only

$$H^2(\mathbf{D}) = \{f(z) = \sum_{k=0}^{\infty} c_k z^k \mid c_k \in \mathbf{C},\ \|f\|_{H^2:=} \sum_{k=0}^{\infty} |c_k|^2 < \infty, |z| < 1\}.$$

$f \in H^2(\mathbf{D})$ implies that f is holomorphic in $\mathbf{D}$, and, as an important property, the partial sum of the infinite series on the unit circle has a $L^2(\partial\mathbf{D})$-limit that equals to the non-tangential boundary limit $\lim_{r\to 1-} f(re^{it})$ almost everywhere. Those boundary limits form a closed subspace of the L^2 space on the boundary circle, denoted by $H^2(\partial\mathbf{D})$. The mapping from $H^2(\mathbf{D})$ to $H^2(\partial\mathbf{D})$ is an isometry. We hence identify the space $H^2(\partial\mathbf{D})$ with $H^2(\mathbf{D})$. If we start from a function f in $L^2(\partial\mathbf{D})$ with Fourier coefficients $c_k, k = 0, \pm1, \pm2, \ldots$, then we have the so-called Hardy spaces decomposition $f = f^+ + f^-$, where

$$f^+(e^{it}) = \sum_{k=0}^{\infty} c_k e^{ikt}, \quad f^-(e^{it}) = \sum_{k=-1}^{-\infty} c_k e^{ikt}.$$

If f is real-valued on the circle, then we have

$$f = 2\mathrm{Re} f^+ - c_0.$$

The last relation shows that approximation of functions in $L^2(\partial\mathbf{D})$ can be reduced to that of functions in the Hardy class. In other contexts we have analogous relations, so the case of Hardy spaces on which we concentrate below will be a prototypical example.

In $H^2(\mathbf{D})$, the normalized reproducing kernels (also known as Szegő kernels) are the rational functions

$$e_a(z) = \frac{\sqrt{1-|a|^2}}{1-\overline{a}z}, \qquad a \in \mathbf{D}. \tag{3.1}$$

Then $\{e_a\}_{a\in\mathbf{D}}$ is a dictionary of $H^2(\mathbf{D})$. Let $A = (a_1, \ldots, a_n) \in \mathbf{D}^n$ be a n-tuple. If the a_k are all distinct, we associate to A the n-tuple $(e_{a_1,\ldots,e_{a_n}})$ of normalized reproducing kernels. More generally, if A consists of $m < n$ distinct points $b_1, \ldots, b_m$ where b_k is repeated l_k times with $l_1 + \cdots + l_m = n$, then we associate to A the n-tuple

$$(E_{\{b_1,1\}}, \ldots, E_{\{b_1,l_1\}}, E_{\{b_2,1\}}, \ldots, E_{\{b_2,l_2\}}, \ldots, E_{\{b_m,1\}}, \ldots, E_{\{b_m,l_m\}}) \tag{3.2}$$

defined as follows. We set $E_{\{0,j\}}(z) = z^j$, and if $a_k \neq 0$ then $E_{\{a_k,j\}}(z) = \frac{c(a_k,j)}{(1-\overline{a}_k z)^j}$, where $c(a_k, j)$ is the constant making $\|E_{\{a_k,j\}}\| = 1$. The orthogonalization of (3.2) is the so-called Takenaka–Malmquist (TM-) system, or orthogonal rational system, $(B_1, \ldots, B_n)$, where

$$B_k(z) = B_{\{a_1,\ldots,a_k\}}(z) = \frac{\sqrt{1-|a_k|^2}}{1-\overline{a}_k z} \prod_{l=1}^{k-1} \frac{z-a_l}{1-\overline{a}_l z}$$

[28, Lecture V]. We call each B_k a *modified Blaschke product*. One recognizes that the rational function $\frac{\sqrt{1-|a_n|^2}}{1-\overline{a}_n z}$ in front is the Szegő kernel and the product thereafter is the Blaschke product with the zeros $a_1, \ldots, a_{k-1}$. The claim is: when studying rational approximation in $H^2(\mathbf{D})$, TM systems are unavoidable. This is no wonder, because every rational function is a linear combination of $E_{\{b_k,j\}}$ for some b_k. Existing studies on TM systems include Laguerre and Kautz systems. The traditional works on TM systems deal with the condition

$$\sum_{k=1}^{\infty}(1-|a_k|)=\infty. \tag{3.3}$$

This is called the *hyperbolic non-separability condition.* For any $p \in [1,\infty)$, it is necessary and sufficient for the corresponding TM system to be complete in $H^p(\mathbf{D})$ ([15]). Also, for $1<p<\infty$ and any sequence $(a_1,\ldots,a_n,\ldots)$, the corresponding TM system is a Schauder basis of the closure of span$\{B_n\}_{n=1}^{\infty}$ in $H^p(\mathbf{D})$ ([33]).

The difference with the current study is that the parameters $a_1,\ldots,a_n,\ldots$ used to approximate a given function are not fixed before hand, nor are they required to satisfy the hyperbolic non-separability condition (3.3), and, correspondingly, the induced TM-system $\{B_n\}$ is not necessarily a basis. Instead, we adaptively select the parameters a_n, as in greedy algorithms, and thus formulate expansions of given signals with fast convergence. Below we will introduce our main algorithm in the unit disc case called *Adaptive Fourier Decomposition*, abbreviated as AFD ([34]).

Let f be any function in $H^2(\mathbf{D})$. Letting $f_1=f$, recursively and for any n complex numbers $a_1,\ldots,a_n$ in $\mathbf{D}$, we have

$$f(z)=\langle f_1,e_{a_1}\rangle e_{a_1}(z)+f_2(z)\frac{z-e_{a_1}}{1-\overline{a}_1z},$$

and

$$f_2(z)=\langle f_2,e_{a_2}\rangle e_{a_2}(z)+f_3(z)\frac{z-e_{a_2}}{1-\overline{a}_2z},$$

etc. so that we arrive at

$$f(z)=\sum_{k=1}^{n}\langle f_k,e_{a_k}\rangle B_{\{a_1,\ldots,a_k\}}(z)+f_{n+1}(z)\prod_{k=1}^{n}\frac{z-a_k}{1-\overline{a}_kz}. \tag{3.4}$$

This identity gives rise to an interpolating rational function to f for the interpolating points $a_1,\ldots,a_n$ where repeated points correspond to interpolation with derivatives of the function.

The identity, furthermore, gives rise to fast approximation in energy if one selects a_k, once $a_1,\ldots,a_{k-1}$ have been fixed, according to the formula:

$$a_k=\arg\max\{|\langle f_k,e_a\rangle|^2 \mid a\in\mathbf{D}\}. \tag{3.5}$$

The following energy relation is to be noted:

$$\|f\|^2=\sum_{k=1}^{n}|\langle f_k,e_{a_k}\rangle|^2+\|f_{n+1}\|^2.$$

Under the selection criterion (3.5) we can show that in the energy sense ([34]), namely in the sense of strong convergence in $H^2(\mathbf{D})$,

$$f = \sum_{k=1}^{\infty} \langle f_k, e_{a_k} \rangle B_k.$$

The above-described AFD, or *Core AFD* algorithm was published in 2012 and lately, in 2015, found to be equivalent with the Pre-Orthogonal Greedy Algorithm ([37]). The motivation of AFD is characterizing positive-instantaneous-frequency decomposition, or mono-component decomposition of signals. There followed two elaborations on AFD of which one is the called unwinding AFD ([27, 30, 17]) and the other is geablack towards n-best rational approximation ([35, 31]).

Unwinding AFD dwells on Core AFD and the principle of energy front-loading for Nevanlinna outer functions in one complex variable. The latter principle addresses the following fact: if $F(z) = \sum_{k=0}^{\infty} c_k z^k$ is a $H^2(\mathbf{D})$-function with the Nevanlinna factorization (see [20], [43]): $F(z) = I(z)O(z)$, where $I(z)$ and $O(z)$ are, respectively, its inner and outer factors, and if we write $O(z) = \sum_{k=0}^{\infty} d_k z^k$, then there holds for any positive integer N that

$$\sum_{k=0}^{N} |c_k|^2 \leq \sum_{k=0}^{N} |d_k|^2.$$

Since

$$\sum_{k=0}^{\infty} |c_k|^2 = \|F\|^2 = \|O\|^2 = \sum_{k=0}^{\infty} |d_k|^2,$$

the above inequality amounts to saying that the outer part of a Hardy function has best polynomial approximation of degree N in $H^2(\mathbf{D})$ that converges faster than that of the original function. This suggests that when decomposing a function in $H^2(\mathbf{D})$, it may be a good strategy to perform the Nevanlinna factorization and then decompose the outer part instead of the original function, to finally multiply back by the inner factor which is a finite Blaschke product, at least when the function is continuous on $\partial\mathbf{D}$. Of course, performing the Nevanlinna factorization is not such an easy business as it essentially involves estimating conjugate functions, and fair judgement should be used in each case.

The related theory is developed in [27, 30, 17]. Experiments show that Unwinding AFD is indeed among the best positive frequency decomposition methods ([32]). The authors became aware late 2015 that the Ph.D. thesis of M. Nahon ([27]) at Yale University, 2000, under the guidance of R. Coifman, develops an analogous unwinding algorithm based on the Nevanlinna factorization ([20, 43]). In a recent paper by Coifman and Steinerberger, theoretical aspects of the algorithm are further developed ([17]).

Cyclic AFD was designed to approach the problem of n-best rational approximation in $H^2(\mathbf{D})$. The problem is formulated as follows. Given $f \in H^2(\mathbf{D})$, find a rational function of the form p/q, with deg$\{p\}$ and deg$\{q\}$ not exceeding n, and q zero-free inside the unit disc, such that $\|f - p/q\|$ is minimum among all possible rational functions of the same kind. The latter are just the rational functions of degree no larger than n in $H^2(\mathbf{D})$. Existence of such minimizing rational functions was proved a long time ago, but a theoretical algorithm for finding p/q to give rise to the minimum is still an open issue. A detailed account of the problem may be found in [4, 11, 5, 12, 14]. Both the RARL2 algorithm (which extends to the matrix-valued case, see `www-sop.inria.fr/apics/RARL2/rarl2.html` for a description and tutorial as well as [7, 9, 19] for further references) and the one through Cyclic AFD ([31]) provide practical algorithms. RARL2 is a descent algorithm using Schur parameters to describe Blaschke products of given degree along with a compactification thereof to ensure convergence to a local optimum. It is used in identification and design of microwave devices, see [29, 39]. The algorithm using Cyclic AFD is parameterized by the zeros of the denominator polynomial, and uses the fact that the expansion as a sum of modified Blaschke products

$$\sum_{k=1}^{n} \langle f_k, e_{a_k} \rangle B_{\{a_1,\ldots,a_k\}}(z)$$

is a rational function of degree no larger than n by construction. The theory and algorithmic scheme of Cyclic AFD are definitely simpler (though the 1-D search over $a \in \mathbf{D}$ iterated at each step to reach a fixed point is nontrivial), but convergence to a local minimum is still an issue. Cyclic AFD corresponds to simultaneous optimal selection of n parameters, while Core AFD corresponds to sequential selection of n parameters ([31]). For some related studies we refer to [5, 8, 14, 13]. Other algorithms based on fixed point heuristics or balanced truncation of Hankel operators to approach rational $H^2(\mathbf{D})$-approximation can be found, *e.g.*, in [41, 22].

The above considerations and algorithms extend to the context of the half-plane rather than the disk, by means of conformal mapping. The reason is that a conformal map from the disk to the half-plane is a rational function of degree 1 (*i.e.*, a Möbius transform) and therefore it preserves rationality and the degree, see [20, Ch. I]. There is also a parallel approach by using TM systems and the corresponding maximal principle in the half-plane. More generally, the relevant extension of what precedes to more general domains is that of best *meromorphic approximation with n poles*, see [10].

For functions of multivalent variables, finding a basis is equivalent to finding a uniqueness set. That is a fundamental task and therefore of great interest in mathematical analysis. It is, however, in many cases difficult to achieve. On the other hand, Szegő kernels are usually simple rational functions, and fast representation of signals as linear combinations of Szegő kernels has great significance in relation to applications. The principles outlined in the last two sections are valid and the results available in a number of contexts for several real and complex variables, and with functions valued in vectors and matrices. We give a brief introduction to this circle of ideas in the following sections.

4. Quaternionic and Clifford contexts for functions of several real variables

Denote by $\mathbf{Q}$ the quaternion algebra, and by $\mathcal{A}_m$ the Clifford algebra of linear dimension 2^m. The analogues of AFD have been formulated for Hardy spaces in the setting of $\mathbf{Q}$ and of $\mathcal{A}_m$ in [36] [42], respectively. The first setting is very much in the spirit of AFD. The second, however, is more in the spirit of General Greedy Algorithm with dictionary consisting of the higher-order Szegő kernels, due to the fact that the inner product of Clifford algebra-valued functions is not necessarily scalar-valued. That prevents the G-S orthogonalization process from being applied.

Define $R_+^{m+1} = \{x = (x_0, \underline{x}) \in R^{m+1}; x_0 > 0, \underline{x} \in R^m\}$. We briefly introduce the AFD (General Greedy Algorithm) in the Hardy space $H^2(R_+^4)$ of the upper half-space (resp. $H^2(R_+^{m+1})$), which consists of $\mathbf{Q}$-valued (resp. $\mathcal{A}_m$-valued) left monogenic functions [21]. For the parallel theory in the Hardy spaces on the unit ball, see [36, 42]. Denote by $\phi_a(x)$ the non-normalized Szegő kernel of $H^2(R_+^4)$ (resp. $H^2(R_+^{m+1})$), *i.e.* $\phi_a(x) = \frac{\overline{x+\overline{a}}}{|x+\overline{a}|^4}$ (resp. $\phi_a(x) = \frac{\overline{x+\overline{a}}}{|x+\overline{a}|^{m+1}}$). Regarding $H^2(R_+^4)$, although $\mathbf{Q}$ is a non-commutative algebra, it is not difficult to apply the P-OGA to $\{\phi_a(x), a, x \in R_+^4\}$ to obtain an orthonormal system $\{\mathcal{B}_n\}_{n=1}^{\infty}$ parameterized by the selected sequence $\{a^{(n)}\}_{n=1}^{\infty}$ in R_+^4. Similar to AFD, one has $\lim_{n\to\infty} ||f - \sum_{k=1}^n \mathcal{B}_k \langle f, \mathcal{B}_k \rangle|| = 0$. We note that the study of AFD in the setting of $\mathbf{Q}$ predates that of P-OGA: the latter is a generalization of the former.

As to $H^2(R_+^{m+1})$, we introduce the completion of the Szegő kernel dictionary given by

$$\widetilde{\mathbf{D}} = \left\{ \frac{\psi_{\beta,a}(x)}{||\psi_{\beta,a}||}, \beta = (\beta_1, \ldots, \beta_{m+1}), a, x \in R_+^{m+1} \right\},$$

where

$$\psi_{\beta,a}(x) = \frac{\partial^{|\beta|}}{\partial a_0^{\beta_1} \cdots \partial a_{m+1}^{\beta_{m+1}}} \phi_a(x).$$

For each $f \in H^2(R_+^{m+1})$, applying the General Greedy Algorithm with $\widetilde{\mathbf{D}}$ one has

$$\lim_{n\to\infty} \left\| f - \sum_{k=1}^{n} \frac{\psi_{\beta^{(k)},a^{(k)}}}{||\psi_{\beta^{(k)},a^{(k)}}||} \left\langle g_k, \frac{\psi_{\beta^{(k)},a^{(k)}}}{||\psi_{\beta^{(k)},a^{(k)}}||} \right\rangle \right\| = 0,$$

where g_k is the standard remainder defined in §2.

The sphere cases are also considered in the quaternionic and Clifford algebra settings ([36, 42]).

In both settings it is shown that a global maximal selection of the parameter is attainable at each step of the recursive process (i.e., $\rho = 1$). In particular, one can obtain rational approximations of functions in $L^2(R^4)$ by applying the corresponding AFD and the well-known Sokhotskyi–Plemelj formula (e.g., [21, 36]).

5. Several complex variables

Similar approximation schemes in settings involving several complex variables have also been studied. In fact, one can consider functions defined on various classical domains, with values in $\mathbf{C}^{\mathbf{N}}$, or $\mathbf{C}^{N\times M}$ matrices, etc.

5.1. Functions defined on n-torus

Denote by T^n the n-torus, where $T = \partial\mathbf{D}$. There are two generalizations of AFD in the Hardy space of the n-torus $H^2(T^n)$ ([37]). One merely consists in processing P-OGA in this context (i.e., it is shown that P-OGA is applicable to $H^2(T^n)$). The other is based on the product-TM system. As to the former, we omit the details as it should be clear from the previous section already how to perform. Below, we give a brief introduction to the latter. For simplicity, we consider only the case $n = 2$.

Denote by $B_k^{\mathbf{a}}(z)$ the modified Blaschke product (a member of the TM system) associated with the sequence $\mathbf{a} = \{a_k\}_{k=1}^{\infty}$ in $\mathbf{D}$. We introduce the *tensor product type modified Blaschke product* $\{B_k^{\mathbf{a}}(z) \otimes B_l^{\mathbf{b}}(w)\}$, where $\mathbf{a}, \mathbf{b} \subset \mathbf{D}$ and $(z, w) \in \mathbf{D^2} = \mathbf{D} \times \mathbf{D}$. For $f \in H^2(T^2)$, we look for a rational approximation of f of separable type given by $f = \lim_{m\to\infty} \sum_{1\le k,l\le m} \langle f, B_k^{\mathbf{a}} \otimes B_l^{\mathbf{b}}\rangle B_k^{\mathbf{a}} \otimes B_l^{\mathbf{b}} = \lim_{m\to\infty} S_m(f)$ in the H^2-norm. Denote by $D_m(f) = S_m(f) - S_{m-1}(f)$ the *m-partial sum difference*. The main step is to select $(a_{m+1}^*, b_{m+1}^*) \in \mathbf{D^2}$ according to the maximal problem

$$(a_{m+1}^*, b_{m+1}^*) := \arg \sup_{(a_{m+1}, b_{m+1})\in\mathbf{D^2}} ||D_{m+1}(f)||^2, \tag{5.1}$$

where $\{a_1, \ldots, a_m\}$ and $\{b_1, \ldots, b_m\}$ are previously fixed.

The existence of (a_{m+1}^*, b_{m+1}^*) is proved in [37] through a technical discussion. In a way similar to previously described AFD, $S_m(f)$ converges to f in the H^2-norm if each (a_k, b_k) is selected according to criterion (5.1). As an application, one can obtain rational approximations of functions in $L^2(T^2)$.

5.2. Functions defined on R^n in the setting of Hardy spaces on tubes

As mentioned in the previous sections, one can obtain rational approximations of functions in L^2 of the boundary of a domain by applying AFD in the domain. Since R^n can be written as the union of 2^n octants, seeking rational approximations of functions in $L^2(R^n)$ motivates the study of AFD in Hardy spaces on tubes over octants ([24]). For the purpose of illustration, it suffices here to investigate AFD in the Hardy space of the tube over the first octant $H^2(T_{\Gamma_1})$, where $T_{\Gamma_1} = \mathbf{C}_+ \times \cdots \times \mathbf{C}_+$.

Denote by $S_z(w)$ the Cauchy–Szegő kernel of $H^2(T_{\Gamma_1})$, i.e.,

$$S_z(w) = \prod_{j=1}^{n} \frac{-1}{2\pi i(w_k - \overline{z}_k)}.$$

By using the methodology P-OGA, one can obtain an orthonormal system $\{\mathcal{B}_k\}_{k=1}^{\infty}$ parameterized by the sequence $\{z^{(k)}\}_{k=1}^{\infty} \subset T_{\Gamma_1}$. Indeed, $\{\mathcal{B}_k\}_{k=1}^{\infty}$ is the G-S orthogonalization of the selected Cauchy–Szegő kernels and, if necessary, their higher-order directional derivatives. It is concluded in [24] that the attainability of a global maximal selection at each step follows from a certain kind of boundary behavior of functions in $H^2(T_{\Gamma_1})$, called "boundary vanishing condition (BVC)" (also see [37]). After verifying the BVC in $H^2(T_{\Gamma_1})$, the convergence follows from the general theory given in [37]. In [24] the P- OGA generalization of AFD in the Hardy spaces on tubes *over regular cones* is also given.

5.3. Matrix-valued signals defined in the unit disc

We denote by $H_2^{p\times q}$ the space of $p \times q$ matrices with entries in $H^2(\mathbf{D})$. In a recent paper of D. Alpay, F. Colombo, T. Qian and I. Sabadini they show that it is possible, as in the scalar case, to decompose those functions as linear combinations of suitably modified matrix-valued Blaschke product, in an adaptive way. The procedure is based on a generalization to the matrix-valued case of the maximum selection principle of 1-D AFD, which involves not only selections of suitable points in the unit disc but also suitable orthogonal projections. It can be shown that the maximum selection principle again gives rise to a convergent algorithm ([1]). The analogous parametrization in terms of Schur analysis and tangential interpolation directions was given earlier in [6], and has been used to design a matrix-valued version of the RARL2 algorithm, see [19].

5.4. Adaptive decomposition: the case of the Drury–Arveson space

Blaschke factors and products have counterparts in the unit ball of C^N, and this fact allows us to extend the maximum selection principle to the case of functions in the Drury–Arveson space of functions analytic in the unit ball of $\mathbf{C}^n$. This gives rise to an algorithm which is a variation of the higher-dimensional AFD. In the corresponding paper of D. Alpay, F. Colombo, T. Qian and I. Sabadini they also introduce infinite Blaschke products in this setting and study their convergence ([2]).

5.5. Matrix-valued signals defined on the polydisc

The polydisc case has been given special attention, due to its connection with image processing. It develops in the context of multi-trigonometric series, like 2-D AFD treated in an earlier subsection. An alternative setting is given in the third paper of D. Alpay, F. Colombo, T. Qian and I. Sabadini where they develop interpolation theory as well as an operator-valued Blaschke product method that offers an adaptive expansion of holomorphic functions in the Hardy space over the polydisc corresponding to signals on the n-torus ([3]).

6. AFD and Aveiro method in reproducing kernel Hilbert spaces

We first note that the Hardy H^2 space is a reproducing kernel Hilbert space, where the reproducing kernel is given by the Szegő kernel. Subsequently, P-OGA was proposed as an expansion algorithm in reproducing kernel Hilbert spaces (see §2) although in the previous sections we restrict ourselves to AFD in various Hardy spaces. The key of AFD (or P-OGA) is the construction of an orthonormal system by applying the G-S orthogonalization process to the selected reproducing kernels and their higher-order derivatives. Such a construction ensures that the approximating function and its derivatives meet interpolation conditions to the approximated function at the selected points.

The study of interpolating functions is closely related to interpolation and sampling problems in reproducing kernel Hilbert spaces, some prototypical aspects of which may be found in [38]. We use the notation $\mathbb{H}_K$ to indicate that $\mathbb{H}$ is a reproducing kernel Hilbert space admitting a reproducing kernel $K(q, \overline{p})$. Suppose that $\mathbb{H}_K$ consists of holomorphic functions defined in an open set $E \subset \mathbf{C}$. Let further $\{p_k\}_{k=1}^{\infty} \subset E$ be a sequence of distinct points. The so-called Aveiro Method, proposed by S. Saitoh et al. in [16], aims at constructing an approximating function to $f \in \mathbb{H}_K$ involving a finite number of sampling points $\{p_1, \ldots, p_n\}$.

Based on this work, the authors of [23] propose the so-called "Aveiro Method under complete dictionary (AMUCD)" by combining the ideas of P-OGA with Aveiro Method. Roughly speaking, AMUCD enhances the power of Aveiro Method in that the approximating function given by AMUCD does not require all elements of $\{p_k\}_{k=1}^{\infty}$ to be distinct. As in AFD, the representation and its derivatives enjoy interpolation properties at $\{p_k\}_{k=1}^{\infty}$. It is shown in [23] that AMUCD is applicable to the classical Hardy spaces and Paley–Wiener spaces. It turns out that AMUCD is, in fact, an alternative representation of AFD. Nevertheless, AMUCD has the advantage not to require working out the related orthonormal system, whereas in many instances of P-OGA one does not know explicit formulas for the related orthonormal system of functions.

References

[1] D. Alpay, F. Colombo, T. Qian, I. Sabadini, *Adaptive orthonormal systems for matrix-valued functions*, preprint, arXiv:1511.07834.

[2] D. Alpay, F. Colombo, T. Qian, I. Sabadini, *Adaptive Decomposition: the Case of the Drury–Arveson Space*, preprint, arXiv:

[3] D. Alpay, F. Colombo, T. Qian, I. Sabadini, *Adaptive Decomposition: the Polydisc Case*, preprint.

[4] L. Baratchart, *Existence and Generic Properties of L^2 Approximants for Linear Systems*, I.M.A. Journal of Math. Control and Information, **3** (1986), 89–101.

[5] L. Baratchart, *A remark on uniqueness of best rational approximation of degree* 1 *in L2 of the circle,* Electronic Transactions on Numerical Analysis, **25** (2006), 54–66.

[6] L. Baratchart and D. Alpay and A. Gombani, *On the differential structure of matrix-valued inner functions*, Operator Theory: Advances and Applications, **73** (1994), 30–68.

[7] L. Baratchart, M. Cardelli, M. Olivi, *Identification and rational L2 approximation, a gradient algorithm*, Automatica, **27**(1991), 413–418.

[8] Laurent Baratchart and Sylvain Chevillard and Tao Qian, *Minimax principle and lower bounds in H2-rational approximation*, Journal of Approximation Theory, **206** (2016), 17–47.

[9] L. Baratchart, M. Olivi, *Critical points and error rank in matrix H2 rational approximation and the strong differential consistency of output error identification from white noise inputs*, Constructive Approximation, **14** (1998), 273–300.

[10] L. Baratchart and F. Mandrea and E.B. Saff and F. Wielonsky, *2-D inverse problems for the Laplacian: A meromorphic approximation approach*, J. de Mathématiques Pures et Appl., **86** (2006), 1–41.

[11] L. Baratchart and E. B. Saff and F. Wielonsky, *A criterion for uniqueness of a critical point in H^2 rational approximation*, J. Analyse Mathématique, **70** (1996), 225–266.

[12] L. Baratchart and H. Stahl and F. Wielonsky, *Asymptotic uniqueness of best rational aproximants of given degree to Markov functions in $\mathcal{L}^2$ of the circle*, Constructive Approximation, **17** (2001), 103–138.

[13] L. Baratchart, H. Stahl and M. Yattselev, *Weighted extremal domains and rational approximation*, Advances in Maths., **229** (2012), 357–407.

[14] L. Baratchart, M. Yattselev, *Asymptotic uniqueness of best rational approximations to complex Cauchy transforms in L2 of the circle*, Recent Treds in Orthogonal Polynomials and Approximation Theory, volume 507 of Contemporary Mathematics, pp. 87–111, Amer. Math. Soc., Providence, RI, 2010.

[15] A. Bultheel, P. Gonzalez-Vera, E. Hendriksen, O. Njastad, *Orthogonal Rational Functions*, Cambridge University Press, 1999.

[16] L.P. Castro, H. Fujiwara, M.M. Rodrigues, and S. Saitoh, *A new discretization method by means of reproducing kernels*, Science and Technology Publication House, 2012, pp. 185–223.

[17] R. Coifman, S. Steinerberger, *Nonlinear phase unwinding of functions*, preprinted on ArXiv.

[18] R.A. DeVore and V.N. Temlyakov, *Some remarks on Greedy Algorithms*, Adv. Comput. Math. **5** (1996), 173–187.

[19] P. Fulcheri and M. Olivi. Matrix rational H^2 approximation: a gradient algorithm based on Schur analysis, *SIAM J. Control Optim.*, **36** (1998), 2103–2127.

[20] J.B. Garnett, Bounded Analytic Functions, Adademic Press, 1987.

[21] J.E. Gilbert and M.A. Murray, *Clifford algebras and Dirac operators in harmonic analysis*, Cambridge Studies in Advanced Mathematics 26, Cambridge University Press, 1991.

[22] S. Gugercin and A. Antoulas and C. Beattie, *H2 model reduction for large-scale linear dynamical systems*, SIAM J. MATRIX ANAL. APPL., **30** (2008), 609–638.

[23] W.-X. Mai and T. Qian, *Aveiro method in reproducing kernel hilbert spaces under complete dictionary*, submitted for publication, arXiv:1604.07592 [math.CV].

[24] W.-X. Mai and T. Qian, *Rational approximation of functions in the Hardy spaces on tubes*, submitted for publication, arXiv:1604.07597 [math.CV].

[25] Y. Mo and T. Qian, *Support vector machine adapted Tikhonov regularization method to solve the Dirichlet problem*, Applied Mathematics and Computation, **245** (2014), 509–519.

[26] Y. Mo, T. Qian and W. Mi, *Sparse Representation in Szegő Kernels through Reproducing Kernel Hilbert Space Theory with Applications*, accepted by International Journal of Wavelet, Multiresolution and Information Processing.

[27] M. Nahon, *Phase Evaluation and Segmentation*, Ph.D. Thesis, Yale University, 2000.

[28] N. Nikolskii, *Treatise on the shift operator*, Springer, 1986.

[29] M. Olivi and F. Seyfert and J.P. Marmorat, *Identification of microwave filters by analytic and rational H^2 approximation*, Automatica, **49** (2013), 317–325.

[30] T. Qian, *Intrinsic mono-component decomposition of functions: An advance of Fourier theory*, Mathematical Methods in Applied Sciences, **33** (2010), 880–891, DOI: 10.1002/mma.1214.

[31] T. Qian, *Cyclic AFD Algorithm for Best Approximation by Rational Functions of Given Order*, Mathematical Methods in the Applied Sciences, **37** (2014), 846–859, DOI: 10.1002/mma.2843.

[32] T. Qian, H. Li, M. Stessin, *Comparison of Adaptive Mono-component Decompositions*, Nonlinear Analysis: Real World Applications, **14** (2013), 1055–1074.

[33] T. Qian, Q.-H. Chen, L.-H. Tan, *Rational Orthogonal Systems are Schauder Bases*, Complex Variables and Elliptic Equations, **59** (2014), 841–846.

[34] T. Qian, Y.-B. Wang, *Adaptive Fourier series – a variation of greedy algorithm*, Advances in Computation Mathematics, **34** (2011), 279–293.

[35] T. Qian and E. Wegert, *Optimal Approximation by Blaschke Forms, Complex Variables and Elliptic Equations*, **58** (2013), 123–133.

[36] T. Qian, W. Sproessig, and J.-X. Wang, *Adaptive Fourier decomposition of functions in quaternionic hardy spaces*, Mathematical Methods in Applied Sciences, **35** (2012), 43–64.

[37] T. Qian, W. Sproessig, *Two-dimensional adaptive Fourier decomposition*, arXiv: 1406.2008 [math. CV].

[38] K. Seip,*Interpolation and sampling in spaces of analytic functions*. University Lecture Series, vol. 33, American Mathematical Society, Providence, RI, 2004.

[39] F. Seyfert and M. Oldoni and M. Olivi and S. Lefteriu and D. Pacaud, *Deembedding of filters in multiplexers via rational approximation and interpolation*, Internat. Jour. of RF and Microwave Computer-Aided Engineering, Wiley, 2015.

[40] V.N. Temlyakov, *Greedy algorithm and m-Term trigonometric approximation*, Constr. Approx., **14** (1998), 569–587.

[41] P. Van Dooren and K.A. Gallivan and P.-A. Absil, *H2 optimal model reduction of MIMO systems*, Applied Mathematics Letters, **21** (2008), 1267–1273.

[42] J. X. Wang and T. Qian, *Approximation of monogenic functions by higher order Szegő kernels on the unit ball and the upper half-space*, Science China: Mathematics, **57** (2014), 1785–1797.

[43] G. Weiss and M. Weiss, *A derivation of the main results of the theory of Hp-spaces*, Rev. Un. Mat. Argentina, **20** (1962), 63–71.

Laurent Baratchart
INRIA, 2004 route des Lucioles, BP 93
F-06902 Sophia-Antipolis Cedex, France
e-mail: `Laurent.Baratchart@inria.fr`

Wei-Xiong Mai and Tao Qian
Department of Mathematics, Faculty of Science and Technology
University of Macau, Macau, China
e-mail: `maiweixiong@gmail.com`, `fsttq@umac.mo`

Modern Trends in Hypercomplex Analysis
Trends in Mathematics, 35–46

A. Kolmogorov and M. Riesz Theorems for Octonion-valued Monogenic Functions

Sultan Catto, Alexander Kheyfits and David Tepper

Abstract. The classical M. Riesz theorem on the boundedness of the conjugation operator for harmonic functions and Kolmogorov's weak-type inequality are proved in the framework of the octonion-valued monogenic functions.

Mathematics Subject Classification (2010). Primary 30G35, 31B05; Secondary 35E99 .

Keywords. Octonion-valued monogenic functions, monogenic conjugates, subharmonicity of powers of monogenic functions.

1. Introduction and statement of results

Monogenic functions of octonion variables, due to their potential, yet not completely fulfilled utility in physics [9], continue to attract attention of researchers. J. Baez in his very informative survey formulates the development of an octonionic analogue of the theory of analytic functions as the first item in his list of 14 important open octonion-related problems [3, p. 201, first bullet]. In this note, we continue the study of this topic; see, e.g., [4, 10, 11, 14, 15, 16] and especially [17] and the references therein.

An octonionic version of M. Riesz theorem, see **MR** below, about conjugate harmonic functions has been recently published [5]; here we give octonionic versions of the classical Kolmogorov's weak-type inequality and shorten the proof of Riesz' theorem [5].

The monogenic functions are analogs of analytic functions in the octonionic framework. Just like the classical theorems of M. Riesz and Kolmogorov, the generalizations of these results are useful in the study of boundary value problems for monogenic functions.

The theorem of Kolmogorov **K2** is valid for all p, $0 < p < 1$. However, due to the fact that powers $|f|^p$ of octonionic monogenic functions are subharmonic only for $6/7 \leq p$ [10], our Theorem 2 below, which is an octonionic analog of the theorem **K2**, is also valid only for $6/7 \leq p$.

First we remind a few standard definitions and classical results, and state our results. The proofs are given in Section 2.

Given a harmonic function $u(x, y)$ in a simply-connected domain of the complex plane, a harmonic function $v(x, y)$ is called its *conjugate harmonic function*, if they satisfy the *Cauchy–Riemann* system of partial differential equations

$$\frac{\partial u}{\partial x} = \frac{\partial v}{\partial y}; \qquad \frac{\partial v}{\partial x} = -\frac{\partial u}{\partial y}.$$

It is clear that a conjugate harmonic function is determined only up to an additive constant; we norm it as $v(0) = 0$.

Theorem MR (M. Riesz, 1927). *Let u be a harmonic function in the unit disc. If*

$$u \in L^p[0, 2\pi]$$

for $1 < p < \infty$, then its conjugate harmonic function $v = \widetilde{u} \in L^p[0, 2\pi]$ as well, and

$$\|v\|_{L^p} \leq A_p \|f\|_{L^p}.$$

The exact value of the constant A_p was found by Pichorides [19], it is $A_p = \tan \frac{\pi}{2p}$ if $1 < p \leq 2$, and its reciprocal $A_p = \cot \frac{\pi}{2p}$ if $2 \leq p < \infty$. We remark that $A_p \approx \frac{2}{\pi(p-1)}$ as $p \to 1$ and $A_p \approx \frac{2}{\pi} p$ as $p \to \infty$.

If $p = 1$, the statement fails, that is, the conjugation operator is not bounded in L^1. However, in this case the operator has the weak-type 1.

Theorem K1 (Kolmogorov, 1925). *Let u be a harmonic function in the unit disk and $v = \widetilde{u}$ its conjugate harmonic function. For any positive $\lambda > 0$, the inequality is valid,*

$$\operatorname{meas}\{x : |v(x)| > \lambda\} \leq A \frac{\|u\|_{L^1}}{\lambda},$$

where meas *stands for the Lebesgue measure and A is an absolute constant.*

Moreover, the following inequality holds good too.

Theorem K2 (Kolmogorov, 1925). *If a harmonic function*

$$u \in L^1[0, 2\pi],$$

then its conjugate harmonic function v, normalized by $v(0) = 0$, is integrable as well, $v = \widetilde{u} \in L^p[0, 2\pi]$ for any $0 < p < 1$, and

$$\|v\|_{L^1} \leq B_p \|u\|_{L^p}$$

where the constant B_p depends on p only.

It should be mentioned that a quaternionic version of the M. Riesz theorem is known, see Avetisyan [2] and the references therein; namely the quaternionic monogenic function is integrable, whenever its vector component is integrable. Conjugate harmonic functions in Clifford algebras were studied by Nolder [18]; our Proposition 1 below is an analog of Lemma 1.6 in [18]. However, the algebra of octonions is not a Clifford algebra due to the non-associativity of the octonions.

The *non-commutative, non-associative, alternative division algebra of octonions* $\mathcal{O}$ is an 8-dimensional vector space with the *basis* elements $\{\mathbf{e}_0 \equiv 1, \mathbf{e}_1, \ldots, \mathbf{e}_7\}$, satisfying the multiplication table

$\mathbf{e}_i \times \mathbf{e}_j$	$\mathbf{e}_1$	$\mathbf{e}_2$	$\mathbf{e}_3$	$\mathbf{e}_4$	$\mathbf{e}_5$	$\mathbf{e}_6$	$\mathbf{e}_7$
$\mathbf{e}_1$	-1	$\mathbf{e}_4$	$\mathbf{e}_7$	$-\mathbf{e}_2$	$\mathbf{e}_6$	$-\mathbf{e}_5$	$-\mathbf{e}_3$
$\mathbf{e}_2$	$-\mathbf{e}_4$	-1	$\mathbf{e}_5$	$\mathbf{e}_1$	$-\mathbf{e}_3$	$\mathbf{e}_7$	$-\mathbf{e}_6$
$\mathbf{e}_3$	$-\mathbf{e}_7$	$-\mathbf{e}_5$	-1	$\mathbf{e}_6$	$\mathbf{e}_2$	$-\mathbf{e}_4$	$\mathbf{e}_1$
$\mathbf{e}_4$	$\mathbf{e}_2$	$-\mathbf{e}_1$	$-\mathbf{e}_6$	-1	$\mathbf{e}_7$	$\mathbf{e}_3$	$-\mathbf{e}_5$
$\mathbf{e}_5$	$-\mathbf{e}_6$	$\mathbf{e}_3$	$-\mathbf{e}_2$	$-\mathbf{e}_7$	-1	$\mathbf{e}_1$	$\mathbf{e}_4$
$\mathbf{e}_6$	$\mathbf{e}_5$	$-\mathbf{e}_7$	$\mathbf{e}_4$	$-\mathbf{e}_3$	$-\mathbf{e}_1$	-1	$\mathbf{e}_2$
$\mathbf{e}_7$	$\mathbf{e}_3$	$\mathbf{e}_6$	$-\mathbf{e}_1$	$\mathbf{e}_5$	$-\mathbf{e}_4$	$-\mathbf{e}_2$	-1

.

The non-commutativity and non-associativity of the algebra $\mathcal{O}$ result in certain difficulties in a study of this algebra. However, as early as in 1933, P. Stein [21] employed the subharmonic functions in his study of M. Riesz' theorem. Since then, this tool has been used by various authors, see [7], [20] and the references therein and above. If the final claim depends only on the modulus $|f|$ of the left- or right-monogenic function f, the use of subharmonicity allows in many problems to fix certain ordering and/or association from the outset and work with this order to the end of the proof, when it can be seen that the result does not depend upon a particular ordering and association.

We use the following notation. Let $\mathbf{x} = (x_0, x_1, \ldots, x_7) \in \mathbb{R}^8$ be a vector of real numbers and

$$\mathbf{oct} = \sum_{j=0}^{7} x_j \mathbf{e}_j$$

be a generic octonion $\mathbf{oct} \in \mathcal{O}$. Consider real-valued continuously differentiable functions

$$f_0(\mathbf{x}), \ldots, f_7(\mathbf{x}), \ \mathbf{x} \in \Omega,$$

in a simply-connected domain $\Omega \subset \mathbb{R}^8$. The octonion-valued left-monogenic functions in Ω are defined as 8-dimensional vector-functions

$$f(\mathbf{x}) = \sum_{j=0}^{7} f_j(\mathbf{x})\mathbf{e}_j, \mathbf{x} \in \Omega, \tag{1.1}$$

satisfying the operator equation

$$D[f] = \mathbf{0}, \tag{1.2}$$

where $D = \sum_{j=0}^{7} \frac{\partial}{\partial x_j}\mathbf{e}_j$ is the Dirac or Cauchy–Riemann–Fueter operator; $\partial/\partial x_j$, $j = 0, 1, \ldots, 7$, are partial derivatives with respect to the coordinates in $\mathbb{R}^8$. Thus, we study functional-theoretical properties of the elements of the kernel of the Dirac operator D.

Solutions of the system $[f]D = \mathbf{0}$ are called right-monogenic functions; the functions, which are both left- and right-monogenic, are called (two-sided) monogenic functions. Hereafter, we always discuss the left-monogenic functions; the proofs go word-by-word for the right- and two-sided monogenic functions. Moreover, since adding a constant to any component f_j of any solution of system (1.2)–(1.3) does not violate the system, we will assume that $f(\mathbf{0}) = \mathbf{0}$.

Combining representations (1.1) and (1.2) and using the linear independence of the basis octonions $\mathbf{e}_0, \ldots, \mathbf{e}_7$, it follows that equation (1.2) is equivalent to a system of eight first-order linear partial differential equations with constant coefficients with respect to the unknown functions $f_0, f_1, \ldots, f_7$. This system can be written down as the matrix equation

$$\begin{bmatrix} \frac{\partial}{\partial x_0} & -\frac{\partial}{\partial x_1} & -\frac{\partial}{\partial x_2} & -\frac{\partial}{\partial x_3} & -\frac{\partial}{\partial x_4} & -\frac{\partial}{\partial x_5} & -\frac{\partial}{\partial x_6} & -\frac{\partial}{\partial x_7} \\ \frac{\partial}{\partial x_1} & \frac{\partial}{\partial x_0} & -\frac{\partial}{\partial x_4} & -\frac{\partial}{\partial x_7} & \frac{\partial}{\partial x_2} & -\frac{\partial}{\partial x_6} & \frac{\partial}{\partial x_5} & \frac{\partial}{\partial x_3} \\ \frac{\partial}{\partial x_2} & \frac{\partial}{\partial x_4} & \frac{\partial}{\partial x_0} & -\frac{\partial}{\partial x_5} & -\frac{\partial}{\partial x_1} & \frac{\partial}{\partial x_3} & -\frac{\partial}{\partial x_7} & \frac{\partial}{\partial x_6} \\ \frac{\partial}{\partial x_3} & \frac{\partial}{\partial x_7} & \frac{\partial}{\partial x_5} & \frac{\partial}{\partial x_0} & -\frac{\partial}{\partial x_6} & -\frac{\partial}{\partial x_2} & \frac{\partial}{\partial x_4} & -\frac{\partial}{\partial x_1} \\ \frac{\partial}{\partial x_4} & -\frac{\partial}{\partial x_2} & \frac{\partial}{\partial x_1} & \frac{\partial}{\partial x_6} & \frac{\partial}{\partial x_0} & -\frac{\partial}{\partial x_7} & -\frac{\partial}{\partial x_3} & \frac{\partial}{\partial x_5} \\ \frac{\partial}{\partial x_5} & \frac{\partial}{\partial x_6} & -\frac{\partial}{\partial x_3} & \frac{\partial}{\partial x_2} & \frac{\partial}{\partial x_7} & \frac{\partial}{\partial x_0} & -\frac{\partial}{\partial x_1} & -\frac{\partial}{\partial x_4} \\ \frac{\partial}{\partial x_6} & -\frac{\partial}{\partial x_5} & \frac{\partial}{\partial x_7} & -\frac{\partial}{\partial x_4} & \frac{\partial}{\partial x_3} & \frac{\partial}{\partial x_1} & \frac{\partial}{\partial x_0} & -\frac{\partial}{\partial x_2} \\ \frac{\partial}{\partial x_7} & -\frac{\partial}{\partial x_3} & -\frac{\partial}{\partial x_6} & \frac{\partial}{\partial x_1} & -\frac{\partial}{\partial x_5} & \frac{\partial}{\partial x_4} & \frac{\partial}{\partial x_2} & \frac{\partial}{\partial x_0} \end{bmatrix} \cdot \begin{bmatrix} f_0 \\ f_1 \\ f_2 \\ f_3 \\ f_4 \\ f_5 \\ f_6 \\ f_7 \end{bmatrix} = \mathbf{0} \tag{1.3}$$

with respect to the scalar real-valued functions $f_0, \ldots, f_7$.

Differentiating the equations of system (1.2)–(1.3) and adding them, one immediately derives the equations

$$\Delta f_0(\mathbf{x}) = \cdots = \Delta f_7(\mathbf{x}) = 0, \ \mathbf{x} \in \Omega,$$

where Δ is the 8-dimensional Laplace operator. Hence, all the components $f_0, \ldots, f_7$ of an octonion-valued monogenic function f are the classical harmonic functions.

Equations (1.2) such that each component f_j, $j \geq 0$, is harmonic, are called the Generalized Cauchy–Riemann systems (GCR) – see Stein and Weiss [20, pp. 231–234]. More general systems

$$\sum_{j=0}^{n} A_j \frac{\partial f}{\partial x_j} + Bf = \mathbf{0}$$

with constant matrices A_j and B were considered, in different context, by Evgrafov [8].

Stein and Weiss have proved that for any GCR system there exists a nonnegative index $p_0 < 1$ such that $|F|^p$ is a subharmonic function for all $p \geq p_0$. It is known (ibid, p. 234) that for the M. Riesz system in $\mathbb{R}^n$,

$$\begin{cases} \frac{\partial f_1}{\partial x_1} + \cdots + \frac{\partial f_n}{\partial x_n} = 0 \\ \frac{\partial f_i}{\partial x_j} = \frac{\partial f_j}{\partial x_i},\ i, j = 1, \ldots, n, \end{cases}$$

the exact value of p_0 is $(n-2)/(n-1)$. Our system (1.2)–(1.3) is not the M. Riesz system, however, the same assertion is valid for system (1.2)–(1.3) in $\mathbf{R}^8$; namely, it has been proven in [10] that for the octonion-valued monogenic functions, that is, for the solutions of system (1.2)–(1.3),

$$p_0 = \left.\frac{n-2}{n-1}\right|_{n=8} = 6/7.$$

As Stein and Weiss have noticed (ibid., p. 233) the inequality $p_0 < 1$ allows one to develop a substantive theory of the Hardy spaces for the corresponding systems (1.2), in our case for octonionic monogenic functions. Certain other properties of the octonion-valued monogenic functions, for instance, the Phragmén–Lindelöf principle, the three-lines theorem, Paley–Wiener theorem, and some others, have been also proven in [4, 14, 15, 16, 17].

For a monogenic function f, the function f_0 is called the *scalar* component of f, the vector-function $f_v = (f_1, \ldots, f_7)$ the *vector* component. It is worth mentioning that even in the case of quaternionic, that is, four-component monogenic functions [2, p. 911, Remark 1.2], in a bound f_0 through f_v from above, the right-hand side must contain all the three components of f_v. That is why in the results below f_0 is estimated from above by f_v.

The gradient of scalar functions $f_0, \ldots, f_7$ is

$$\nabla f_j(\mathbf{x}) = \left(\frac{\partial f_j(\mathbf{x})}{\partial x_0}, \frac{\partial f_j(\mathbf{x})}{\partial x_1}, \ldots, \frac{\partial f_j(\mathbf{x})}{\partial x_7} \right),$$

thus

$$\left|\nabla f_j(\mathbf{x})\right|^2 = \left|\frac{\partial f_j(\mathbf{x})}{\partial x_0}\right|^2 + \left|\frac{\partial f_j(x)}{\partial x_1}\right|^2 + \cdots + \left|\frac{\partial f_j(x)}{\partial x_7}\right|^2.$$

We introduce also the 64-tuple function

$$\nabla f(\mathbf{x}) = (\nabla f_0(\mathbf{x}), \nabla f_1(\mathbf{x}), \ldots, \nabla f_7(\mathbf{x})),$$

called the *second gradient* of f [20, Chap. VI, Sect. 5.9], and the 56-tuple function ∇f_v,

$$\nabla f_v(\mathbf{x}) = (\nabla f_1(\mathbf{x}), \nabla f_2(\mathbf{x}), \ldots, \nabla f_7(\mathbf{x})), \tag{1.4}$$

hence

$$\left|\nabla f\right|^2 = \left|\nabla f_0\right|^2 + \left|\nabla f_v\right|^2. \tag{1.5}$$

To prove the boundedness of the conjugation operator for the octonion-valued monogenic functions in L^p, $1 < p < \infty$, firstly we notice that due to the positivity of the subharmonic function $|f(\mathbf{x})|^p$ when $p \geq 6/7$, if f is integrable over the boundary of the domain, then f and all its harmonic components f_i, $i = 0, 1, \dots, 7$, have non-tangential boundary values almost everywhere on the boundary. When $\mathbf{x}$ is at the boundary of the domain, by $f(\mathbf{x})$ we mean these non-tangential boundary values. We let B denote the unit ball in $\mathbb{R}^8$, $S = \partial B$, and $d\sigma$ the surface area measure on S. Now we state our results. Theorem 1.1 is an analog of M. Riesz theorem in the octonionic framework.

Theorem 1.1. *Let f be monogenic in B and $0 \leq r < 1$. Then for any finite $p > 1$ there exists the constant C_p depending on p only such that*

$$\int_S |f(r,\theta)|^p d\sigma(\theta) \leq C \int_S |f_v(r,\theta)|^p d\sigma(\theta)$$

for any constant $C \geq C_p$.

If $1 < p \leq 2$, then we can take

$$C_p = \frac{7}{p-1},$$

and for $2 \leq p < \infty$,

$$C_p = 7(p-1),$$

hence the constants have the same asymptotic behavior when $p \to 1$ or $p \to \infty$ as in Pichorides' result.

Our proofs are based on the following inequality, which may be of independent interest.

Proposition 1.2. *If f is an octonionic monogenic function, then*

$$\left|\nabla f_0\right|^2 \leq 7 \left|\nabla f_v\right|^2, \tag{1.6}$$

therefore

$$\left|\nabla f\right|^2 \leq 8 \left|\nabla f_v\right|^2, \tag{1.7}$$

where ∇f_0 is the gradient of the scalar function f_0 and the second gradients ∇f and ∇f_v are defined above, see (1.4) and the paragraph before it.

The next corollary follows immediately.

Corollary 1.3. *Under the conditions of Theorem 1.1, for $1 < p < \infty$,*

$$\int_S |f_0(r,\theta)|^p d\sigma(\theta) \leq \left(C_p^{1/p} + 1\right)^p \int_S |f_v(r,\theta)|^p d\sigma(\theta).$$

In the case $0 < p \leq 1$, when the Riesz theorem is known to fail, some relevant results were established by Kolmogorov and Zygmund. The following statements are similar to Kolmogorov's Theorem **K2** if $p < 1$ and his weak-type estimate, Theorem **K1** if $p = 1$.

Proposition 1.4. *If f is octonion-monogenic in the unit ball B and integrable on the unit sphere, $f \in L(S)$, then for any p, $6/7 \leq p \leq 1$, and $0 < r < 1$,*

$$\left(\int_S |f(r,\theta)|^p d\sigma(\theta) \right)^{1/p} \leq (\sigma_8)^{\frac{1-p}{p}} \int_S |f(\theta)| d\sigma(\theta),$$

where σ_8 is the surface area of the unit sphere S in $\mathbb{R}^8$.

As a corollary, we straightforwardly derive the weak-type inequality for the octonion-valued monogenic functions, which can be thought of as a (weak) substitution for the weak-type inequality in the case of monogenic functions.

Theorem 1.5. *Assuming the conditions of Proposition 1.4 and setting $p = 6/7$ in it, we have*

$$meas\left\{ \mathbf{x} \in B \middle| \, |f_0(\mathbf{x})| > \lambda \right\} \leq \frac{7}{8} \frac{||f_v||^2_{L^2(S)}}{\lambda^2}$$

where meas stands for the Lebesgue measure in $\mathbb{R}^8$.

Remark 1.6. We prove the results for the functions in a ball, however, similar assertions are valid for functions in the half-spaces as well – cf. for example, [13, Lecture 19].

Remark 1.7. It is worth mentioning that Theorem 1.5 claims a stronger rate of decay of the measure as $\lambda \to \infty$, namely λ^{-2} rather than the classical rate λ^{-1}, however, the bound involves the L^2 norm rather than L^1 norm.

Remark 1.8. As a corollary of Theorem 1.1, it is possible to derive a proof of the Paley–Wiener theorem for the monogenic functions, Cf. [17].

2. Proofs

As usual, derivatives of the vector-functions $f(\mathbf{x})$ and $f_v(\mathbf{x})$ are computed component-wise, that is,

$$\frac{\partial f(\mathbf{x})}{\partial x_k} = \sum_{j=0}^{7} \frac{\partial f_j(\mathbf{x})}{\partial x_k} \mathbf{e}_j \quad \text{for } k = 0, 1, \ldots, 7.$$

The following proofs essentially use properties of the superharmonic functions, see, for example, [1] or [13]. We mention here only that for smooth functions $u(\mathbf{x})$, $\mathbf{x} \in \Omega$, which is the case in our work, the superharmonicity is equivalent to the inequality $\Delta u(\mathbf{x}) \leq 0$ for all $\mathbf{x}$ in the domain Ω. In turn, the latter is equivalent to the mean-value inequality, namely, the average of a superharmonic function over any sphere (or ball) within its domain does not exceed the value of the function at the center of the sphere (or ball).

This "superharmonic" approach was developed by M. Essen [7].

Proof of Proposition 1.2. The first equation of system (1.2)–(1.3) implies that

$$\frac{\partial f_0(\mathbf{x})}{\partial x_0}=\frac{\partial f_1(\mathbf{x})}{\partial x_1}+\cdots+\frac{\partial f_7(\mathbf{x})}{\partial x_7},$$

the sequel equations of (1.3) give similar expressions for the next partial derivatives $\frac{\partial f_0(\mathbf{x})}{\partial x_j}$, $j=1,2,\ldots,7$. Squaring each of these equations and using the elementary inequality

$$(a_1+a_2+\cdots+a_7)^2\le 7\left(a_1^2+\cdots+a_7^2\right), \tag{2.1}$$

where a_j are real numbers, we derive the estimates

$$\left(\frac{\partial f_0(\mathbf{x})}{\partial x_0}\right)^2\le 7\left(\left(\frac{\partial f_1(\mathbf{x})}{\partial x_1}\right)^2+\left(\frac{\partial f_2(\mathbf{x})}{\partial x_2}\right)^2+\cdots+\left(\frac{\partial f_7(\mathbf{x})}{\partial x_7}\right)^2\right),$$

$$\left(\frac{\partial f_0(\mathbf{x})}{\partial x_1}\right)^2\le 7\left(\left(\frac{\partial f_1(\mathbf{x})}{\partial x_0}\right)^2+\left(\frac{\partial f_2(\mathbf{x})}{\partial x_4}\right)^2+\cdots+\left(\frac{\partial f_7(\mathbf{x})}{\partial x_3}\right)^2\right),$$

$$\left(\frac{\partial f_0(\mathbf{x})}{\partial x_2}\right)^2\le 7\left(\left(\frac{\partial f_1(\mathbf{x})}{\partial x_4}\right)^2+\left(\frac{\partial f_2(\mathbf{x})}{\partial x_0}\right)^2+\cdots+\left(\frac{\partial f_7(\mathbf{x})}{\partial x_6}\right)^2\right),$$

$$\cdots\quad\cdots\quad\cdots$$

$$\left(\frac{\partial f_0(\mathbf{x})}{\partial x_7}\right)^2\le 7\left(\left(\frac{\partial f_1(\mathbf{x})}{\partial x_3}\right)^2+\left(\frac{\partial f_2(\mathbf{x})}{\partial x_6}\right)^2+\cdots+\left(\frac{\partial f_7(\mathbf{x})}{\partial x_0}\right)^2\right).$$

Keeping in mind the matrix in (1.3), we notice that every first-order partial derivative of each component f_j occurs in these inequalities exactly once. Therefore, adding up these eight inequalities for $(\partial f_0/\partial x_0)^2$, …, $(\partial f_0/\partial x_7)^2$, we get inequality (1.6) of Proposition 1.2, namely,

$$|\nabla f_0(\mathbf{x})|^2\le 7\left(|\nabla f_1(\mathbf{x})|^2+\cdots+|\nabla f_7(\mathbf{x})|^2\right)=7|\nabla f_v(\mathbf{x})|^2.$$

Combining the latter with (1.5), we deduce inequality (1.7), thus completing the proof of the proposition. □

Remark 2.1. A simple example of the 7-vector $(1,1,1,1,1,1,1)$ shows that the factor 7 in (2.1) cannot be decreased.

The next lemma can be traced back to Kuran [12] (subharmonic version) and Essen [7].

Lemma 2.2. *If f is a monogenic function and $1<p\le 2$, then the function*

$$g(\mathbf{x})=|f(\mathbf{x})|^p-C|f_v(\mathbf{x})|^p$$

is superharmonic in B, whenever a constant $C\ge C_p=\frac{7}{p-1}$.

Proof. Since the components $f_0, f_1, \dots, f_7$ of the monogenic function f are harmonic functions, it suffices to prove that the Laplacian Δg is nonpositive, $\Delta g(\mathbf{x}) \leq 0$. The proof is based on the following well-known equation, which can be deduced by a direct computation,

$$\Delta|f(\mathbf{x})|^p = p|f(\mathbf{x})|^{p-4}\left\{(p-2)\sum_{j=0}^{7}\left(f(\mathbf{x})\cdot\frac{\partial f(\mathbf{x})}{\partial x_j}\right)^2 + |f(\mathbf{x})|^2\sum_{j=0}^{7}\left|\frac{\partial f(\mathbf{x})}{\partial x_j}\right|^2\right\},$$

where $f(\mathbf{x})\cdot\frac{\partial f(\mathbf{x})}{\partial x_j} = \sum_{k=0}^{7} f_k(\mathbf{x})\cdot\frac{\partial f_k(\mathbf{x})}{\partial x_j}$ – see [20, Chap. VI, Proof of Theor. 4.9]. Since $p \leq 2$, the latter implies the inequality

$$\Delta|f(\mathbf{x})|^p \leq p|f(\mathbf{x})|^{p-2}\sum_{j=0}^{7}\left|\frac{\partial f(\mathbf{x})}{\partial x_j}\right|^2. \tag{2.2}$$

Similarly,

$$\Delta|f_v(\mathbf{x})|^p = p|f_v(\mathbf{x})|^{p-4}\left\{(p-2)\sum_{j=0}^{7}\left(f_v(\mathbf{x})\cdot\frac{\partial f_v(\mathbf{x})}{\partial x_j}\right)^2 + |f_v(\mathbf{x})|^2\sum_{j=0}^{7}\left|\frac{\partial f_v(\mathbf{x})}{\partial x_j}\right|^2\right\}.$$

Estimating the scalar product by the Schwarz inequality, we compute

$$\sum_{j=0}^{7}\left(f_v\cdot\frac{\partial f_v}{\partial x_j}\right)^2 \leq \sum_{j=0}^{7}|f_v|^2\left|\frac{\partial f_v}{\partial x_j}\right|^2 = |f_v|^2|\nabla f_v|^2.$$

Obviously $|f_v| \leq |f|$, hence

$$|f|^{p-2} \leq |f_v|^{p-2},$$

for we consider here $p \leq 2$. Since

$$|\nabla f_v|^2 = \sum_{j=0}^{7}\left|\frac{\partial f_v}{\partial x_j}\right|^2,$$

we derive

$$\begin{aligned}\Delta|f_v|^p &\geq p|f_v|^{p-4}\left\{(p-2)|f_v|^2|\nabla f_v|^2 + |f_v|^2|\nabla f_v|^2\right\}\\ &= p(p-1)|f_v|^{p-2}|\nabla f_v|^2.\end{aligned} \tag{2.3}$$

Inserting (2.2) and (2.3) into the equation for $\Delta g = \Delta|f|^p - C\Delta|f_v|^p$, we complete the proof. □

Proof of Theorem 1.1. As in [7] or [13], we first consider the case $1 < p \leq 2$. Let

$$f(\mathbf{x}) = \sum_{j=0}^{7} f_j(\mathbf{x})\mathbf{e}_j$$

be a monogenic octonion-valued function, f_0 its scalar component and

$$f_v = \sum_{j=1}^{7} f_j\mathbf{e}_j \equiv (f_1, \dots, f_7)$$

its vector component. As was mentioned above, without loss of generality we assume that $f(\mathbf{0}) = 0$.

By Lemma 1.1, the function $|f|^p - C|f_v|^p$ with $C \geq 7/(p-1)$ is superharmonic, thus its spherical mean does not exceed its value at the center of the sphere. Due to the assumption $f(\mathbf{0}) = 0$, we arrive at the inequality

$$\int_S |f(x)|^p d\sigma \leq C \int_S |f_v(x)|^p d\sigma, \tag{2.4}$$

we sought for. The non-commutativity and non-associativity of the octonions are bypassed here, for we integrate only scalar-valued functions.

For $2 \leq p < \infty$, we proceed by duality. Set $q = p/(p-1)$, thus $1 < q \leq 2$, in which case the conclusion has already been proven. Let $S(r)$ be the sphere of radius r centered at the origin of $\mathbb{R}^8$. Writing down the norm in $L^p(S(1-\epsilon))$, $0 < \epsilon < 1$, as the supremum over the unit ball in the dual space L^q, using the symmetry of the involved bilinear form (cf. [13, p. 144]), then estimating this form by the Hölder inequality, and finally applying the result proven above for the case $q \in (1, 2]$, we finish the proof. □

Proof of Proposition 1.4. The function

$$h(r) = \left(\int_S |f(r,\theta)|^p d\sigma(\theta)\right)^{1/p}$$

is increasing on $(0,1)$ [1, Cor. 3.2.6] and upper-bounded there by the condition; therefore, application of the Hölder inequality with $1/p > 1$ and $1/(1-p)$ completes the proof. □

Proof of Theorem 1.5. From inequality (2.4) with $p = 2$ and $C = C_2 = 7$ we get

$$\frac{1}{\sigma_8} \int_S |f(\mathbf{x})|^2 d\sigma \leq \frac{7}{\sigma_8} \int_S |f_v(\mathbf{x})|^2 d\sigma,$$

and since the volume average is dominated by the surface average [1, Cor. 3.2.6], the latter implies the inequality

$$\frac{1}{\omega_8} \int_B |f(\mathbf{x})|^2 dx \leq \frac{7}{\sigma_8} \int_S |f_v(\mathbf{x})|^2 d\sigma,$$

where ω_8 is the volume of the 8-dimensional unit ball. The conclusion follows if we drop in the integral on the left a positive part where $|f| \leq \lambda$ and note that $\sigma_8 = 8\omega_8$. □

Acknowledgment

This is an expanded version of the talk given by the second author at the 10th Congress of ISAAC at Macau University, August 3–8, 2015. A.K. is grateful to the hosts and organizers of the Congress for invitation and excellent organization of the meeting.

The work was started when the second author was on leave from Bronx Community College / CUNY.

The authors are thankful to Professor Andrew McInerney and Professor Mohamed Messaoudene for useful stimulating discussions, and to the anonymous reviewers, whose careful reading and thoughtful remarks helped us to improve the paper.

References

[1] D.H. Armitage, S.J. Gardiner, *Classical Potential Theory*, Springer-Verlag, 2001.

[2] K. Avetisyan, *Subharmonicity and a version of Riesz theorem on harmonic conjugates*, Adv. Appl. Clifford Algebras, **24**(2014), 909–919.

[3] J.C. Baez, *The octonions*, Bull. Amer. Math. Soc., **39** (2002), 145–205.

[4] S. Catto, A. Kheyfits, *Asymptotic behavior of octonion-valued monogenic functions in tube domains*, in *Group Theoretical Methods in Physics,* 2006, Proc. of the XXVI International Colloquium on Group Theoretical Methods in Physics (Group 26), New York, USA, 26–30 June 2006. Canopus Publishing Limited, 2009, 140–145.

[5] S. Catto, A. Kheyfits, D. Tepper, *M. Riesz Theorem on Conjugate Harmonic Functions for Octonion-Valued Monogenic Functions*, XXIII International Conference on Integrable Systems and Quantum Symmetries (ISQS-23). J. of Physics: Conference Series **670** (2016) 012017, 8 pp.

[6] R.R. Coifman, G. Weiss, *On subharmonicity inequalities involving solutions of generalized Cauchy-Riemann equations*, Studia Math., **36** (1970), 77–83.

[7] M. Essén, *A superharmonic proof of the M. Riesz conjugate function theorem*, Ark. Mat., **22**, 1984, No. 2, 241–249.

[8] M.A. Evgrafov, *Generalization of the Phragmén–Lindelöf theorems for analytic functions to solutions of different elliptic systems*, Izv. Akad. Nauk SSSR, Ser. Mat., **27** (1963), 843–854.

[9] F. Gürsey, C.-H. Tze, *On the Role of Division, Jordan, and Related Algebras in Particle Physics*, World Scientific, 1996.

[10] A. Kheyfits, D. Tepper, *Subharmonicity of powers of octonion-valued monogenic functions and some applications*, Bull. of the Belgian Math. Soc. – Simon Stevin, **13** (2006), 609–617.

[11] A. Kheyfits, *Phragmén–Lindelöf principle for Clifford monogenic functions*, in *Complex Analysis and its Applications*, Proc. of the 15th International Conference on Finite or Infinite-Dimensional Complex Analysis and Applications. Osaka City Univ., Osaka (2007), 237–240.

[12] Ü. Kuran, *n-dimensional extensions of theorems on conjugate functions*, Proc. London Math. Soc., (3) **15** (1965), 713–730.

[13] B.Ya. Levin, *Lectures on Entire Functions*, Amer. Math. Soc., Providence, RI, 1996.

[14] X. Li, L. Peng, *On Stein–Weiss conjugate harmonic function and octonion analytic function*, Approx. Theory and its Appl. **16** (2000), 28–36.

[15] X. Li, L. Peng, *The Cauchy integral formulas on the octonions*, Bull. Math. Soc. Belgique, **9** (2002), 47–64.

[16] X. Li, L. Peng, *Three-line theorem on the octonions*, Acta Math. Sinica, English Ser., **20** (2004), 483–490.

[17] X. Li, L. Peng, T. Qian, *The Paley–Wiener theorem in the non-commutative and non-associative octonions*, Sci. China Ser. A **52** (2009), 129–141.

[18] C. Nolder, *Conjugate harmonic functions and Clifford algebras*, J. Math. Analysis Appl., **302** (2005), 137–142.

[19] S.K. Pichorides, *On the best values of constants in the theorems of M. Riesz, Zygmund and Kolmogorov*, Studia Math., **44** (1972), 165–179.

[20] E.M. Stein, G. Weiss, *Introduction to Fourier Analysis on Euclidean Spaces*, Princeton Univ. Press, 1971.

[21] P. Stein, *On a theorem of M. Riesz*, J. London Math. Soc., **8** (1933), 242–247.

Graduate School and Baruch College/CUNY
One Bernard Baruch Way
New York, NY 10010, USA
e-mail: scatto@gc.cuny.edu

Graduate School and Bronx Community College/CUNY
2155 University Avenue
Bronx, NY 10453, USA
e-mail: akheyfits@gc.cuny.edu

Temple University
1801 N. Broad St.
Philadelphia, PA 19122, USA
e-mail: detepper@temple.edu

Modern Trends in Hypercomplex Analysis
Trends in Mathematics, 47–77

Compressed Sensing with Nonlinear Fourier Atoms

Paula Cerejeiras, Qiuhui Chen, Narciso Gomes and Stefan Hartmann

Abstract. We study the problem of compressed sensing for nonlinear Fourier atoms and Takenaka–Malmquist systems. We show that reconstruction by means of a ℓ_1-minimization is possible with high probability.

Mathematics Subject Classification (2010). Primary 94A20; Secondary 30D50, 15A52.

Keywords. Nonlinear Fourier atoms; compressed sensing; interpolation; Takenaka–Malmquist system.

1. Introduction

In the last decade a new paradigm has taken hold in signal and image processing: compressed sensing. The possibility of reconstructing a signal by only a few measurements under the condition that the representation in a given basis or frame is sparse has allowed to look at new methods and algorithms. Although sparsity constraints are directly connected only with non-convex optimization the uniqueness property shown by Candès, Rhomberg, and Tao [11] allows the application of simple convex algorithms, such as linear programming. This has been applied to a variety of situations, but here we are interested in applying it to a generalization of the standard Fourier basis, the case of so-called nonlinear Fourier atoms. Decomposition algorithms for this kind of atoms were thoroughly investigated by the group of T. Qian in Macau during the last decade ([14]). Although the original starting point for T. Qian was the investigation into a mathematical justification of the Hilbert–Huang transform and the empirical mode decomposition, the underlying structure is much older. The whole approach is in fact based on the question of decomposing a function on the unit circle in terms of Blaschke products, that means in terms of the so-called Takenaka–Malmquist system. Despite being one of the classic topics in Complex Analysis this system is rather unknown in the signal processing community which has its focus on Wavelet and Gabor decompositions. This is the main reason why investigations from the point of signal processing into this system basically restarted within the last decade, not only by T. Qian and his co-authors, but also by M. Pap and her collaborators in the framework of the study of the so-called Voice transform ([26] and [27]). Due to its close connection with the group of Möbius transformations they can also be used to describe dilated functions on the unit circle which was used in the definition of hyperbolic wavelets. While the adaptive Fourier decompositions of T. Qian showed its capacities in a variety of examples, principally in linear systems theory, it still constitutes a greedy algorithm and it is a priori not clear that in applications the number of atoms can be kept sufficiently small as not to be affected by the exponentially rising costs. In fact, in the recent PhD-thesis by L. Shuang (cf. [31]) a comparison between the AFD (Adaptive Fourier Decomposition)-method and Basis Pursuit where made showing that there are indeed situations where a Basis Pursuit has an advantage. The mathematical justification for the applicability of Basis Pursuit was given only by an asymptotic analysis and, therefore, is only valid for large scale matrices. Here, we will use a compressed sensing approach to the reconstruction of a given signal in terms of Takenaka–Malmquist systems. The direct approach to compressed sensing involves the checking of the null space property. Traditionally, this is done by verifying the RIP condition, that is to say, that the sampling matrix behaves almost as an isometry for sparse vectors since RIP implies the null space property (see [17]). Unfortunately, a direct verification of this condition for the kind of matrices we are dealing with is extremely difficult. Therefore, we will use the approach by Rauhut [30] for the case of the classic Fourier basis to give a general answer. But a direct adaptation represents some additional problems. For instance, the calculation of the expectation value turns out to be much more demanding due to the lack of structure (no easy multiplication rule). Furthermore, in the last section we will make a comparison between our approach and the results in [31]. We are not going to make a comparison with the AFD method of T. Qian since that comparison was already done in [31].

2. Non-linear Fourier atoms

Nonlinear Fourier atoms are a family of nonlinear Fourier bases, seen as an extension of the classical Fourier basis, that have been constructed and applied to signal processing [13, 15]). For any complex number $a = re^{it_a}$, $r = |a| < 1$, the nonlinear phase function $\theta_a(t)$ is defined by the non-tangential boundary value of the Möbius transformation

$$\tau_a(z) = \frac{z-a}{1-\bar{a}z},$$

that is, the nonlinear Fourier atom is given by

$$e^{i\theta_a(t)} := \tau_a(e^{it}) = \frac{e^{it}-a}{1-\bar{a}e^{it}}.$$

Note that $\theta_a(t+2\pi) = \theta_a(t) + 2\pi$ and its derivative is the Poisson kernel

$$\theta_a'(t) = \frac{1-r^2}{1+r^2-2r\cos(t_a-t)} = \mathrm{Re}\left(\frac{e^{it}+re^{it_a}}{e^{it}-re^{it_a}}\right), \tag{2.1}$$

which satisfies

$$0 < \frac{1-r}{1+r} \le \theta_a'(t) \le \frac{1+r}{1-r}. \tag{2.2}$$

For any sequence $(c_k)_{k\in\mathbb{Z}}$ of finite nonzero terms it holds

$$\sum_{k\in\mathbb{Z}} |c_k|^2 = \frac{1}{2\pi}\int_{\mathbb{T}} \left|c_k e^{ikx}\right|^2 dx = \frac{1}{2\pi}\int_{\mathbb{T}} \left|c_k e^{ik\theta_a(t)}\right|^2 \theta_a'(t)dt,$$

which by combining it with (2.2) implies that we can consider the so-called nonlinear Fourier basis $\{e^{in\theta_a(t)}, n\in\mathbb{Z}\}$ of $\mathcal{L}^2(\mathbb{T})$ with $\mathbb{T} = \{z\in\mathbb{C} : |z| = 1\}$ denoting the unit circle ([13], [29]). Note that if $a = 0$, $\{e^{in\theta_a(t)}, n\in\mathbb{Z}\}$ reduces to the classic Fourier basis $\{e^{int}, n\in\mathbb{Z}\}$. These atoms are star-like functions, convex with positive phase derivative on the boundary and they are linked to nonharmonic Fourier series and TM systems (see, for instance, [28]).

2.1. Hardy spaces

We consider the following function spaces:

- $\mathcal{L}^2(\mathbb{T})$ as the Hilbert space of square integrable functions over the unit circle.
- For $1\le p<\infty$ the Hardy space $\mathcal{H}^p$ is defined as the space of all analytic functions f in $\mathbb{D}$ for which the norm

$$\|f\|_p = \sup_{0\le r<1}\left(\frac{1}{2\pi}\int_{[0,2\pi]} |f(re^{it})|^p dt\right)^{1/p}$$

  if finite ([18]). The space $\mathcal{H}^\infty$ consists of all bounded analytic functions f in $\mathbb{D}$ with norm given by

$$\|f\|_\infty = \sup_{|z|<1} |f(z)|.$$

For functions in $\mathcal{H}^p(\mathbb{D})$, $1\le p\le\infty$, the radial limit

$$\tilde{f}(e^{it}) = \lim_{r\to 1} f(re^{it})$$

exists almost everywhere in t (Fatou's Theorem), and indeed $\tilde{f}\in\mathcal{L}^p(\mathbb{T})$. Moreover

$$\|f\|_p = \sup_{0\le r<1}\left(\frac{1}{2\pi}\int_{[0,2\pi]} |f(re^{it})|^p dt\right)^{1/p} =: \|\tilde{f}\|_{\mathcal{L}^p(\mathbb{T})}.$$

We normally identify f with $\tilde{f}$ and can regard $\mathcal{H}^p$ as the subspace of those functions in $\mathcal{L}^p(\mathbb{T})$ for which the negative Fourier coefficients vanish, that is:

$$\frac{1}{2\pi}\int_{[0,2\pi]} \tilde{f}(e^{it})e^{-int}dt = 0$$

for all $n < 0$. Then a function $\tilde{f} \sim \sum_{n=0}^{\infty} a_n z^n$ can be naturally identified with the power series $f(z) = \sum_{n=0}^{\infty} a_n z^n$ defining an analytic function f in $\mathbb{D} = \{z \in \mathbb{C} : |z| < 1\}$.

One can also obtain the extension from $\tilde{f}$ to f by convolving it with the Poisson kernel K_r, namely

$$f(re^{it}) = \frac{1}{2\pi} \int_{[0,2\pi]} K_{r,t}(t_a - t)\tilde{f}(e^{it})dt$$

where $K_{r,t}(t_a - t)$ is Poisson kernel from (2.1).

The case $p = 2$ is simpler since for a function $f : z \mapsto \sum_{n=0}^{\infty} a_n z^n$ we have

$$\|f\|_2 = \left(\sum_{n=0}^{\infty} |a_n|^2 \right)^{1/2}.$$

We have the following inclusions:

$$\mathcal{H}^{\infty} \subset \mathcal{H}^p \subset \mathcal{H}^q \subset \mathcal{H}^1,$$

for $1 < q \leq p < \infty$.

2.2. Takenaka–Malmquist system as non-linear Fourier atoms

The Takenaka–Malmquist (TM) system belongs to the families of unit analytic signals with nonlinear phase and is closely linked to non-linear Fourier atoms.

Given a sequence $(a_k)_{k=1}^{\infty}$ of points in $\mathbb{D}$ we associate with it the modified Blaschke products $B_1, B_2, \ldots$, defined by

$$\mathcal{B}_n(z) = \frac{\sqrt{1 - |a_n|^2}}{1 - \bar{a}_n z} \varphi_n(z) \quad \text{and} \quad \mathcal{B}_{-n}(z) = \overline{\mathcal{B}_n\left(1/\bar{z}\right)},$$

for $n = 0, 1, 2, \ldots$. Hereby the finite Blaschke product is given by

$$\varphi_p(z) = \prod_{k=0}^{p-1} \frac{z - a_k}{1 - \bar{a}_k z}, \tag{2.3}$$

where $a_k \in \mathbb{D}$ for all $k \geq 0$. Although the usual orthonormal Fourier atoms $\{e^{in\omega}, n \in \mathbb{Z}\}$ form an orthogonal basis in $\mathcal{L}^2(\mathbb{T})$ in general the nonlinear Fourier atoms $\{e^{in\theta(t)}, n \in \mathbb{Z}\}$ are not orthogonal. The Gram–Schmidt orthonormalization process leads to the TM basis $\{\mathcal{B}_n : n \geq 0\}$ which is known to be complete and orthogonal in $\mathcal{H}^2(\mathbb{D})$ and to the basis $\{\mathcal{B}_n : n \in \mathbb{Z}\}$ which forms an orthogonal basis in $\mathcal{L}^2(\mathbb{T})$ if and only if it satisfies the condition

$$\sum_{n=1}^{\infty} (1 - |a_n|) = \infty. \tag{2.4}$$

Condition (2.4) implies that the parameters a_k in (2.3) converge to 0. We recall that if $a_k = 0$ we obtain the classical Fourier basis as a usual case (see, for instance, [8]).

The inner product of two complex functions g_1 and g_2 in $\mathbb{T}$ is defined as

$$\langle g_1, g_2 \rangle = \frac{1}{2\pi i} \int_{\mathbb{T}} \overline{g_1(z)} g_2(z) \frac{dz}{z}. \tag{2.5}$$

Alternatively from (2.5) the inner product can be written as

$$\langle g_1, g_2 \rangle = \frac{1}{2\pi} \int_0^{2\pi} \overline{g_1(e^{ix})} g_2(e^{ix}) dx$$

and the induced norm will be denoted by $\| \cdot \|_2$.

To give some examples when $a_n = b$ ($n \in \mathbb{N}_0$), we have $\mathcal{B}_n = L_n^b$ ($n \in \mathbb{N}_0$) forming the discrete Laguerre system, and in case of $a_{2k-1} = a, a_{2k} = b$ ($k \in \mathbb{N}_0$) we get $\{\mathcal{B}_n, n \in \mathbb{N}_0\}$ as the Kautz system investigated in [3].

3. Sparse sampling in Takenaka–Malmquist system

Since we want to follow the approach of Rauhut [30] we have to look at his setting with trigonometric polynomials being replaced by atoms of the form

$$\mathcal{B}_{a_n}(x) = \frac{\sqrt{1 - |a_n|^2}}{1 - \overline{a}_n e^{ix}} \prod_{k=0}^{n-1} \frac{e^{ix} - a_k}{1 - \overline{a}_k e^{ix}},$$

where $a_n \in \mathbb{C}$ is such that $|a_n| < 1$. These atoms are star-like functions, convex with positive phase derivative on the boundary and they are linked to TM systems insofar as they represent elements of the orthogonal basis generated by $(a_0, a_1, \ldots, a_n, \ldots)$. Hereby, we denote by $\prod_d$ the space spanned by at most d of its elements. In other words an element $f \in \prod_d$ is of the form

$$f(x) = \sum_{n=1}^{d} c_n \frac{\sqrt{1 - |a_n|^2}}{1 - \overline{a}_n e^{ix}} \prod_{k=0}^{n-1} \frac{e^{ix} - a_k}{1 - \overline{a}_k e^{ix}}, \qquad x \in [0, 2\pi],\ |a_k| < 1,\ c_n \in \mathbb{C}. \tag{3.1}$$

We assume that the sequence of coefficients $c = (c_k)$ is supported on a set T which is much smaller than the dimension of $\prod_d$, that is to say, the finite combination in (3.1) is sparse. However, a priori nothing is known about T apart from its maximum size. Thus, it is useful to introduce the set (not a linear space) $\prod_d(M) \subset \prod_d$ of all polynomials of type (3.1) such that their sequence of coefficients $c = (c_n)$ have support on a set $T \subset \{1, \ldots, d\}$ satisfying $|T| \leq M$, i.e., $f \in \prod_d(M)$ is of the form

$$f(x) = \sum_{n \in T} c_n \frac{\sqrt{1 - |a_n|^2}}{1 - \overline{a}_n e^{ix}} \prod_{k=0}^{n-1} \frac{e^{ix} - a_k}{1 - \overline{a}_k e^{ix}}.$$

Again, the objective is, given a sampling set $X := \{x_1, x_2, \ldots, x_N\}$ of independent random variables having uniform distribution on $[0, 2\pi]$, to reconstruct $f \in \prod_d(M)$ from the samples $f(x_j)$ at those N randomly chosen points.

3.1. Description of the main results

In this paper we will prove the following theorems.

Theorem 3.1. *Assume $f \in \prod_q(M)$ with some sparsity $M \in \mathbb{N}$. Let $x_1, x_2, \dots, x_N \in [0, 2\pi]$ be independent random variables having the uniform distribution on $[0, 2\pi]$. Choose $n \in \mathbb{N}$, $\beta > 0$, $\kappa > 0$ and $K_1, \dots, K_n \in \mathbb{N}$ such that*

$$a := \sum_{m=1}^{n} \beta^{n/K_m} < 1 \quad \text{and} \quad \frac{\kappa}{1-\kappa} \leq \frac{1-a}{1+a} M^{-3/2}. \tag{3.2}$$

Set $\theta := (N\mathcal{K})/M$. Then with probability at least

$$1 - \left(\mathcal{C}_d \left[d\beta^{-2n} \sum_{m=1}^{n} G_{2mK_m}(\theta) + \kappa^{-2} \mathcal{K}^{2n}\, M\, N^{2n} G_{2n}(\theta) \right] \right), \tag{3.3}$$

where $d := \dim(\prod_d)$, $G_n(\theta) = \theta^{-n} \sum_{k=1}^{\lfloor \frac{n}{2} \rfloor} S_2(n,k)\theta^k$, and $S_2(n,k)$ denote the Stirling numbers of the second kind, f can be reconstructed exactly from its sample values $f(x_1), \dots, f(x_N)$ by solving the minimization problem

$$\min \|(c_n)\|_1 := \sum_{n=1}^{d} |c_n|,$$

$$s.a. \quad f(x_j) := \sum_{n=1}^{d} c_n \frac{\sqrt{1-|a_n|^2}}{1-\overline{a_n}e^{ix_j}} \prod_{k=0}^{n-1} \frac{e^{ix_j} - a_k}{1-\overline{a_k}e^{ix_j}}, j = 1, \dots, N. \tag{3.4}$$

For a given n it is reasonable to take $K_m \approx m/n$, $m = 1, \dots, n$, rounding m/n to the nearest integer. Then we can choose β quite close to the maximal value such that $a = \sum_{m=1}^{n} \beta^{n/K_m} < 1$. By our choice of K_m we approximately have

$$\sum_{m=1}^{n} \beta^{n/K_m} \approx \sum_{m=1}^{n} \beta^m \approx \frac{\beta}{1-\beta}.$$

Thus, the optimal β will always be close to $1/2$.

Although we are not going to prove them in this paper the following theorems can be easily obtained by adapting the proof of the previous theorem. Their proofs are straightforward adaptation of the corresponding proofs in [30] with the necessary modifications coming from the proof of Theorem 3.1. For more details we refer to the thesis [20]. To make it more clear the principal difficulty resides in the proof of Theorem 3.1 since the adaptations for the proofs of the other two theorems do not depend on the actual choice of the system, i.e., the proof of the first theorem provides the necessary basis for the proof of the other two theorems without the need for further modifications.

Theorem 3.2. *There exists an absolute constant $C > 0$ such that the following is true. Assume $f \in \prod_d(M)$ for some sparsity $M \in \mathbb{N}$. Let $x_1, x_2, \dots, x_N \in [0, 2\pi]$ be independent random variables having the uniform distribution on $[0, 2\pi]$. If for some $\epsilon > 0$ it holds*

$$N \geq CM \log(d\epsilon) \tag{3.5}$$

then with probability at least $1-\epsilon$ *the function* f *can be recovered from its sample values* $f(x_j)$, $j = 1, \dots, N$, *by solving the* ℓ_1*-minimization problem* (3.4).

Theorem 3.3. *Let* $x_1, x_2, \dots, x_N \in [0, 2\pi]$ *be independent random variables having the uniform distribution on* $[0, 2\pi]$. *Further assume that* T *is a random subset of* $[0, 2\pi]$ *modeled by*

$$a := \sum_{m=1}^{n} \beta^{n/K_m} < 1 \quad \text{and} \quad \frac{k}{1-k} \le \frac{1-a}{1+a}((\alpha+1)\mathbb{E}|T|)^{-3/2}. \tag{3.6}$$

Then with probability at least

$$1-\left(\kappa^{-2}W(n,N,\mathbb{E}|T|,d)+\beta^{-2n}d\sum_{m=1}^{n} Z(K_m,m,N,\mathbb{E}|T|,d)+\exp\left(-\frac{3\alpha^2}{6+2\alpha}\mathbb{E}|T|\right)\right) \tag{3.7}$$

any $f \in \prod_T \subset \prod_q(|T|)$ *can be reconstructed exactly from its sample values* $f(x_1), \dots, f(x_N)$ *by solving the minimization problem* (3.4).

3.2. Proof of the main result

To prove our theorem we have to introduce some auxiliary notations: $\ell_2(D)$, $\ell_2(T)$, $\ell_2(X)$ will denote the ℓ_2-spaces of sequences indexed by $D = \{1, 2, \dots, d\}$, T, and X, respectively, all endowed with the usual Euclidean norm. Moreover, we introduce the operator $\mathcal{F}_X : \ell_2(D) \to \ell_2(X)$ given as

$$\mathcal{F}_X := \left[\frac{\sqrt{1-|a_n|^2}}{1-\overline{a_n}e^{ix_j}} \prod_{k=0}^{n-1} \frac{e^{ix_j}-a_k}{1-\overline{a_k}e^{ix_j}}\right]_{j=1,\dots,N,\ n=1,\dots,d}.$$

We recall that $|a_k| < 1$, for all $k = 1, \dots, d$.

By $\mathcal{F}_{TX}$ we represent the restriction of $\mathcal{F}_X$ to sequences supported only on T, thus, an operator acting from $\ell_2(T)$ in $\ell_2(X)$. Furthermore, the adjoint operators are given by $\mathcal{F}_X^* : \ell_2(X) \to \ell_2(D)$ and $\mathcal{F}_{TX}^* : \ell_2(X) \to \ell_2(T)$.

Our problem is to reconstructing a sequence $c \in \ell_2(D)$ from $\beta = \mathcal{F}_X c \in \ell_2(X)$ by solving the problem $\min \|c\|_1$ subject to $\mathcal{F}_X c = \beta$. Obviously, if $k \notin \operatorname{supp} c$ then $\operatorname{sgn}(c)_k = 0$ while $|\operatorname{sgn}(c)_k| = 1$ for all $k \in \operatorname{supp} c$.

Our proof is based on the following lemma (see also [30] and [11]).

Lemma 3.4. *Let* $c \in \ell_2(D)$ *and* $T := \operatorname{supp} c$. *Assume* $\mathcal{F}_{TX} : \ell_2(T) \to \ell_2(X)$ *to be injective. Suppose that there exists a vector* $P \in \ell_2(D)$ *with the following properties:*

(i) $P_k = \operatorname{sgn}(c)$ *for all* $k \in T$,
(ii) $|P_k| < 1$ *for all* $k \notin T$,
(iii) *there exists a vector* $\lambda \in \ell_2(X)$ *such that* $P = \mathcal{F}_X^* \lambda$.

Then c *is the unique minimizer to problem* (3.4).

Proof. Let us assume $X \neq \emptyset$ and $c \neq 0$ to exclude the trivial cases. Furthermore, let us suppose that the vector P exists. Let b be any vector different to c with

$\mathcal{F}_X b = \mathcal{F}_X c$. Consider $h := b - c$, then $\mathcal{F}_X q$ vanishes on X. This means that for b_k, $k \in T$, we have the following estimate

$$\begin{aligned}|b_k| &= |c_k + d_k| = |(c_k + h_k)\overline{\operatorname{sgn} c_k}\operatorname{sgn} c_k| \\ &= |\,(c_k\,\overline{\operatorname{sgn} c_k} + h_k\,\overline{\operatorname{sgn} c_k})\operatorname{sgn} c_k| = |\,|c_k| + h_k\,\overline{\operatorname{sgn} c_k}||\overline{\operatorname{sgn} c_k}| \\ &= |\,|c_k| + h_k\,\overline{\operatorname{sgn} c_k}| \geq |c_k| + \operatorname{Re}(h_k\,\overline{\operatorname{sgn} c_k}) = |c_k| + \operatorname{Re}\left(h_k\,\overline{P_k}\right).\end{aligned}$$

Thus, for any $k \in T$ we have $|c_k| + \operatorname{Re}\left(h_k\,\overline{P_k}\right) \leq |b_k|$. Otherwise, for $k \notin T$ we have $\operatorname{Re}\left(h_k\,\overline{P_k}\right) \leq |h_k| = |h_k|$ since $|P_k| < 1$. Thus

$$\|b\|_{\ell_1} \geq \|c\|_{\ell_1} + \sum_{k\in[-q,q]\cap\mathbb{Z}} \operatorname{Re}\left(h_k\,\overline{P_k}\right).$$

Now, from condition (iii) we can conclude

$$\begin{aligned}\sum_{k\in[-q,q]\cap\mathbb{Z}} \operatorname{Re}\left(h_k\,\overline{P_k}\right) &= \operatorname{Re}\left(\sum_{k\in[-q,q]\cap\mathbb{Z}} h_k\,\overline{(\mathcal{F}_X^*\lambda)_k}\right) \\ &= \operatorname{Re}\left(\sum_{i=1}^{N} (\mathcal{F}_X\,h)(x_i)\overline{\lambda(x_i)}\right) = 0\end{aligned}$$

whereas $\mathcal{F}_X\,h$ vanishes. Thus, $\|b\|_{\ell_1} \geq \|c\|_{\ell_1}$. The equality holds when $\|h_k\| = \operatorname{Re}\left(h_k\,\overline{P_k}\right)$ for all $k \notin T$. Since $\|P_k\| < 1$, this forces h to vanish outside of T. Taking in account the injectivity of $\mathcal{F}_{TX}$ we have that since $\mathcal{F}_X h$ vanishes on X, h vanishes identically and we have $b = c$. Thus, this shows that c is the unique minimizer to the problem (3.4). □

Before we can start our proof we need an additional lemma about the maximum of the Blaschke product. Let us point out that the usual estimates on the maximum of a Blaschke product which can be found in the literature are not good enough in this context since we want to estimate a probability.

Lemma 3.5. *Let be $\epsilon > 0$ and $a_n, z \in \mathbb{C}$ such that $a_n = \alpha + i\beta$ and $z = x + iy$, $n = 1, \ldots, d$. If $|a_n| < 1$, $|z| = 1$, and either $|\pm 1 - a_n| = \epsilon$ or $|\pm i - a_n| = \epsilon$ then the set of functions*

$$f_{a_n}(z) = \frac{1 - |a_n|^2}{|1 - \overline{a}_n z|^2}, \tag{3.8}$$

has a uniform maximum which can be estimated by

$$\max\nolimits_{a_n\in\mathbb{D}}\left(\max\nolimits_{z_A\in\mathbb{T}} \frac{1 - |a_n|^2}{|1 - \overline{a}_n z|^2}\right) \leq \frac{2}{\epsilon} - 1.$$

Since this proof consists of lengthy, but straightforward calculations we refer to [20].

For the proof of our main theorem we have to remark that if $N \geq |T|$ then $\mathcal{F}_{TX}$ is injective almost surely. This means that we need to show now that with high probability there exists a vector P with the properties assumed in Lemma 3.4.

To do this we introduce the restriction operator $R_T : \ell_2(D) \to \ell_2(T)$ given by $R_T c_k = c_k$ for $k \in T$ and its adjoint $R_T^* = E_T : \ell_2(T) \to \ell_2(D)$ which is the operator that extends a vector outside T by zero, i.e., $(E_T d)_k = d_k$ for $k \in T$ and $(E_T d)_k = 0$ otherwise.

Now assume for the moment that $\mathcal{F}_{TX}^* \mathcal{F}_{TX} : \ell_2(T) \to \ell_2(T)$ is invertible. In this case we can construct P explicitly by

$$P := \mathcal{F}_X^* \mathcal{F}_{TX} (\mathcal{F}_{TX}^* \mathcal{F}_{TX})^{-1} R_T \mathrm{sgn}(c),$$

where as before $T := \mathrm{supp} c$. Then clearly P has property (i) and property (iii) in Lemma 3.4 with

$$\lambda := \mathcal{F}_{TX} (\mathcal{F}_{TX}^* \mathcal{F}_{TX})^{-1} R_T \mathrm{sgn}(c) \in \ell_2(X).$$

We are left with proving that P has property (ii) of Lemma 3.4 with high probability.

To this end we follow [30] and introduce the auxiliary operators

$$H : \ell_2(T) \to \ell_2(D), \qquad H := DE_T - \mathcal{F}_X^* \mathcal{F}_{TX}$$

and

$$H_0 : \ell_2(T) \to \ell_2(T), \qquad H_0 := R_T H = DI_T - \mathcal{F}_{TX}^* \mathcal{F}_{TX},$$

where I_T denotes the identity on $l^2(T)$ and the diagonal matrix D has entries

$$D_{mm} = \sum_{j=1}^{N} \frac{\sqrt{1-|a_m|^2}}{1-a_m e^{-ix_j}} \prod_{\ell=0}^{m-1} \frac{e^{-ix_j} - \overline{a_\ell}}{1-a_\ell e^{-ix_j}} \frac{\sqrt{1-|a_m|^2}}{1-\overline{a_m} e^{ix_j}} \prod_{\ell=0}^{m-1} \frac{e^{ix_j} - a_\ell}{1-\overline{a_\ell} e^{ix_j}}.$$

Obviously, H_0 is self-adjoint, and $H = [h_{mn}]$, where

$$h_{mn} := D_{mn}\delta_{mn} - \sum_{j=1}^{N} \frac{\sqrt{1-|a_m|^2}}{1-a_m e^{-ix_j}} \prod_{\ell=0}^{m-1} \frac{e^{-ix_j} - \overline{a_\ell}}{1-a_\ell e^{-ix_j}} \frac{\sqrt{1-|a_n|^2}}{1-\overline{a_n} e^{ix_j}} \prod_{k=0}^{n-1} \frac{e^{ix_j} - a_k}{1-\overline{a_k} e^{ix_j}}, \tag{3.9}$$

acts on a vector as

$$(Hc)_m = - \sum_{\substack{n=1,\\ n\neq m}}^{d} \sum_{j=1}^{N} \frac{\sqrt{1-|a_m|^2}}{1-a_m e^{-ix_j}} \prod_{\ell=0}^{m-1} \frac{e^{-ix_j} - \overline{a_\ell}}{1-a_\ell e^{-ix_j}} \frac{\sqrt{1-|a_n|^2}}{1-\overline{a_n} e^{ix_j}} \prod_{k=0}^{n-1} \frac{e^{ix_j} - a_k}{1-\overline{a_k} e^{ix_j}} c_n. \tag{3.10}$$

Now we can write

$$P = (DE_T - H)(DI_T - H_0)^{-1} R_T \mathrm{sgn}(c).$$

As we are interested in property (ii) in Lemma 3.4 we consider only values of P on $T^c = D \setminus T$. Since $R_{T^c} E_T = 0$ we have

$$P_k = -D^{-1} R_{T^c} H \left(I_T - D^{-1} H_0 \right)^{-1} R_T \mathrm{sgn}(c) \qquad \text{for all} \quad k \in T^c.$$

We can study the term $\left(I_T - D^{-1}H_0\right)^{-1}$ via the von Neumann series (see, for instance, [22]) of $\left(I_T - (D^{-1}H_0)^n\right)^{-1} = I_T + A_n$ with

$$A_n := \sum_{r=1}^{\infty} \left(D^{-1}H_0\right)^{rn}, \qquad n \in \mathbb{N}. \tag{3.11}$$

Here, we get

$$\left(I_T - (D^{-1}H_0)\right)^{-1} = (I_T + A_n) \sum_{m=0}^{n-1} \left(D^{-1}H_0\right)^m.$$

Therefore, we can write on the complement of T

$$R_{T^c}P = H(I_T + A_n)\Big(\sum_{m=0}^{n-1} \left(D^{-1}H_0\right)^m\Big)D^{-1}R_T\mathrm{sgn}(c) = -\big(P^{(1)} + P^{(2)}\big),$$

where

$$P^{(1)} = DS_nD^{-1}\mathrm{sgn}(c) \quad \text{and} \quad P^{(2)} = HA_nR_T(I + S_{n-1})D^{-1}\mathrm{sgn}(c),$$

with $S_n := \sum_{m=0}^{n-1} \left(D^{-1}HR_T\right)^m$.

Since our goal is to estimate $\mathbb{P}(\sup_{k\in T^c}|P_k| \geq 1)$ we consider $a_1, a_2 > 0$ to be numbers satisfying $a_1 + a_2 = 1$ and we have

$$\mathbb{P}\big(\sup_{k\in T^c}|P_k| \geq 1\big) \leq \mathbb{P}\big(\big\{\sup_{k\in T^c}|P_k^{(1)}| \geq a_1\big\} \cup \big\{\sup_{k\in T^c}|P_k^{(2)}| \geq a_2\big\}\big). \tag{3.12}$$

This leads to the estimates

$$\begin{aligned}\mathbb{P}\big(|P_k^{(1)}| \geq a_1\big) &= \mathbb{P}\left(\left|\left(DS_nD^{-1}\mathrm{sgn}(c)\right)_k\right| \geq a_1\right)\\ &\leq \mathbb{P}(E_k) := \mathbb{P}\Big(\sum_{m=1}^{n} |(D(D^{-1}HR_T)^m D^{-1}\mathrm{sgn}(c))_k| \geq a_1\Big).\end{aligned}$$

and

$$\begin{aligned}\sup_{k\in T^c}|P_k^{(2)}| &\leq \|P^{(2)}\|_\infty \\ &\leq \|HA_nD^{-1}\|_{\ell^\infty(T)\to\ell^\infty(D)}\big(1 + \|R_TDS_{n-1}D^{-1}\mathrm{sgn}(c)\|_{\ell^\infty(T)}\big)\end{aligned} \tag{3.13}$$

where $\ell^\infty(D)$ denotes the space of bounded sequences indexed by D.

For the term $\|R_TDS_{n-1}D^{-1}\mathrm{sgn}(c)\|_{\ell^\infty(T)}$ similarly as in (3.13) we get

$$\begin{aligned}\mathbb{P}\big(|(DS_{n-1}D^{-1}\mathrm{sgn}(c))_k| \geq a_1\big) &\leq \mathbb{P}(E_k)\\ &= \mathbb{P}\Big(\sum_{m=1}^{n} |(D(D^{-1}HR_T)^m D^{-1}\mathrm{sgn}(c))_k| \geq a_1\Big)\\ &= \mathbb{P}\Big(\sum_{m=1}^{n} |((HD^{-1}R_T)^m D^{-1}\mathrm{sgn}(c))_k| \geq a_1\Big).\end{aligned}$$

Now, we know as well that

$$\left\|HA_nD^{-1}\right\|_\infty \le \left\|HD^{-1}DA_nD^{-1}\right\|_\infty \le \left\|HD^{-1}\right\|_\infty \left\|DA_nD^{-1}\right\|_\infty. \tag{3.14}$$

with $DA_nD^{-1} = \sum_{r=1}^{\infty} D\left(D^{-1}H_0\right)^{rn} D^{-1}$.

For the estimation of the term $\left\|DA_nD^{-1}\right\|_\infty$ we can use the Frobenius norm $\|A\|_F^2 = Tr(AA^*)$. Note that the trace of a product of a matrix with a diagonal matrix D with same dimension is commutative, i.e., $Tr(AD) = Tr(DA)$. Therefore, in our case $\left\|DA_nD^{-1}\right\|_F^2 = Tr(A_nA_n^*)$. Thus, for now, we suppose that

$$\left\|D\left(D^{-1}H_0\right)^n D^{-1}\right\|_F \le \kappa < 1. \tag{3.15}$$

From the definition (3.11) of A_n, it follows that

$$\begin{aligned}\left\|DA_nD^{-1}\right\|_F &= \left\|D\sum_{r=1}^{\infty}\left(D^{-1}H_0\right)^{rn} D^{-1}\right\|_F \le \sum_{r=1}^{\infty}\left\|D(D^{-1}H_0)^n D^{-1}\right\|_F^r \\ &\le \sum_{r=1}^{\infty}\kappa^r = \frac{\kappa}{1-\kappa}.\end{aligned} \tag{3.16}$$

For the term $\left\|HD^{-1}\right\|_\infty$, we have to remind us that the matrix H is given by

$$H = [h_{\ell k}] := D_{\ell k} - \sum_{j=1}^{N} \frac{\sqrt{1-|a_\ell|^2}}{1-a_\ell e^{-ix_j}} \prod_{s=0}^{\ell-1} \frac{e^{ix_j}-a_s}{1-\overline{a_s}e^{ix_j}} \frac{\sqrt{1-|a_k|^2}}{1-\overline{a_k}e^{ix_j}} \prod_{m=0}^{k-1} \frac{e^{-ix_j}-\overline{a_m}}{1-a_m e^{-ix_j}},$$

such that by (3.9) and $\left|\prod_{s=0}^{\ell-1} \frac{e^{ix}-a_s}{1-\overline{a_s}e^{ix}}\right| = 1$ and $\left|\prod_{m=0}^{k-1} \frac{e^{-ix}-\overline{a_m}}{1-a_m e^{-ix}}\right| = 1$ we obtain

$$\left\|HD^{-1}\right\|_\infty \le \sup_\ell \sum_{\ell\neq k\in T}\sum_{j=1}^{N} \left|\frac{\sqrt{1-|a_\ell|^2}}{1-\overline{a_\ell}e^{i\cdot x_j}}\right| \left|\frac{\sqrt{1-|a_k|^2}}{1-a_k e^{-i\cdot x_j}}\right| \left|D_{kk}^{-1}\right|. \tag{3.17}$$

Using the estimate from Lemma 3.5 in the last expression, we can further estimate

$$\begin{aligned}&\sup_\ell \sum_{\ell\neq k\in T}\sum_{j=1}^{N} \sqrt{\frac{2}{\epsilon}-1}\sqrt{\frac{2}{\epsilon}-1}\left|D_{kk}^{-1}\right| \le \sup_\ell \sum_{\ell\neq k\in T} N\left(\frac{2}{\epsilon}-1\right)\left|D_{kk}^{-1}\right| \\ &= N\left(\frac{2}{\epsilon}-1\right)\sup_\ell \sum_{\ell\neq k\in T}\left|D_{kk}^{-1}\right| \le N\frac{2}{\epsilon}\sup_\ell \sum_{\ell\neq k\in T}\left|D_{kk}^{-1}\right|.\end{aligned}$$

We recall that

$$\begin{aligned}|D_{kk}| &= \sum_{j=1}^{N}\frac{1-|a_\ell|^2}{\left|1-\overline{a_\ell}e^{i\cdot x_j}\right|^2} \ge N\cdot \min_j \frac{1-|a_\ell|^2}{\left|1-\overline{a_\ell}e^{i\cdot x_j}\right|^2} \ge N\frac{1-|a_\ell|^2}{(1+|a_\ell|)^2} \\ &= N\frac{1-|a_\ell|}{1+|a_\ell|} \ge N\frac{\epsilon}{2}.\end{aligned} \tag{3.18}$$

Then, taking into account (3.18) expression (3.17) be further simplified to

$$\sup_\ell \sum_{\ell\neq k\in T}\sum_{j=1}^{N}\left|\frac{\sqrt{1-|a_\ell|^2}}{1-\overline{a_\ell}e^{i\cdot x_j}}\right|\left|\frac{\sqrt{1-|a_k|^2}}{1-a_k e^{-i\cdot x_j}}\right| |D_{kk}^{-1}| \leq N\frac{2}{\epsilon}\sup_\ell \sum_{\ell\neq k\in T}|D_{kk}^{-1}|$$
$$\frac{N\epsilon}{2}\frac{2}{N\epsilon}(|T|-1) < |T|. \tag{3.19}$$

Now, from (3.14), (3.16) and (3.19), we get the estimate

$$\begin{aligned}\left\|HA_nD^{-1}\right\|_\infty &\leq \left\|HD^{-1}DA_nD^{-1}\right\|_\infty \leq \left\|HD^{-1}\right\|_\infty \left\|DA_nD^{-1}\right\|_\infty \\ &\leq |T|\frac{\kappa}{1-\kappa}.\end{aligned} \tag{3.20}$$

Thus, from (3.15) and $\|S_{n-1}\mathrm{sgn}(c)\|_\infty < a_1$ it follows

$$\sup_{k\in T^c}|P_k^{(2)}| \leq (1+a_1)\frac{\kappa}{1-\kappa}|T|^{\frac{3}{2}}. \tag{3.21}$$

Therefore, taking into account the estimate (3.18) from (3.21) and if

$$\frac{\kappa}{1-\kappa} \leq \frac{a_2}{1+a_1}|T|^{-\frac{3}{2}} \tag{3.22}$$

then $\sup_{k\in T^c}|P_k^{(2)}| \geq a_2$ as intended.

Also it follows from (3.22) that $\kappa < 1$ and $|T| \geq 1$ (note that if $T = \emptyset$ then $c = 0$) and ℓ_1-minimization will clearly recover f. Furthermore, we have

$$\mathbb{P}\big(\sup_{k\in T^c}|P_k| \geq 1\big) \leq \sum_{k\in D}\mathbb{P}(E_k) + \mathbb{P}\big(\left\|(D^{-1}H_0)^n\right\|_F \geq \kappa\big). \tag{3.23}$$

Thus, we need to estimate $\mathbb{P}(E_k)$ and $\mathbb{P}\big(\left\|(D^{-1}H_0)^n\right\|_F \geq \kappa\big)$.

3.3. Analysis of powers of G_0

Our above considerations mean that we need to estimate the powers of the random matrix $G_0 = D^{-1}H_0$ in Frobenius norm. In fact we will need the expectation value $\mathbb{E}_X$ of it to estimate the probability.

Lemma 3.6. *It holds*

$$\mathbb{E}_X[\|G_0^n\|_F^2] \leq \sum_{t=1}^{\min\{n,N\}}\frac{N!}{(N-t)!}\sum_{\mathcal{A}\in P(2n,t)} \mathcal{C}_d\,\mathcal{C}(\mathcal{A},T),$$

with

$$\mathcal{C}_d := \left(\prod_{s=1}^{2n}\sum_{l_s=1}^{N}\left|\mathcal{B}_{k_s}(e^{ix_{l_s}})\right|^2\right)^{-1} \tag{3.24}$$

and

$$\mathcal{C}(\mathcal{A},T) := \sum_{k_1,k_2,\dots,k_{2n}\in T, k_r\neq k_{r+1}}\prod_{A\in\mathcal{A}}\left(\frac{2}{\epsilon}-1\right)^t \tag{3.25}$$

for a small ϵ defined as in Lemma 3.5.

Proof. As first step we have to remark that it will be a quite hard problem to calculate the expectation value with respect to a Blaschke product and, therefore, we are more interested to get an estimate.

Let us start by taking a closer look at the self-adjoint matrix G_0. We need to estimate $\|G_0^n\|_F^2 = \mathrm{Tr}(G_0^{2n})$. Since we have

$$\begin{aligned}\mathcal{F}_X &:= \left[\mathcal{B}_k(e^{ix_j})\right]_{j=1,\ldots,N,k=1,\ldots,d} \\ &= \left[\frac{\sqrt{1-|a_k|^2}}{1-\overline{a_k}e^{ix_j}}\prod_{t=0}^{k-1}\frac{e^{ix_j}-a_t}{1-\overline{a_t}e^{ix_j}}\right]_{j=1,\ldots,N,k=1,\ldots,d},\end{aligned} \tag{3.26}$$

we know that the diagonal matrix D and its inverse D^{-1} are given by

$$D := \left[\sum_{j=1}^{N}\left|\mathcal{B}_k(e^{ix_j})\right|^2\delta_{k\ell}\right]_{k,\ell\in T}, \quad D^{-1} := \left[\sum_{j=1}^{N}\left|\mathcal{B}_k(e^{ix_j})\right|^2\delta_{k\ell}\right]^{-1}_{k,\ell\in T},$$

and, additionally, we have the matrix H_0 given by

$$H_0 := \left[(1-\delta_{lk})\sum_{j=1}^{N}\overline{\mathcal{B}_l(e^{ix_j})}\mathcal{B}_k(e^{ix_j})\right]_{k,l\in T}.$$

Consider now the entry (l,k) of the matrix $G_0 = D^{-1}H_0$. Here we get

$$G_0(l,k) := (D^{-1}H_0)_{lk} = (1-\delta_{lk})\frac{\mathcal{B}_l \bullet \mathcal{B}_k}{\|\mathcal{B}_l\|^2}, \tag{3.27}$$

where we used the discrete scalar product

$$\mathcal{B}_l \bullet \mathcal{B}_k = \sum_{j=1}^{N}\overline{\mathcal{B}_l(e^{ix_j})}\mathcal{B}_k(e^{ix_j})$$

and the norm

$$\|\mathcal{B}_k\|^2 = \sum_{j=1}^{N}\left|\mathcal{B}_k(e^{ix_j})\right|^2.$$

We aim to estimate the square of Frobenius norm of (3.27), i.e.,

$$\left\|(D^{-1}H_0)^n\right\|_F^2 = \mathrm{Tr}\left[(D^{-1}H_0)^{2n}\right] = \mathrm{Tr}\left[G_0^{2n}\right].$$

We need to calculate $G_0^{2n}(k_1,k_1)$. From now on, and in order to simplify the notation, we always assume $k_j \neq k_{j+1}$ without explicitly mentioning it. The general form of $G_0^{2n}(k_1,k_1)$ is given by

$$G_0^{2n}(k_1,k_1) = \sum_{k_2,\dots,k_{2n}\in T} G_0(k_1,k_2)G_0(k_2,k_3)\cdots G_0(k_{2n},k_1)$$
$$= \sum_{k_2,\dots,k_{2n}\in T}\prod_{r=1}^{2n} G_0(k_r,k_{r+1}).$$

In a more detailed form we have

$$G_0^{2n}(k_1,k_1) = \sum_{k_2,\dots,k_{2n}\in T} \frac{\mathcal{B}_{k_1}\bullet\mathcal{B}_{k_2}}{\|\mathcal{B}_{k_1}\|^2}\frac{\mathcal{B}_{k_2}\bullet\mathcal{B}_{k_3}}{\|\mathcal{B}_{k_2}\|^2}\cdots\frac{\mathcal{B}_{k_{2n}}\bullet\mathcal{B}_{k_1}}{\|\mathcal{B}_{k_{2n}}\|^2}$$
$$= \sum_{k_2,\dots,k_{2n}\in T}\sum_{j_1,\dots,j_{2n}=1}^{N} \frac{\overline{\mathcal{B}_{k_1}(e^{ix_{j_1}})}\mathcal{B}_{k_2}(e^{ix_{j_1}})}{\sum_{l_1=1}^N\left|\mathcal{B}_{k_1}(e^{ix_{l_1}})\right|^2}\frac{\overline{\mathcal{B}_{k_2}(e^{ix_{j_2}})}\mathcal{B}_{k_3}(e^{ix_{j_2}})}{\sum_{l_2=1}^N\left|\mathcal{B}_{k_2}(e^{ix_{l_2}})\right|^2}\cdots$$
$$\cdots\frac{\overline{\mathcal{B}_{k_{2n}}(e^{ix_{j_{2n}}})}\mathcal{B}_{k_1}(e^{ix_{j_{2n}}})}{\sum_{l_{2n}=1}^N\left|\mathcal{B}_{k_{2n}}(e^{ix_{l_{2n}}})\right|^2}.$$

As in [30] we have to switch to partitions $\mathcal{A}$ since some indices i_r might be the same which means that we cannot use directly the product rule for the expectation value since it is only valid for independent random variables. We associate a partition $\mathcal{A}=(A_1,A_2,\dots,A_t)$ of $\{1,\dots,2n\}$ to a certain vector $i_1,\dots,i_{2n}$ such that $i_r=i_{r'}$ if and only if r and r' are contained in the same set $A_i\in\mathcal{A}$. This is allows us to unambiguously write i_A instead of i_r if $r\in A$. Furthermore, if $A\in\mathcal{A}$ contains only a pair of elements $\overline{\mathcal{B}_{a_{t_r}}(z_A)}\mathcal{B}_{a_{t_{r+1}}}(z_A)$ then the term will vanish due to the condition $k_r\neq k_{r+1}$. Thus, we only need to consider partitions $\mathcal{A}$ satisfying $|A|\geq 2$ for all $A\in\mathcal{A}$, i.e., partitions in $P(2n,t)$ with $t>1$. Additionally, we need to remember that the number of vectors $(\ell_{A_1},\dots,\ell_{A_t})\in\{1,\dots,N\}^t$ with different entries is exactly $N\cdots(N-t+1)=N!/(N-t)!$ if $N\geq t$ and 0 if $N\leq t$. Thus, we obtain the expectation value of the trace of the matrix

$$\mathbb{E}_X\left[\mathrm{Tr}\left(G_0^{2n}\right)\right] = \sum_{k_1,k_2,\dots,k_{2n}\in T}\sum_{j_1,\dots,j_{2n}=1}^{N}\prod_{A\in\mathcal{A}}\left(\prod_{s=1}^{2n}\sum_{l_s=1}^{N}\left|\mathcal{B}_{k_s}(e^{ix_{l_s}})\right|^2\right)^{-1}$$
$$\times\,\mathbb{E}_X\left[\prod_{r\in A}\overline{\mathcal{B}_{k_r}(e^{ix_{j_A}})}\mathcal{B}_{k_{r+1}}(e^{ix_{j_A}})\right]. \tag{3.28}$$

This description is possible since the terms $\sum_{l_s=1}^N\left|\mathcal{B}_{k_s}(e^{ix_{l_s}})\right|^2$ are actually fixed constants. The complete calculation of (3.28) can be found in the appendix.

Since x_{j_A} has uniform distribution on $[0,2\pi]$ from (3.28) we have to look for the expectation value

$$\mathbb{E}_X\left[\prod_{r\in A}\overline{\mathcal{B}_{k_r}(e^{ix_{j_A}})}\mathcal{B}_{k_{r+1}}(e^{ix_{j_A}})\right].$$

For $|A| = 2$, (note that $k_r \neq k_{r+1}$, $k_s \neq k_{s+1}$ and $k_r = k_{s+1}$) we obtain

$$\mathbb{E}_X \left[\overline{\mathcal{B}_{k_r}(e^{ix_{j_A}})}\mathcal{B}_{k_{r+1}}(e^{ix_{j_A}})\overline{\mathcal{B}_{k_s}(e^{ix_{j_A}})}\mathcal{B}_{k_{s+1}}(e^{ix_{j_A}})\right]$$
$$= \mathbb{E}_X \left[|\mathcal{B}_{k_r}(e^{ix_{j_A}})|^2\mathcal{B}_{k_{r+1}}(e^{ix_{j_A}})\overline{\mathcal{B}_{k_s}(e^{ix_{j_A}})}\right]$$
$$= \frac{1}{2\pi i}\int_{|z_A|=1} \frac{1-|a_{k_r}|^2}{|1-a_{k_r}\overline{z_A}|^2}\frac{\sqrt{1-|a_{k_{r+1}}|^2}}{1-\overline{a_{k_{r+1}}}z_A}\prod_{s=0}^{k_{r+1}-1}\frac{z_A - a_s}{1-\overline{a_s}z_A}$$
$$\times \frac{\sqrt{1-|a_{k_s}|^2}}{1-a_{k_s}\overline{z_A}}\prod_{u=0}^{k_s-1}\frac{\overline{z_A}-\overline{a_u}}{1-a_u\overline{z_A}}\frac{\sqrt{1-|a_{k_{s+1}}|^2}}{1-\overline{a_{k_{s+1}}}z_A}\prod_{n=0}^{k_{s+1}-1}\frac{z_A-a_n}{1-\overline{a_n}z_A}\frac{dz_A}{z_A}. \tag{3.29}$$

On the one hand, if $k_s = k_{r+1}$ from expression (3.29) we obtain

$$\mathbb{E}_X \left[|\mathcal{B}_{k_r}(e^{ix_{j_A}})|^2|\mathcal{B}_{k_s}(e^{ix_{j_A}})|^2\right] = \frac{1}{2\pi i}\int_{|z_A|=1}\frac{1-|a_{k_r}|^2}{|1-a_{k_r}\overline{z_A}|^2}\frac{1-|a_{k_r}|^2}{|1-a_{k_r}\overline{z_A}|^2}\frac{dz_A}{z_A}$$
$$= \frac{1-|a_{k_r}|^2}{1-\overline{a_{k_r}}a_{k_s}}\frac{a_{k_s}}{a_{k_s}-a_{k_r}} + \frac{1-|a_{k_s}|^2}{1-\overline{a_{k_s}}a_{k_r}}\frac{a_{k_r}}{a_{k_r}-a_{k_s}}.$$

To estimate the last value we will consider the estimate by Lemma 3.5 and introduce the quantities $\epsilon_2 = |a_r - a_s|$ and ϵ_1 keeping in mind that $|a| < 1$. Thereby, we get

$$\frac{1-|a_{k_r}|^2}{1-\overline{a_{k_r}}a_{k_s}}\frac{a_{k_s}}{a_{k_s}-a_{k_r}} + \frac{1-|a_{k_s}|^2}{1-\overline{a_{k_s}}a_{k_r}}\frac{a_{k_r}}{a_{k_r}-a_{k_s}}$$
$$\leq \sqrt{\frac{2}{\epsilon_1}-1}\frac{|a_1|}{\epsilon_2} + \sqrt{\frac{2}{\epsilon_1}-1}\frac{|a_2|}{\epsilon_2} \leq 2\sqrt{\frac{2}{\epsilon_1}-1}\frac{1}{\epsilon_2} = \frac{2}{\epsilon_2}\sqrt{\frac{2}{\epsilon_1}-1}.$$

On the other hand, if $k_s \neq k_{r+1}$ from expression (3.29) we obtain

$$\mathbb{E}_X \left[\left|\mathcal{B}_{k_r}(e^{ix_{j_A}})\right|^2\overline{\mathcal{B}_{k_{r+1}}(e^{ix_{j_A}})}\mathcal{B}_{k_s}(e^{ix_{j_A}})\right]$$
$$= \frac{1}{2\pi}\int_{|z_A|=1}\frac{1-|a_{k_r}|^2}{\left|1-a_{k_r}\overline{z_A}\right|^2}\frac{\sqrt{1-|a_{k_{r+1}}|^2}}{1-a_{k_{r+1}}\overline{z_A}}$$
$$\times \prod_{t=0}^{k_{r+1}-1}\frac{\overline{z_A}-\overline{a_t}}{1-a_t\overline{z_A}}\frac{\sqrt{1-|a_{k_s}|^2}}{1-\overline{a_{k_s}}z_A}\prod_{u=0}^{k_s-1}\frac{z_A-a_u}{1-\overline{a_u}z_A}\frac{dz_A}{z_A}$$
$$= \frac{1}{2\pi}\int_{|z_A|=1}\frac{1-|a_{k_r}|^2}{1-\overline{a}_{k_r}z_A}\frac{z_A}{z_A-a_{k_r}}\frac{\sqrt{1-|a_{k_{r+1}}|^2}}{z_A-a_{k_{r+1}}}$$
$$\times \prod_{t=0}^{k_{r+1}-1}\frac{\overline{z_A}-\overline{a_t}}{1-a_t\overline{z_A}}\frac{\sqrt{1-|a_{k_s}|^2}}{1-\overline{a_{k_s}}z_A}\prod_{u=0}^{k_s-1}\frac{z_A-a_u}{1-\overline{a_u}z_A}dz_A. \tag{3.30}$$

From the last expression by assuming $k_s > k_{r+1}$ we get

$$\begin{aligned}
&\mathbb{E}_X\left[|\mathcal{B}_{k_r}(e^{ix_{j_A}})|^2|\mathcal{B}_{k_s}(e^{ix_{j_A}})|^2\right]\\
&= \frac{1}{2\pi}\int_{|z_A|=1} \frac{1-|a_{k_r}|^2}{1-\overline{a_{k_r}}z_A}\frac{z_A}{z_A-a_{k_r}}\frac{\sqrt{1-|a_{k_{r+1}}|^2}}{z_A-a_{k_{r+1}}}\frac{\sqrt{1-|a_{k_s}|^2}}{1-\overline{a_{k_s}}z_A}\prod_{u=k_{r+1}}^{k_s-1}\frac{z_A-a_u}{1-\overline{a_u}z_A}dz_A\\
&= a_{k_r}\frac{\sqrt{1-|a_{k_{r+1}}|^2}}{1-\overline{a_{k_{r+1}}}a_{k_r}}\frac{\sqrt{1-|a_{k_s}|^2}}{1-\overline{a_{k_s}}a_{k_r}}\prod_{u=k_{r+1}+1}^{k_s-1}\frac{a_{k_r}-a_u}{1-\overline{a_u}a_{k_r}}. \qquad (3.31)
\end{aligned}$$

We can estimate the absolute value of the last result in (3.31) as

$$\begin{aligned}
&|a_{k_r}|\left|\frac{\sqrt{1-|a_{k_{r+1}}|^2}}{1-\overline{a_{k_{r+1}}}a_{k_r}}\frac{\sqrt{1-|a_{k_s}|^2}}{1-\overline{a_{k_s}}a_{k_r}}\prod_{u=k_{r+1}+1}^{k_s-1}\frac{a_{k_r}-a_u}{1-\overline{a_u}a_{k_r}}\right|\\
&\le \sqrt{\frac{1-|a_{k_{r+1}}|^2}{|1-\overline{a_{k_{r+1}}}a_{k_r}|^2}}\sqrt{\frac{1-|a_{k_s}|^2}{|1-\overline{a_{k_s}}a_{k_r}|^2}}\prod_{u=k_{r+1}+1}^{k_s-1}\left|\frac{a_{k_r}-a_u}{1-\overline{a_u}a_{k_r}}\right| \le \frac{2}{\epsilon}-1,
\end{aligned}$$

such that

$$|a_{k_r}|<1 \quad \text{and} \quad d(a_{k_r},a_u) = \left|\frac{a_{k_r}-a_u}{1-\overline{a_u}a_{k_r}}\right| < 1,$$

where $d(a_{k_r},a_u)$ is called *pseudo-hyperbolic distance* between a_u and a_{k_r}.

Note that if $k_s < k_{r+1}$, we obtain basically the same result. For $|A| = n$, we shall obtain in the same way

$$\mathbb{E}\left[\prod_{s=1}^{n}\overline{\mathcal{B}_{k_{r_s}}(e^{ix_{j_s}})}\mathcal{B}_{k_{r_s+1}}(e^{ix_{j_s}})\right] \le \left(\frac{2}{\epsilon}-1\right)^{n-1}.$$

Before we show this we are going to give two remarks. First of all we can easily check that if the components are all different, i.e.,

$$\mathbb{E}\left[\prod_{s=1}^{n}\overline{\mathcal{B}_{k_{r_s}}(e^{ix_{j_s}})}\mathcal{B}_{k_{r_s+1}}(e^{ix_{j_s}})\right]$$

with $k_{r_s}\neq k_{r_t}$ when $s\neq t$ we get zero as the result. Furthermore, it is also easy to see that the highest value in the estimates happen when the pair $\overline{\mathcal{B}_{k_r}(e^{ix_{j_A}})\mathcal{B}_{k_s}(e^{ix_{j_A}})}$ repeats itself n times with $r_s \neq r_s+1$.

This means that, although keeping $r\neq t$, the main problem is to estimate the values of the integral for the expectation value

$$\mathbb{E}_X\left[\left(\overline{\mathcal{B}_{a_r}(x_{s_A})}\mathcal{B}_{a_t}(x_{s_A})\right)^n\right] = \frac{1}{2\pi}\int_{|z_A|=1}\left(\overline{\mathcal{B}_{a_r}(x_{s_A})}\mathcal{B}_{a_t}(x_{s_A})\right)^n\frac{dz}{iz_A} \qquad (3.32)$$

which leads to

$$\begin{aligned}
&\mathbb{E}_X\left[\left(\mathcal{B}_{a_r}(x_{s_A})\overline{\mathcal{B}_{a_t}(x_{s_A})}\right)^n\right]\\
&= \frac{1}{2\pi}\int_{|z_A|=1}\left(\frac{\sqrt{1-|a_r|^2}}{1-\overline{a_r}z_A}\frac{\sqrt{1-|a_t|^2}}{1-a_t\overline{z_A}}\right)^n\frac{dz_A}{iz_A}, \quad \text{with } r\neq t.
\end{aligned} \qquad (3.33)$$

Let us start our estimate as follows

$$\left|\frac{1}{2\pi}\int_{|z|=1}\left(\frac{\sqrt{1-|a_r|^2}}{1-\overline{a_r}z}\frac{\sqrt{1-|a_t|^2}}{1-a_t\overline{z}}\right)^n\frac{dz}{iz}\right|$$

$$\leq \frac{1}{|2\pi i|}\int_{|z|=1}\left|\sqrt{\frac{1-|a_r|^2}{|1-\overline{a_r}z|^2}}\sqrt{\frac{1-|a_t|^2}{|1-a_t\overline{z}|^2}}\right|^n |dz|.$$

$$\leq \left(\max_{a_n\in\mathbb{D}}\max_{z\in\mathbb{T}}\frac{1-|a_n|^2}{|1-\overline{a_n}z|^2}\right)^n \leq \left(\frac{2}{\epsilon}-1\right)^n.$$

From this using Lemma 3.5 we get

$$\mathcal{C}(\mathcal{A},T) := \sum_{r_1,r_2,\ldots,r_{2n}\in T, r_j\neq r_{j+1}} \prod_{A\in\mathcal{A}}\left(\frac{2}{\epsilon}-1\right)^{|A|} = \prod_{A\in\mathcal{A}}\left(\frac{2}{\epsilon}-1\right)^{|A|} T^{2n-|A|+1}$$

$$= \mathcal{K}\times\#\Big\{(r_1,r_2,\ldots,r_{2n})\in T^{2n} : r_j\neq r_{j+1}$$

$$\wedge \sum_{r\in A}\int_{|z_A|=1}\prod_{r=1}^{2n}\overline{\mathcal{B}_{a_r}(z_A)}\mathcal{B}_{a_{r+1}}(z_A)\frac{dz_A}{iz_A}\neq 0,\ \forall A\in\mathcal{A}\Big\}, \tag{3.34}$$

where

$$\mathcal{K} := \prod_{A\in\mathcal{A}}\left(\frac{2}{\epsilon}-1\right)^{|A|}. \tag{3.35}$$

Let us remark that in the above considerations everything depends on the result of the integral

$$\int_{|z_A|=1}\prod_{r=1}^{2n}\overline{\mathcal{B}_{a_r}(z_A)}\mathcal{B}_{a_{r+1}}(z_A)\frac{dz_A}{iz_A} \tag{3.36}$$

for $\mathcal{A}\in P(2n,t)$. Here, the indices $(t_1,t_2,\ldots,t_{2n})\in T^{2n}$ are subjected to the $|A| = t$ constraints where the above integral is different of zero. This also is the principal point if one wants to extend the proof to more general cases like frames. □

3.4. Analysis of $\mathbb{P}(E_k)$

We have to study the term $\mathbb{P}(E_k)$ in (3.23). In the usual manner, let $\beta_m = \beta^{n/K_m}$, $m = 1,\ldots,n$ be positive numbers satisfying

$$\sum_{m=1}^{n}\beta_m = a_1$$

and $K_m \in \mathbb{N}$, $m = 1, \dots, n$, some natural numbers. For $k \in D$ as before we have

$$\mathbb{P}(E_k) = \mathbb{P}\Big(\sum_{m=1}^{n} |((D^{-1}HR_T)^m \operatorname{sgn}(c))_k| \geq a_1 \Big)$$

$$= \sum_{m=1}^{n} \mathbb{P}(|((D^{-1}HR_T)^m \operatorname{sgn}(c))_k|^{2K_m} \beta_m^{-2K_m} \geq 1) \tag{3.37}$$

$$\leq \sum_{m=1}^{n} \mathbb{E}[|((D^{-1}HR_T)^m \operatorname{sgn}(c))_k|^{2K_m}] \beta_m^{-2K_m}, \tag{3.38}$$

where we now have $\beta_m^{-2K_m} = \beta^{-2n}$ for all m. We remark that (3.37) and (3.38) are obtained from (even when the expectation is infinite) $\mathbb{E}(X) = \sum_{i=1}^{\infty} \mathbb{P}(X \geq i)$.

Thus, we obtain

$$\mathbb{P}(E_k) \leq \beta^{-2n} \sum_{m=1}^{n} \mathbb{E}[|((D^{-1}HR_T)^m \operatorname{sgn}(c))_k|^{2K_m}] \tag{3.39}$$

and the condition $a_1 = \sum_{m=1}^{n} \beta_m$ reads as

$$a_1 = a = \sum_{m=1}^{n} \beta^{n/K_m} < 1.$$

The above consideration means that we have to study the expectation value appearing in (3.39). The following proof is similar to the one of Lemma 3.6.

Lemma 3.7. *For $k \in D$ and $c \in \ell_2(D)$ with* $\operatorname{supp} c = T$ *we have*

$$\mathbb{E}\left[\left|\left((D^{-1}HR_T)^m \operatorname{sgn} c\right)_k\right|^{2K}\right] \leq \sum_{t=1}^{\min\{Km,N\}} \frac{N!}{(N-t)!} \sum_{\mathcal{A} \in P(2Km,t)} \mathcal{C}_d \times \mathcal{B}(\mathcal{A}, T).$$

Hereby, we identify partitions of $[2Km]$ in $P(2Km,t)$ with partitions of $[2K] \times [m]$ with fixed k_r and k_{r+1} such that $k_r \neq k_{r+1}$,

$$\mathcal{C}_d = \left(\prod_{q=1}^{2K} \prod_{s=1}^{m} \sum_{l_s=1}^{N} \left| \mathcal{B}_{k_s^{(q)}}(e^{ix_{l_s}^{(q)}}) \right|^2 \right)^{-1},$$

and

$$\mathcal{B}(\mathcal{A}, T) = \sum_{\substack{k_1^{(1)}, k_2^{(1)}, \dots, k_m^{(1)} \in T \\ \vdots \\ k_1^{(2K)}, k_2^{(2K)}, \dots, k_m^{(2K)} \in T \\ k_j^{(p)} \neq k_{j+1}^{(p)},\ j \in [m]}} \prod_{A \in \mathcal{A}} \left(\frac{2}{\epsilon} - 1 \right)^t$$

for a small ϵ defined as in Lemma 3.5.

For the proof of this lemma, we reefer to the appendix.

Now the objective is to complete the proof of Theorem 3.1 by joining all the established facts. Consider the absolute value of the quantity $C(\mathcal{A},T)$ defined in (3.25) for A $\in P(2n,t)$. Here the indices $(k_1,\dots,k_{2n}) \in T^{2n}$ are subjected to the $|A| = t$ linear constraints (3.36) for all $A \in \mathcal{A}$. These constraints are independent except when the sum of the integrals is zero. Thus, from (3.34) and (3.35) we can estimate

$$|C(\mathcal{A},T)| \le \mathcal{K}^t\, |T|^{2n-t+1} \le \mathcal{K}^t M^{2n-t+1}. \tag{3.40}$$

By Lemma 3.6 we obtain

$$\begin{aligned}\mathbb{E}\left[\left\|D^{-1}H_0^n\right\|_F^2\right] &\le C_d \sum_{t=1}^{\min\{n,N\}} \frac{N!}{(N-t)!} \sum_{\mathcal{A}\in P(2n,t)} \mathcal{K}^t\, |T|^{2n-t+1}\\ &\le C_d M^{2n+1} \sum_{t=1}^{n} \left(\frac{N\mathcal{K}}{M}\right)^t S_2(2n,t),\end{aligned}$$

where $S_2(n,t) = |P(2n,t)|$ are the associated Stirling numbers of the second kind.

Set $\theta = \frac{N\mathcal{K}}{M}$. Markov's inequality now yields

$$\begin{aligned}\mathbb{P}\left(\left\|\left(D^{-1}H_0\right)^n\right\|_F \ge \kappa\right) = \mathbb{P}\left(\left\|\left(D^{-1}H_0\right)^n\right\|_F^2 \ge \kappa^2\right) &\le \kappa^{-2}\,\mathbb{E}\left[\left\|\left(D^{-1}H_0\right)^n\right\|_F^2\right]\\ &\le \kappa^{-2} C_d \mathcal{K}^{2n}\, M\, N^{2n} G_{2n}(\theta).\end{aligned}$$

Let us note that from (3.15) we have $\kappa < 1$. So, we have $\left\|D\left(D^{-1}H_0\right)^n D^{-1}\right\|_F \le \kappa$ which implies $\left(I_T - \left(D^{-1}H_0\right)^n\right)$ is invertible by a von Neumann series. In the same way, $\mathcal{F}_{TX}^{*}\mathcal{F}_{TX} = D_T\left(I_T - D^{-1}H_0\right)$ is invertible. Hence, also $\mathcal{F}_{TX}$ is injective.

Let us now consider $P(E_k)$. By Lemma 3.7 we need to bound $B(\mathcal{A},T)$ defined in (A.2), i.e., the number of vectors $(k_j^{(p)}) \in T^{2Km}$ satisfying $k_j^{(p)} = k_{j+1}^{(p)}$ for all $A \in \mathcal{A}$ with $\mathcal{A} \in P(2Km,t)$. These are t independent linear constraints. Therefore, the number of these indices is bounded from above by $|T|^{2Km-t} \le M^{2Km-t}$. Thus, in same way, by taking $\theta = \frac{N\mathcal{K}}{M}$, we obtain

$$\mathbb{E}_X\left[\left|\left(\left(D^{-1}HR_T\right)^m \operatorname{sgn} c\right)_k\right|^{2K}\right] \le C_d \sum_{t=1}^{Km} (N\mathcal{K})^t\, S_2(2Km,t) M^{2Km-t}.$$

Let us consider $\mathbb{P}(\text{failure})$ as the probability that the exact reconstruction of f by ℓ_1-minimization fails.

By Lemma 3.4 and (3.23) we again obtain

$$\begin{aligned}\mathbb{P}(\text{failure}) &\le \mathbb{P}\left(\{\mathcal{F}_{TX}\text{is not injective}\} \cup \{\sup\nolimits_{k\in T^c}|P_k| \ge 1\}\right)\\ &\le C_d \sum_{k\in D} \mathbb{P}(E_k) + \mathbb{P}\left(\left\|\left(D^{-1}H_0\right)^n\right\| \ge k\right)\end{aligned}$$

$$\leq \mathcal{C}_d\, d\beta^{-2n} \sum_{m=1}^{n} G_{2mK_m}(\theta) + \kappa^{-2}\, \mathcal{C}_d \mathcal{K}^{2n}\, M\, N^{2n} G_{2n}(\theta)$$
$$= \mathcal{C}_d \left[d\beta^{-2n} \sum_{m=1}^{n} G_{2mK_m}(\theta) + \kappa^{-2} \mathcal{K}^{2n}\, M\, N^{2n} G_{2n}(\theta) \right]$$

under the conditions

$$a_1 = a = \sum_{m=1}^{n} \beta^{n/K_m} < 1, \quad a_2 + a_1 = 1, \text{ i.e., } a_2 = 1 - a,$$
$$\frac{\kappa}{1-\kappa} \leq \frac{a_2}{1+a_1} M^{-3/2} = \frac{1-a}{1+a} M^{-3/2}.$$

4. Applications

The main goal of this section is to present some numerical experiments using Takenaka–Malmquist systems. Since their main field of application is in the study of transfer functions in systems identification we choose some examples of such functions. To allow for comparison we use the following two examples of transfer functions from a recent thesis [31],

$$E(z) = e^{e^z} \quad \text{and} \quad F(z) = \frac{0.247z^4 + 0.0355z^3}{0.3329z^2 - 1.2727z + 1}.$$

But before this we take the Poisson kernel

$$P(\zeta, z_0) = \frac{r^2 - |z_0|^2}{2\pi r |z_0 - \zeta|^2}$$

as an example. This function is of particular interest to us due to the fact that whenever $|z_0| \to 1$ we get a singularity at the point z_0 whose influence can be studied for different parameters z_0.

For the numerical calculations we use the Matlab toolbox $\ell 1$-Magic [10] which adopts a Linear Programming to minimize the ℓ_1-norm of our coefficients x subject to $y = Ax$ using the primal-dual interior point method (see, for instance, [34]) with A being our sampling matrix.

Since $\ell 1$-Magic works with real-valued vectors we need to modify our (complex-valued) system. We rewrite our complex multiplication $(\alpha + \mathbf{i}\beta)(v + \mathbf{i}w) = a + \mathbf{i}b$ as a matrix-vector multiplication, i.e.,

$$\begin{pmatrix} \alpha & -\beta \\ \beta & \alpha \end{pmatrix} \begin{pmatrix} v \\ w \end{pmatrix} = \begin{pmatrix} a \\ b \end{pmatrix}.$$

This allows us to rewrite the complex linear system in the form $\tilde{y} = \mathcal{M}\tilde{x}$ with

$$M = \begin{pmatrix} \operatorname{Re}(A) & -\operatorname{Im}(A) \\ \operatorname{Im}(A) & \operatorname{Re}(A) \end{pmatrix} \quad \text{and} \quad \tilde{x} = \begin{pmatrix} \operatorname{Re}(x) \\ \operatorname{Im}(x) \end{pmatrix}, \quad \tilde{y} = \begin{pmatrix} \operatorname{Re}(y) \\ \operatorname{Im}(y) \end{pmatrix}.$$

We need to choose our points a_i for the Blaschke products. To this end we take a grid given by the points

$$\{z_{k\ell} = r_k e^{\mathbf{i}\frac{2\pi\ell}{2^{2k}}},\ \ell = 0, 1, \dots, 2^{2k} - 1, k = 0, \dots, m\}$$

where $r_k = \frac{2^k - 2^{-k}}{2^k + 2^{-k}}$ denotes the radius of the concentric circles such that on the circle with radius r_k we take 2^{2k} equidistant points (see for instance [25]). From this grid we take N randomly chosen points, i.e., a vector $\mathbf{a} = (a_1, \dots, a_N)$.

For our examples we made the simulation using Matlab 8.5.0(R2015a) running on a laptop with Intel(R) Core(TM) i3-4010U CPU 1.70 GHz, RAM 4GB, Windows 10, OS 64-bit(win64).

Example. Consider the Poisson kernel over the unit circle (see [21])

$$g(\zeta, z_0) = \frac{1 - |z_0|^2}{2\pi|z_0 - \zeta|^2}.$$

In this example we choose z_0 to be near to zero, in this case $z_0 = -0.1$ and the function is sampled by $N = 1300$ measurements. In Figure 1 we can see the reconstruction of the function by using only $M = 55$ samples. This corresponds to 4.23% of our total measurements.

When we choose $z_0 = -0.5$ with the same number of measurements ($N = 1300$) we can see a similar level of reconstruction (cf. Figure 1) with a slight increase in the number of used samples ($M = 70$).

For values of z_0 near to the unit circle, i.e., $|z_0| \cong 1$ (in our case we choose $z_0 = -0.8$), with the same number of measurements ($N = 1300$) we can see that again with a slight increase in the number of taken sampling points ($M = 130$) we get a similar quality of reconstruction while using the same percentage (10%) for a smaller number of sampling points ($M = 20$) in the case of $z_0 = -0.8$ we still can get a decent approximation with a dramatically smaller running time.

From this example we redraw the following observations:

1. Within the same number of measurements, when $|z_0|$ is near to zero we have the best reconstruction in the least time.
2. When the modulus of the parameter z_0 is close to 1 it requires more samples to reconstruct the signal.
3. The reconstruction is better in case when $|a_j - r| < \epsilon$ with ϵ relatively small and the parameter a_j being randomly chosen.

We will use the next examples to compare our results with the results from the PhD-thesis of L. Shuang [31]. Note that in his case he chooses the parameter a_j from an a-priori given grid while in our case we use randomly chosen parameters.

Example. Consider the example of the transfer function (from [31])

$$F(z) = \frac{0.247z^4 + 0.0355z^3}{0.3329z^2 - 1.2727z + 1}.$$

For this example we sample the above function using 1000 samples (same as in [31]).

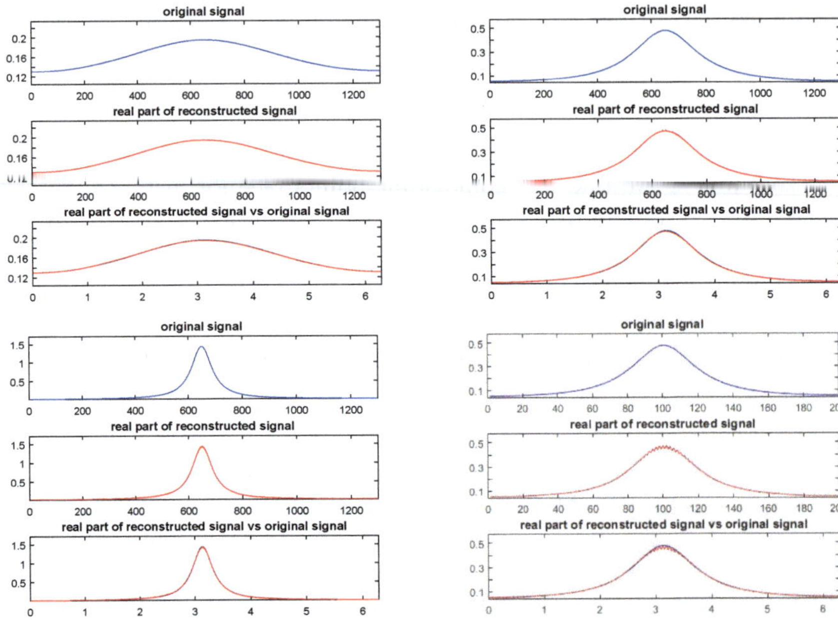

FIGURE 1. From left to right, top to bottom: $z_0 = -0.1$, 55 sampling points corresponds to $\approx 4.23\%$ of total measurements. Time $= 329.85$ s, relative error $= 0.0045$; $z_0 = -0.5$, 70 sampling points corresponds to $\approx 5.38\%$ of total measurements. Time $= 302.72$ s, relative error $= 0.0057$; $z_0 = -0.8, M = 130$, 130 sampling points equivalent to 10% of total measurements. Time $= 324.75$ s, relative error $= 0.0094$; $z_0 = -0.8, M = 20$, 20 sampling points corresponds to 10% of total measurements. Time $= 2.0242$ s, relative error $= 0.0145$

In Shuang's work the relative error is 0.022 compared to the relative error of approximately 0.0004 in our case. Additionally, let us take a look to what happens if we take a bigger number of samples of the original function ($M = 110$). The original signal and the reconstructed signal can be seen in Figure 2. We can point out that the relative error is less than 0.0002 compared to a relative error of 0.004 in Shuang's work.

Unfortunately, since the author in [31] did not provide any information on the used hardware any comparison of runtimes is pure speculation.

Example. Consider the function

$$E(z) = e^{e^z}.$$

As we did in the previous example we sample our function in the same way (1000 samples) as in [31].

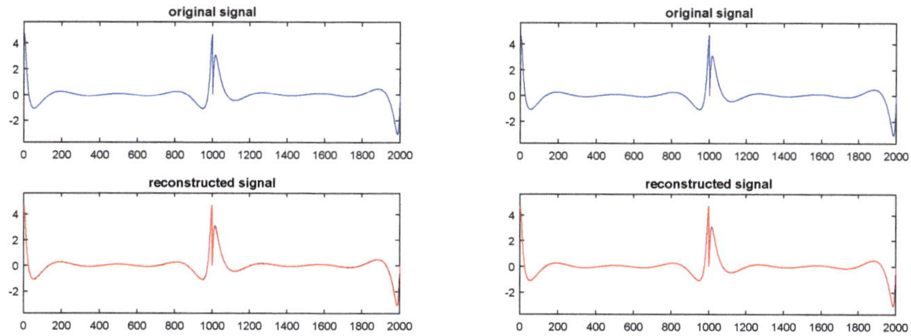

FIGURE 2. Left: the original image and the reconstructed image obtained from 55 random samples (5, 5%), relative error = 3.2271e-04. Right: The original signal and the reconstructed signal obtained from 110 samples (11%), relative error = 1.8086e-04.

In Figure 3 – Top Left we can observe that with only $M = 12$ sampling points we can reconstruct our function with relative error = 0.00046. In the work of Shuang the reconstruction was done with a relative error of 0.0004 but using a larger number of sampling points ($M = 55$).

In Figure 3 – Top Right we can observe that if we use the same number of sampling points $M = 55$ as in [31] then we get a relative error of 0.00002.

For $M = 110$ (Figure 3 – Bottom) the relative error is 0.000004 in contrast to the relative error from Shuang's example which is 0.00003.

Again, since the author in [31] did not provide any information on the used hardware any comparison of run times is pure speculation.

Taking into account these two last examples we can make the following observations:

1. Using the same number of measurements our method provides a better approximation than the approach in the thesis of Shuang [31];
2. Moreover, the same relative error is attained with our method by using a smaller number of sampling points.

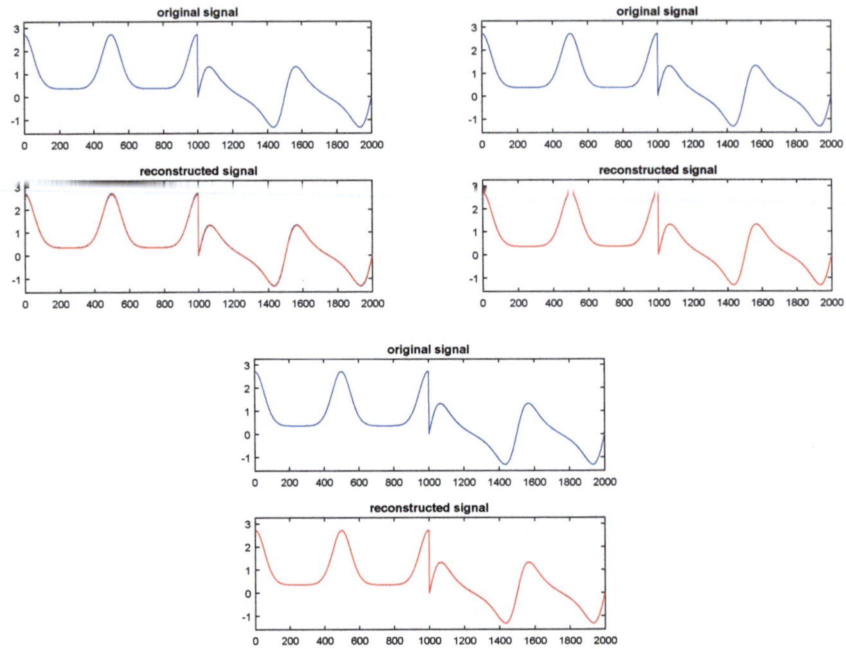

FIGURE 3. Top Left: reconstruction using 12 sampling points (corresponding to 1.2% of total measurements). Relative error = 4.5945e-04. Top Right: 55 sampling points corresponds to 5, 5% of total measurements. Relative error = 1.8646e-05. Bottom: 110 sampling points corresponds to 11% of total measurements. Relative error = 3.8346e-06.

Appendix: Additional calculi

A.1. Estimation of the term $\left\|HD^{-1}\right\|_\infty$

$$
\begin{aligned}
&\left\|HD^{-1}\right\|_\infty \\
&= \left\| \left(\sum_{j=1}^{N} \frac{\sqrt{1-|a_\ell|^2}}{1-a_\ell e^{-ix_j}} \prod_{s=0}^{\ell-1} \frac{e^{ix_j}-a_s}{1-\overline{a_s}e^{ix_j}} \frac{\sqrt{1-|a_k|^2}}{1-\overline{a_k}e^{ix_j}} \prod_{m=0}^{k-1} \frac{e^{-ix_j}-\overline{a_m}}{1-a_m e^{-ix_j}} D_{kk}^{-1} \right)_{\ell k} \right\|_\infty \\
&= \sup_\ell \sum_{\ell \neq k \in T} \left| \sum_{j=1}^{N} \frac{\sqrt{1-|a_\ell|^2}}{1-a_\ell e^{-ix_j}} \prod_{s=0}^{\ell-1} \frac{e^{ix_j}-a_s}{1-\overline{a_s}e^{ix_j}} \frac{\sqrt{1-|a_k|^2}}{1-\overline{a_k}e^{ix_j}} \prod_{m=0}^{k-1} \frac{e^{-ix_j}-\overline{a_m}}{1-a_m e^{-ix_j}} D_{kk}^{-1} \right| \\
&\leq \sup_\ell \sum_{\ell \neq k \in T} \left(\sum_{j=1}^{N} \left| \frac{\sqrt{1-|a_\ell|^2}}{1-\overline{a_\ell} e^{i\cdot x_j}} \right| \left| \prod_{s=0}^{\ell-1} \frac{e^{ix_j}-a_s}{1-\overline{a_s}e^{ix_j}} \right| \right. \\
&\qquad\qquad \left. \times \left| \frac{\sqrt{1-|a_k|^2}}{1-a_k e^{-i\cdot x_j}} \right| \left| \prod_{m=0}^{k-1} \frac{e^{-ix_j}-\overline{a_m}}{1-a_m e^{-ix_j}} D_{kk}^{-1} \right| \right) \\
&\leq \sup_\ell \sum_{\ell \neq k \in T} \sum_{j=1}^{N} \left| \frac{\sqrt{1-|a_\ell|^2}}{1-\overline{a_\ell} e^{i\cdot x_j}} \frac{\sqrt{1-|a_k|^2}}{1-a_k e^{-i\cdot x_j}} D_{kk}^{-1} \right| \\
&\leq \sup_\ell \sum_{\ell \neq k \in T} \sum_{j=1}^{N} \left| \frac{\sqrt{1-|a_\ell|^2}}{1-\overline{a_\ell} e^{i\cdot x_j}} \right| \left| \frac{\sqrt{1-|a_k|^2}}{1-a_k e^{-i\cdot x_j}} \right| \left| D_{kk}^{-1} \right|.
\end{aligned}
$$

A.2. Expectation value of the trace of G_0^{2n}

$$
\begin{aligned}
\mathbb{E}_X\left[\operatorname{Tr}\left(G_0^{2n}\right)\right] &= \sum_{k_1,k_2,\dots,k_{2n}\in T} \sum_{j_1,\dots,j_{2n}=1}^{N} \mathbb{E}_X \left[\frac{\prod_{r=1}^{2n} \overline{\mathcal{B}_{k_r}(e^{ix_{j_r}})} \mathcal{B}_{k_{r+1}}(e^{ix_{j_r}})}{\prod_{s=1}^{2n} \sum_{l_s=1}^{N} \left|\mathcal{B}_{k_s}(e^{ix_{l_s}})\right|^2} \right] \\
&= \sum_{k_1,k_2,\dots,k_{2n}\in T} \sum_{j_1,\dots,j_{2n}=1}^{N} \prod_{A\in\mathcal{A}} \mathbb{E}_X \left[\frac{\prod_{r\in A} \overline{\mathcal{B}_{k_r}(e^{ix_{j_A}})} \mathcal{B}_{k_{r+1}}(e^{ix_{j_A}})}{\prod_{s=1}^{2n} \sum_{l_s=1}^{N} \left|\mathcal{B}_{k_s}(e^{ix_{l_s}})\right|^2} \right] \\
&= \sum_{k_1,k_2,\dots,k_{2n}\in T} \sum_{j_1,\dots,j_{2n}=1}^{N} \prod_{A\in\mathcal{A}} \left(1 \Big/ \prod_{s=1}^{2n} \sum_{l_s=1}^{N} \left|\mathcal{B}_{k_s}(e^{ix_{l_s}})\right|^2 \right) \\
&\qquad \times \mathbb{E}_X \left[\prod_{r\in A} \overline{\mathcal{B}_{k_r}(e^{ix_{j_A}})} \mathcal{B}_{k_{r+1}}(e^{ix_{j_A}}) \right].
\end{aligned}
$$

We recall that the terms $\sum_{l_s=1}^{N} \left|\mathcal{B}_{k_s}(e^{ix_{l_s}})\right|^2$ are fixed constants.

A.3. Proof of Lemma 3.7

Proof. Let be $\sigma := \operatorname{sgn}(c)$ and

$$\left((D^{-1}HR_T)^m \operatorname{sgn} c\right)_k = \sum_{\ell_1,\dots,\ell_m=1}^{N} \sum_{\substack{k_1,\dots,k_m\in T,\\ k_j\neq k_{j+1},\\ j=1,\dots,m}} \sigma(k_m) \frac{\prod_{r=0}^{m} \overline{\mathcal{B}_{k_r}(x_{l_r})}\mathcal{B}_{k_{r+1}}(x_{l_r})}{\prod_{s=1}^{2m}\sum_{l_s=1}^{N} \left|\mathcal{B}_{k_s}(e^{ix_{l_s}})\right|^2}$$

with $k_0 := k$. From this we obtain

$$\begin{aligned}
&\left|\left((D^{-1}HR_T)^m \operatorname{sgn} c\right)_{k_0}\right|^2 \\
&= \sum_{\ell_1^{(1)},\dots,\ell_m^{(1)}=1}^{N} \sum_{\ell_1^{(2)},\dots,\ell_m^{(2)}=1}^{N} \sum_{\substack{k_1^{(1)},k_2^{(1)},\dots,k_m^{(1)}\in T\\ k_1^{(2)},k_2^{(2)},\dots,k_m^{(2)}\in T\\ k_j^{(p)}\neq k_{j+1}^{(p)},\ j\in[m],\ p=1,2}} \overline{\sigma(k_m^{(1)})}\sigma(k_m^{(2)}) \\
&\quad\times \frac{\prod_{r=1}^{m}\overline{\mathcal{B}_{k_r^{(1)}}\left(x_{l_r}^{(1)}\right)}\mathcal{B}_{k_{r+1}^{(1)}}\left(x_{l_r}^{(1)}\right)}{\prod_{s=1}^{m}\sum_{l_s=1}^{N}\left|\mathcal{B}_{k_s^{(1)}}(e^{ix_{l_s}^{(1)}})\right|^2}\,\frac{\prod_{r=1}^{m}\overline{\mathcal{B}_{k_r^{(2)}}\left(x_{l_r}^{(2)}\right)}\mathcal{B}_{k_{r+1}^{(2)}}\left(x_{l_r}^{(2)}\right)}{\prod_{s=1}^{m}\sum_{l_s=1}^{N}\left|\mathcal{B}_{k_s^{(2)}}(e^{ix_{l_s}^{(2)}})\right|^2} \\
&= \sum_{\ell_1^{(1)},\dots,\ell_m^{(1)}=1}^{N} \sum_{\ell_1^{(2)},\dots,\ell_m^{(2)}=1}^{N} \sum_{\substack{k_1^{(1)},k_2^{(1)},\dots,k_m^{(1)}\in T\\ k_1^{(2)},k_2^{(2)},\dots,k_m^{(2)}\in T\\ k_j^{(p)}\neq k_{j+1}^{(p)}\ j\in[m]\ p=1,2}} \overline{\sigma(k_m^{(1)})}\sigma(k_m^{(2)}) \\
&\quad\times \frac{\prod_{r=1}^{m}\frac{\sqrt{1-|a_{k_r^{(1)}}|^2}}{1-a_{k_r^{(1)}}e^{-ix_{l_r}^{(1)}}}\prod_{q_1^{(1)}=0}^{k_r^{(1)}-1}\frac{e^{-ix_{l_r}^{(1)}}-\overline{a_{q_1^{(1)}}}}{1-a_{q_1^{(1)}}e^{-ix_{l_r}^{(1)}}}\frac{\sqrt{1-|a_{k_{r+1}^{(1)}}|^2}}{1-\overline{a_{k_{r+1}^{(1)}}}e^{ix_{l_r}^{(1)}}}\prod_{q_2^{(1)}=0}^{k_{r+1}^{(1)}-1}\frac{e^{ix_{l_r}^{(1)}}-a_{q_2^{(1)}}}{1-\overline{a_{q_2^{(1)}}}e^{ix_{l_r}^{(1)}}}}{\prod_{s=1}^{m}\sum_{l_s=1}^{N}\left|\mathcal{B}_{k_s^{(1)}}(e^{ix_{l_s}^{(1)}})\right|^2} \\
&\quad\times \frac{\prod_{r=1}^{m}\frac{\sqrt{1-|a_{k_r^{(2)}}|^2}}{1-a_{k_r^{(2)}}e^{-ix_{l_r}^{(2)}}}\prod_{s_1^{(2)}=0}^{k_r^{(2)}-1}\frac{e^{-ix_{l_r}^{(2)}}-\overline{a_{s_1^{(2)}}}}{1-a_{s_1^{(2)}}e^{-ix_{l_r}^{(2)}}}\frac{\sqrt{1-|a_{k_{r+1}^{(2)}}|^2}}{1-\overline{a_{k_{r+1}^{(2)}}}e^{ix_{l_r}^{(2)}}}\prod_{s_2^{(2)}=0}^{k_{r+1}^{(2)}-1}\frac{e^{ix_{l_r}^{(2)}}-a_{s_2^{(2)}}}{1-\overline{a_{s_2^{(2)}}}e^{ix_{l_r}^{(2)}}}}{\prod_{s=1}^{m}\sum_{l_s=1}^{N}\left|\mathcal{B}_{k_s^{(2)}}(e^{ix_{l_s}^{(2)}})\right|^2}
\end{aligned}$$

where $k_0^{(1)} = k_0^{(2)} = k_0 = k$.

Taking a $2K$th power yields

$$\left|\left(\left(D^{-1}HR_T\right)^m \operatorname{sgnc}\right)_{k_0}\right|^{2K} = \sum_{\substack{\ell_1^{(1)},\ldots,\ell_m^{(1)}=1\\ \vdots \\ \ell_1^{(2K)},\ldots,\ell_m^{(2K)}=1}} \sum_{\substack{k_1^{(1)},k_2^{(1)},\ldots,k_m^{(1)}\in T\\ \vdots\\ k_1^{(2K)},k_2^{(2K)},\ldots,k_m^{(2K)}\in T\\ k_j^{(p)}\neq k_{j+1}^{(p)}}} \overline{\sigma(k_m^{(1)})}\sigma(k_m^{(2)})\cdots\overline{\sigma(k_m^{(2K-1)})}\sigma(k_m^{(2K)})$$

$$\times\left(\prod_{p=1}^{2K}\prod_{r=1}^{m}\mathcal{B}_{k_r^{(p)}}\left(x_{l_r}^{(p)}\right)\mathcal{B}_{k_{r+1}^{(p)}}\left(x_{l_r}^{(p)}\right)\right)\Big/\left(\prod_{q=1}^{2K}\prod_{s=1}^{m}\sum_{l_s=1}^{N}\left|\mathcal{B}_{k_s^{(q)}}(e^{ix_{l_s}^{(q)}})\right|^2\right)$$

$$= \sum_{\substack{\ell_1^{(1)},\ldots,\ell_m^{(1)}=1\\ \vdots \\ \ell_1^{(2K)},\ldots,\ell_m^{(2K)}=1}} \sum_{\substack{k_1^{(1)},k_2^{(1)},\ldots,k_m^{(1)}\in T\\ \vdots\\ k_1^{(2K)},k_2^{(2K)},\ldots,k_m^{(2K)}\in T\\ k_j^{(p)}\neq k_{j+1}^{(p)}}} \overline{\sigma(k_m^{(1)})}\sigma(k_m^{(2)})\cdots\overline{\sigma(k_m^{(2K-1)})}\sigma(k_m^{(2K)})$$

$$\times\frac{\displaystyle\prod_{p=1}^{2K}\prod_{r=1}^{m}\frac{\sqrt{1-|a_{k_r^{(p)}}|^2}}{1-a_{k_r^{(p)}}e^{-ix_{l_r}^{(p)}}}\prod_{q_1^{(p)}=0}^{k_r^{(p)}-1}\frac{e^{-ix_{l_r}^{(p)}}-\overline{a}_{q_1^{(p)}}}{1-a_{q_1^{(p)}}e^{-ix_{l_r}^{(p)}}}\frac{\sqrt{1-|a_{k_{r+1}^{(p)}}|^2}}{1-\overline{a}_{k_{r+1}^{(p)}}e^{ix_{l_r}^{(p)}}}\prod_{q_2^{(p)}=0}^{k_{r+1}^{(p)}-1}\frac{e^{ix_{l_r}^{(p)}}-a_{q_2^{(p)}}}{1-\overline{a}_{q_2^{(p)}}e^{ix_{l_r}^{(1)}}}}{\displaystyle\prod_{q=1}^{2K}\prod_{s=1}^{m}\sum_{l_s=1}^{N}\left|\mathcal{B}_{k_s^{(q)}}(e^{ix_{l_s}^{(q)}})\right|^2}$$

where $k_0^{(p)}=k$, $p=1,\ldots,2K$. Further, recall that $|\sigma(k)|=1$ on T.

Taking the expectation value $\mathbb{E}$ yields

$$\mathbb{E}\left[\left|((HR_T)^m \operatorname{sgnc})_{k_0}\right|^{2K}\right]\leq \sum_{\substack{\ell_1^{(1)},\ldots,\ell_m^{(1)}=1\\ \vdots \\ \ell_1^{(2K)},\ldots,\ell_m^{(2K)}=1}} \sum_{\substack{k_1^{(1)},k_2^{(1)},\ldots,k_m^{(1)}\in T\\ \vdots\\ k_1^{(2K)},k_2^{(2K)},\ldots,k_m^{(2K)}\in T\\ k_j^{(p)}\neq k_{j+1}^{(p)}}}$$

$$\times\mathbb{E}\left[\frac{\displaystyle\prod_{p=1}^{2K}\prod_{r=1}^{m}\frac{\sqrt{1-|a_{k_r^{(p)}}|^2}}{1-a_{k_r^{(p)}}e^{-ix_{l_r}^{(p)}}}\prod_{q_1^{(p)}=0}^{k_r^{(p)}-1}\frac{e^{-ix_{l_r}^{(p)}}-\overline{a}_{q_1^{(p)}}}{1-a_{q_1^{(p)}}e^{-ix_{l_r}^{(p)}}}\frac{\sqrt{1-|a_{k_{r+1}^{(p)}}|^2}}{1-\overline{a}_{k_{r+1}^{(p)}}e^{ix_{l_r}^{(p)}}}\prod_{q_2^{(p)}=0}^{k_{r+1}^{(p)}-1}\frac{e^{ix_{l_r}^{(p)}}-a_{q_2^{(p)}}}{1-\overline{a}_{q_2^{(p)}}e^{ix_{l_r}^{(1)}}}}{\displaystyle\prod_{q=1}^{2K}\prod_{s=1}^{m}\sum_{l_s=1}^{N}\left|\mathcal{B}_{k_s^{(q)}}(e^{ix_{l_s}^{(q)}})\right|^2}\right].$$

$$= \sum_{\substack{\ell_1^{(1)},\dots,\ell_m^{(1)}=1 \\ \vdots \\ \ell_1^{(2K)},\dots,\ell_m^{(2K)}=1}} \sum_{\substack{k_1^{(1)},k_2^{(1)},\dots,k_m^{(1)}\in T \\ \vdots \\ k_1^{(2K)},k_2^{(2K)},\dots,k_m^{(2K)}\in T \\ k_j^{(p)}\neq k_{j+1}^{(p)}}} \left(\prod_{q=1}^{2K}\prod_{s=1}^{m}\sum_{l_s=1}^{N}\left|\mathcal{B}_{k_s^{(q)}}(e^{ix_{l_s}^{(q)}})\right|^2\right)^{-1} \tag{A.1}$$

$$\times\mathbb{E}\left[\prod_{p=1}^{2K}\prod_{r=1}^{m}\frac{\sqrt{1-|a_{k_r^{(p)}}|^2}}{1-a_{k_r^{(p)}}e^{-ix_{l_r}^{(p)}}}\prod_{q_1^{(p)}=0}^{k_r^{(p)}-1}\frac{e^{-ix_{l_r}^{(p)}}-\overline{a_{q_1^{(p)}}}}{1-a_{q_1^{(p)}}e^{-ix_{l_r}^{(p)}}}\frac{\sqrt{1-|a_{k_{r+1}^{(p)}}|^2}}{1-\overline{a_{k_{r+1}^{(p)}}}e^{ix_{l_r}^{(p)}}}\prod_{q_2^{(p)}=0}^{k_{r+1}^{(p)}-1}\frac{e^{ix_{l_r}^{(p)}}-a_{q_2^{(p)}}}{1-\overline{a_{q_2^{(p)}}}e^{ix_{l_r}^{(1)}}}\right].$$

(with equality if all the entries of σ are equal on T). Let us consider the expectation value appearing in the sum. As in the proof of Lemma 3.6 we have to take into account that some of the indices $\ell_r^{(p)}$ might coincide. This requires to introduce some additional notation. Let $\left(\ell_r^{(p)}\right)_{r=1,\dots,m}^{p=1,\dots,2K} \subset \{1,\dots,N\}^{2Km}$ be some vector of indices and let $\mathcal{A}=(A_1,\dots,A_t)\,, A_i\subset\{1,\dots,m\}\times\{1,\dots,2K\}$ be a corresponding partition such that (r,p) and (r',p') are contained in the same block if and only if $\ell_r^{(p)}=\ell_{r'}^{(p')}$ may unambiguously be written ℓ_A instead of $\ell_r^{(p)}$ if $(r,p)\in A$. Using that all ℓ_A for $A\in\mathcal{A}$ are different and that the x_{ℓ_A} are independent we may write the expectation value in the sum in (A.1) as

$$\mathbb{E}_X\left[\prod_{p=1}^{2K}\prod_{r=1}^{m}\frac{\sqrt{1-|a_{k_r^{(p)}}|^2}}{1-a_{k_r^{(p)}}e^{-ix_{l_r}^{(p)}}}\right.$$
$$\left.\times\prod_{q_1^{(p)}=0}^{k_r^{(p)}-1}\frac{e^{-ix_{l_r}^{(p)}}-\overline{a_{q_1^{(p)}}}}{1-a_{q_1^{(p)}}e^{-ix_{l_r}^{(p)}}}\frac{\sqrt{1-|a_{k_{r+1}^{(p)}}|^2}}{1-\overline{a_{k_{r+1}^{(p)}}}e^{ix_{l_r}^{(p)}}}\prod_{q_2^{(p)}=0}^{k_{r+1}^{(p)}-1}\frac{e^{ix_{l_r}^{(p)}}-a_{q_2^{(p)}}}{1-\overline{a_{q_2^{(p)}}}e^{ix_{l_r}^{(1)}}}\right]$$
$$=\prod_{A\in\mathcal{A}}\mathbb{E}_X\left[\prod_{(r,p)\in A}\frac{\sqrt{1-|a_{k_r^{(p)}}|^2}}{1-a_{k_r^{(p)}}e^{-ix_{l_A}^{(p)}}}\right.$$
$$\left.\times\prod_{q_1^{(p)}=0}^{k_r^{(p)}-1}\frac{e^{-ix_{l_A}^{(p)}}-\overline{a_{q_1^{(p)}}}}{1-a_{q_1^{(p)}}e^{-ix_{l_A}^{(p)}}}\frac{\sqrt{1-|a_{k_{r+1}^{(p)}}|^2}}{1-\overline{a_{k_{r+1}^{(p)}}}e^{ix_{l_A}^{(p)}}}\prod_{q_2^{(p)}=0}^{k_{r+1}^{(p)}-1}\frac{e^{ix_{l_A}^{(p)}}-a_{q_2^{(p)}}}{1-\overline{a_{q_2^{(p)}}}e^{ix_{l_A}^{(1)}}}\right].$$

Note that if $A\in\mathcal{A}$ contains only one element then the last expression vanishes due to the condition $k_r^{(p)}\neq k_{r+1}^{(p)}$ Thus, we only need to consider partitions $\mathcal{A}$ in $P(2Km,t)$. Now we are able to rewrite the inequality in (A.1) as

$$\mathbb{E}_X\left[\left|\left(\left(D^{-1}HR_T\right)^m\operatorname{sgn}c\right)_k\right|^{2K}\right]\leq\sum_{t=1}^{\min\{Km,N\}}\frac{N!}{(N-t)!}\sum_{\mathcal{A}\in P(2Km,t)}\mathcal{C}_d\,\mathcal{B}(\mathcal{A},T),$$

with

$$\mathcal{C}_d = \left(\prod_{q=1}^{2K} \prod_{s=1}^{m} \sum_{l_s=1}^{N} \left| \mathcal{B}_{k_s^{(q)}}(e^{ix_{l_s}^{(q)}}) \right|^2 \right)^{-1}$$

and

$$\mathcal{B}(\mathcal{A}, T) = \sum_{\substack{k_1^{(1)}, k_2^{(1)}, \ldots, k_m^{(1)} \in T \\ \vdots \\ k_1^{(2K)}, k_2^{(2K)}, \ldots, k_m^{(2K)} \in T \\ k_j^{(p)} \neq k_{j+1}^{(p)} \; j \in [m]}} \prod_{A \in \mathcal{A}} \left(\frac{2}{\epsilon} - 1 \right)^t \tag{A.2}$$

for a small ϵ such that $|(\pm 1, 0) - a_n| = \epsilon$ or $|(0, \pm 1) - a_n| = \epsilon$.

This proves the lemma. □

References

[1] R. Baraniuk, M. Davenport, R. DeVore, M. Wakin, *A simple proof of the restricted isometry property for random matrices*, Constr. Approx., **28 (3)**, (2008), 253–263.

[2] R.F. Bass, K. Gröchenig, *Random sampling of multivariate trigonometric polynomials*, SIAM J. Math. Anal., **36 (3)** (2004), 773–795.

[3] J. Bokor, F. Schipp, *Approximate linear H^∞ identification in Laguerre and Kautz basis*, Automatica, A Journal of IFAC, **34** (1998), 463–468.

[4] J. Bokor, F. Schipp, *L^∞ system approximation algorithms generated by ϑ summations*, Automatica, A Journal of IFAC, **33 (11)** (1997), 2019–2024.

[5] M. Bóna, *Combinatorics of Permutations*, 2nd Edition, Chapman and Hall, CRC Press, Boca Raton, 2012.

[6] S. Boyd, L. Vandenberghe, *Convex Optimization*, Cambridge University Press, Cambridge, 2004.

[7] S.A. Broughton, K. Bryan, *Discrete Fourier Analysis and Wavelets – Applications to Signal and Image Processing*, John Wiley & Sons, 2009.

[8] A. Bultheel, P. Carrette, *Fourier analysis and the Takenaka–Malmquist basis*, in: Conference on Decision and Control Maui, Hawaii, USA, December 1, 2003, pp. 486–491.

[9] E. Candès, *A restricted isometry property and it implications for Compressed Sensing*. Comptes Rendus Mathématique, 2008.

[10] E. Candès, J. Romberg, *ℓ1-MAGIC: Recovery of Sparse Signals via Convex Programming*, 2005. Available at http://www.acm.caltech.edu/l1magic.

[11] E. Candès, J. Romberg, T. Tao, *Robust uncertainty principles: Exact signal reconstruction from highly incomplete frequency information*, IEEE Trans. Inform. Theory, **52 (2)** (2006), 489–509.

[12] S. Chen, D.L. Donoho, *Basis Pursuit: Technical report*, Department of Statistics, Stanford University (1994), available at
http://statweb.stanford.edu/ donoho/Reports/1994/asilomar.pdf

[13] Q. Chen, L.Q. Li, T. Qian, *Two families of unit analytic signals with nonlinear phase*, Physica D, **221 (1)**, (2006) 1–12.

[14] Q. Chen, L. Li, T. Qian, *Time-frequency Aspects of Nonlinear Fourier Atoms*, Wavelet analysis and applications. Selected papers based on the presentations at the conference (WAA 2005), Macau, China, 29th November–2nd December 2005, Basel – Birkhäuser, Applied and Numerical Harmonic Analysis, (2007), pp. 287–297.

[15] R.A. DeVore, *Nonlinear approximation*, Acta Numer., **7**, (1998), 51–150.

[16] D.L. Donoho, *Compressed Sensing*, IEEE Trans. Inform. Theory, **52 (4)** (2006), 1289–1306.

[17] S. Foucart, H. Rauhut. *A Mathematical Introduction to Compressive Sensing*. Birkhäuser, 2013.

[18] J.B. Garnett, *Bounded Analytic Functions*, Academic Press, Inc., 1981.

[19] I. Gohberg, *Continuous and Discrete Fourier Transforms, Extension Problems and Wiener–Hopf Equations*, Vol. Operator Theory: Advances and Applications, Birkhäuser Basel, 2012.

[20] N.R. Gomes, *Compressed Sensing in Clifford Analysis*, PhD thesis, Universidade de Aveiro, 2015.

[21] Y. Katznelson, *An introduction to Harmonic Analysis*, Dover, 1976.

[22] V.I. Lebedev, *An Introduction to Functional Analysis in Computational Mathematics: An Introduction*, Birkhäuser, Boston, 2012.

[23] S. Mallat, Z. Zhang, *Matching Pursuit With Time-Frequency Dictionaries*, IEEE Trans. Signal Process., **41 (12)**, (1993), 3397–3415.

[24] W. Mi, T. Qian, F. Wan, *A fast adaptive model reduction method based on Takenaka–Malmquist systems*, Syst. Control Lett., 61 (1), (2012), 223–230.

[25] M. Pap, *Hyperbolic Wavelets and Multiresolution in* $H^2(\mathbb{T})$, J. Fourier Anal. appl., **17 (5)** (2011), 755–766.

[26] M. Pap, *The voice transform generated by a representation of the Blaschke group on the weighted Bergman spaces*, Ann. Univ. Sci. Budapest. Sect. Comput., **33** (2010), 321–342.

[27] M. Pap, F. Schipp, *The voice transform on the Blaschke group III*, Publ. Math. Debrecen, **75 (1-2)**, (2009), 263–283.

[28] T. Qian, Y.B. Wang, *Adaptive Fourier Series – A Variation of Greedy Algorithm*, Adv. Comput. Math., **34 (3)** (2011), 279–293.

[29] T. Qian, E. Wegert, *Optimal approximation by Blaschke forms*, Complex Var. Elliptic Equ., **58 (1)**, (2013), 123–133.

[30] H. Rauhut, *Random sampling of sparse trigonometric polynomials*, Appl. Comput. Harmon. Anal. **22 (1)** (2007), 16–42.

[31] L. Shuang, *Adaptive Signal Decomposition of Hardy Space.* PhD Thesis, Macau University, 2013.

[32] G. Teschke, *Multi-frame representations in linear inverse problems with mixed multiconstraints*, Appl. Computat. Harmon. Anal. **22 (1)**, (2007), 43–60

[33] J.A. Tropp, *Greed is good: Algorithmic results for sparse approximation*, IEEE Trans. Inform. Theor. **50 (10)**, (2004), 2231–2242.

[34] S.J. Wright, *Primal-Dual Interior-Point Methods.* Philadelphia, PA: SIAM, 1997

Acknowledgment

The work of the first, third and fourth authors was partially supported by Portuguese funds through the CIDMA – Center for Research and Development in Mathematics and Applications, and the Portuguese Foundation for Science and Technology ("FCT – Fundação para a Ciência e a Tecnologia"), within project UID/MAT/ 0416/2013. The second author is supported in part by NSFC grant 11571083. The work of the third author was supported by the PhD grant Programa Ciência Global – SFRH/BD/51485/2011. The work of the fourth author was supported by the ERASMUS program.

Paula Cerejeiras
CIDMA, Departamento de Matemática
Universidade de Aveiro
Campus de Santiago
P-3810-193, Aveiro, Portugal
e-mail: pceres@ua.pt

Qiuhui Chen
Department of Mathematics
College of Mathematics and Informatics
South China Agricultural University
Guangzhou, China
e-mail: chenqiuhui@hotmail.com

Narciso Gomes
Departamento Matemática
Faculdade Ciência e Tecnologia
Universidade de Cabo Verde
Campus de Palmarejo
Box 279, Praia, Cabo Verde
e-mail: narciso.gomes@docente.unicv.edu.cv

Stefan Hartmann
former member of CIDMA, Departamento de Matemática
Universidade de Aveiro
Campus de Santiago
P-3810-193, Aveiro, Portugal
e-mail: Stefan.Hartmann@ua.pt

Modern Trends in Hypercomplex Analysis
Trends in Mathematics, 79–110

Script Geometry

Paula Cerejeiras, Uwe Kähler, Franciscus Sommen and Adrian Vajiac

Abstract. In this paper we describe the foundation of a new kind of discrete geometry and calculus called Script Geometry. It allows to work with more general meshes than classic simplicial complexes. We provide the basic definitions as well as several examples, like the Klein bottle and the projective plane. Furthermore, we also introduce the corresponding Dirac and Laplace operators which should lay the groundwork for the development of the corresponding discrete function theory.

Mathematics Subject Classification (2010). 30G35, 65N30, 58A14.

Keywords. Script, simplicial topology, discrete exterior derivative, Hodge–Dirac operator, discrete Laplacian, cohomology.

1. Introduction

In the last two decades one can observe an ever increasing interest in the analysis of discrete structures. On one hand the fact that nowadays everybody can harness large computational power, but the computer is restricted to work with discrete values only, created an increased interest in working with discrete structures. This is true even for persons who are originally unrelated to the field. An outstanding example can be seen in the change of the philosophy of the Finite Element Method.

From the classical point of view the finite element method is essentially a method for discretization of partial differential equations via a variational formulation, i.e., one first establishes the variational formulation and discretizes the problem by creating ansatz spaces via introducing a mesh (normally by triangularization) and (spline) functions defined over the mesh. One of the major problems with this approach is that there is no a priori connection between the choice of the mesh and the variational formulation. The modern approach lifts the problem and, therefore, the finite element modelation directly on to the mesh, resulting in the so-called Finite Element Exterior Calculus [1, 14]. Hereby, one chooses first the mesh and introduces a boundary operator given by the mesh which induces the corresponding discrete variational formulation. From a practical point of view this is even more interesting since finite element meshes are also widely being applied in other fields, such as computer graphics [13, 14]. In this framework notions of discrete vector fields and operators acting on them, e.g., discrete divergence and curl, appear in a rather canonical way instead of being introduced artificially by additional discretizing a continuous formulation. This also leads to immediate applications such as the problem of discrete Hodge decompositions of 3D vector fields on irregular grids. In this context one can also study the notion of a Dirac operator [27].

Yet, if we look at the literature the existing theory is based on working with simplicial complexes and triangularizations [13].

But the meshes in FEM or in computer graphics are not just restricted to meshes coming from triangularization and representing simplicial complexes. Already discretizations based on quadrilaterals hexagons are not in this class. Therefore, a more general geometrical approach than the one based on simplicial complexes is needed. Furthermore, there are problems in other fields (like physics) which are traditionally modeled in a continuous ways. Nowadays, such problems are more and more studied directly on the discrete level, the principal example being the Ising model from statistical physics as opposed to the continuous Heisenberg model. But also here one is not just limited to classic lattices or triangularizations, yet for more general lattices which are not just simplicial complexes a corresponding geometrical theory is missing. These models require a discrete function theory to work with them, similar to the 2D-case where discrete complex analysis plays a major role. In fact most of the recent advances on the 2D-Ising model by S. Smirnov and his collaborators are based on a clever interaction between classic and discrete complex analysis [29]. This is possible since discrete complex analysis is under (more or less) constant development since the forties [25, 28].

Unfortunately, the same cannot be said about the higher-dimensional case. While lately one can observe several approaches to create a discrete function theory in higher dimensions based on lattice discretizations of the Dirac operator (see [32, 26, 21, 18, 19, 4, 8]) they are closer in spirit to finite difference methods than finite element methods ([5, 22, 23, 2, 6]). Nevertheless, these approaches lead to a well-established function theory [17, 9, 10, 11, 12, 20, 7]. For a function theory in connection with the above-mentioned finite element exterior calculus we do not want to be restricted to meshes coming from simplicial complexes. Therefore, one needs a new kind of geometry which allows to work directly with general meshes.

In this paper we are going to lay the foundations of a new type of discrete geometry called script geometry which is not restricted to simplicial complexes. After a short review of simplicial topology we define the principal objects as well as introduce the corresponding Dirac and Laplace operators as discrete versions of the abstract Hodge–Dirac operator. Furthermore, to give a more clear understanding of what we are aiming at we are going to present several examples, such as the Möbius strip, the Klein bottle, the torus, and the projective plane. It is our modest hope that the presented framework will be interesting enough to be explored by many mathematicians in the future.

2. Brief review of simplicial topology

An *abstract simplicial complex* is a collection $\mathcal{S}$ of finite non-empty sets, such that if A is an element of $\mathcal{S}$, then every non-empty subset of A is also an element of $\mathcal{S}$. An element A of $\mathcal{S}$ is called a *simplex* of $\mathcal{S}$; its dimension is one less than the number of its elements, and each non-empty subset of A is called a *face* of A. Vertices of $\mathcal{S}$ are the one-point elements v of $\mathcal{S}$, and $\{v\}$ is by definition a 0-simplex.

If K is a topological simplicial complex and V its vertex set, then the collection of all subsets $\{a_0, \ldots, a_n\}$ of V such that the vertices $a_0, \ldots, a_n$ span a simplex of K, is called the *vertex scheme* of K. The vertex scheme of a topological simplicial complex is an example of abstract simplicial complex. In fact, every abstract complex $\mathcal{S}$ is isomorphic to the vertex scheme for some simplicial complex K, called also the *geometric realization* of $\mathcal{S}$, uniquely determined up to a linear isomorphism.

Let σ be an abstract simplex. Two orderings of its vertex set are equivalent if they differ by an even permutation. There are two equivalence classes (in dimensions bigger than 1), each one of them called an *orientation* of σ. For 0-simplexes, there is only one orientation.

If K is a simplicial complex, then a p-chain on K is a function c from the set of oriented p-simplices of K to $\mathbb{Z}$, such that: (a) $c(\sigma) = -c(-\sigma)$; and (b) $c(\sigma) = 0$ for all but finitely many oriented p-simplices σ. Addition of oriented p-chains is done by adding their integer values. The resulting group is denoted by $C_p(K)$.

If σ is an oriented simplex, the *elementary chain* c corresponding to σ is the function defined as follows: (a) $c(\sigma) = 1$, (b) $c(-\sigma) = -1$, and (c) $c(\tau) = 0$ for all other oriented simplices. The usual convention denotes by σ both the oriented simplex and its elementary p-chain c. This allows the notation $-\sigma$ for the simplex with opposite orientation than σ.

A well-known result is that $C_p(K)$ is a free Abelian group, a basis is obtained by orienting each p-simplex and using the corresponding elementary chains as a basis. Therefore, with the exception of $C_0(K)$, the groups $C_p(K)$ have no natural basis, as one must orient the p-simplices in K in an arbitrary fashion to obtain a basis.

The homomorphism of groups:

$$\partial_p : C_p(K) \to C_{p-1}(K)$$

is called the *boundary operator*, defined by

$$\partial_p[v_0, \ldots, v_p] := \sum_{i=0}^{p} (-1)^i [v_0, \ldots, \widehat{v_i}, \ldots, v_p],$$

where the hat means deletion from the array. The operator ∂_p is well defined and it has the property

$$\partial_p(-\sigma) = -\partial_p(\sigma),$$

for all simplices σ. For example:

$$\partial_1[v_0, v_1] = v_1 - v_0, \qquad \partial_2[v_0, v_1, v_2] = [v_1, v_2] - [v_0, v_2] + [v_0, v_1].$$

It can be proved that

$$\partial_{p-1} \circ \partial_p = 0,$$

so the kernel of ∂_p, denoted by $Z_p(K)$ is the group of p-cycles, and the image of ∂_{p+1}, denoted by $B_p(K)$, is the group of p-boundaries. The pth homology group of K is then defined as

$$H_p(K) := Z_p(K)/B_p(K).$$

Cohomology is usually defined using the Hom functor. That makes cocycles to be "picket fences" inside triangularizations of manifolds.

3. Script geometry

Let us start with the definition of our most basic object, the notion of a *script*.

Definition 3.1. A *script* is a collection

$$\mathfrak{S} := \{\mathfrak{S}_{-1}, \mathfrak{S}_0, \mathfrak{S}_1, \dots, \mathfrak{S}_k, \dots, \mathfrak{S}_m\} \tag{3.1}$$

of sets $\mathfrak{S}_k$, the elements of which are called *k-cells*. In particular,

$$\begin{aligned} &\mathfrak{S}_{-1} := \{\infty\}, \qquad \mathfrak{S}_0 := \{p_1, \dots, p_j, \dots, p_{n_0}\}, \\ &\mathfrak{S}_1 := \{l_1, \dots, l_j, \dots, l_{n_1}\}, \qquad \mathfrak{S}_2 := \{v_1, \dots, v_j, \dots, v_{n_2}\}, \dots, \\ &\mathfrak{S}_k := \{c_1^k, \dots, c_j^k, \dots, c_{n_k}^k\}. \end{aligned}$$

Traditionally $0, 1$ and 2-cells are called *points*, *lines* and *planes*, respectively.

Definition 3.2. A linear combination over $\mathbb{Z}$ of k-cells is called a *k-chain*:

$$C_k := \sum\nolimits_j \lambda_j^k c_j^k, \qquad \lambda_j^k \in \mathbb{Z}, \tag{3.2}$$

and we denote the module of k-chains by $\mathfrak{C}_k$. The support of a k-chain C_k is the set of k-cells c_j^k that are involved in the linear combination (3.2), i.e., for which $\lambda_j^k \neq 0$.

Definition 3.3. The *boundary map* ∂ from $\mathfrak{S}_k$ into $\mathfrak{C}_{k-1}$, the module of $(k-1)$-chains, is defined by:

$$\partial c_j^k := \sum\nolimits_s \mu_j^{k,s} c_s^{k-1}, \qquad \mu_j^{k,s} \in \mathbb{Z} \tag{3.3}$$

which naturally extends to the module $\mathfrak{C}_k$, and it is subject to $\partial^2 = 0$.

Let us remark that the coefficients $\mu_j^{k,s}$ in (3.3) are uniquely determining the boundary operator ∂. For example, if P_0 is a 0-chain, then if $P_0 = \sum_j \lambda_j^0 p_j$, we have:

$$\partial p_j = 1 \cdot \infty, \qquad \partial P_0 = \left(\sum\nolimits_j \lambda_j^0\right) \cdot \infty,$$

therefore $\partial P_0 = 0$ if and only if $\sum_j \lambda_j^0 = 0$. For a generic k-cell c_j^k, since $\partial^2 = 0$ by definition, we have:

$$0 = \partial^2 c_j^k = \partial \left(\sum_s \mu_j^{k,s} c_s^{k-1} \right) = \sum_\ell \left(\sum_s \mu_j^{k,s} \mu_s^{k-1,\ell} \right) c_\ell^{k-2}$$

therefore:

$$\sum_s \mu_j^{k,s} \mu_s^{k-1,\ell} = 0,$$

for all ℓ.

Definition 3.4. A k-chain C_k for which $\partial C_k = 0$, it is called a *k-cycle.* A k-chain

$$C_k = \sum_j \lambda_j^k c_j^k$$

for which $\lambda_j^k = \pm 1$ is called an *oriented surface*, or simply a *surface.* A surface C_k for which $\partial C_k = 0$ is a *closed* surface.

Definition 3.5. A script $\mathfrak{S}$ for which every cell boundary ∂c_j^k is a closed surface is called a *geoscript.*

Definition 3.6. A closed surface C_k is called *tight* if and only if for every closed surface C'_k with $\operatorname{supp} C'_k \subset \operatorname{supp} C_k$, it follows that $C'_k = \pm C_k$, i.e., C_k is the only closed surface, up to sign, with support inside $\operatorname{supp} C_k$.

A *tight* cell c is a cell for which ∂c is a closed tight surface. A geoscript is called *tight* if all its cells have a boundary which is a tight surface, i.e., all its cells are tight cells.

Any point p_j is obviously tight. A line l is tight if and only if $\partial l = p_j - p_k$, i.e., every tight line connects two points. Every plane v which is tight has a boundary

$$\partial v = \sum_{j=1}^{t} \lambda_j l_j, \qquad \lambda_j = \pm 1,$$

which forms a polygon, i.e., $\lambda_j \partial l_j = p_j - p_{j+1}$ whereby $p_{t+1} = p_1$, and all points $p_1, \ldots, p_t$ are different.

In Figure 1, we have drawn two examples of tight scripts and the far right one is a non-tight script. Please note that the "loop" script in Figure 1 is defined by:

$$\mathfrak{S}_0 = \{p_0, p_1\}, \qquad \mathfrak{S}_1 = \{l_1, l_2\}, \qquad \mathfrak{S}_2 = \{v\},$$
$$\partial l_1 = p_1 - p_0, \quad \partial l_2 = p_0 - p_1, \quad \partial v = l_1 + l_2.$$

Note that a tight geoscript of dimension ≤ 2 is always topologically equivalent to a CW-complex. For higher-dimensional geoscripts the situation can be more general than CW-complexes.

The cells in a geoscript are oriented cells and can each come in two states of orientation that are determined by the boundary map $\partial(c_j^k)$, i.e., if one replaces c_j^k by $d_j^k = -c_j^k$ then also $\partial(d_j^k) = -\partial(c_j^k)$. But in general there could be more than two orientations on (closed surfaces inside) $\operatorname{supp} \partial(c_j^k)$ and so the mere knowledge

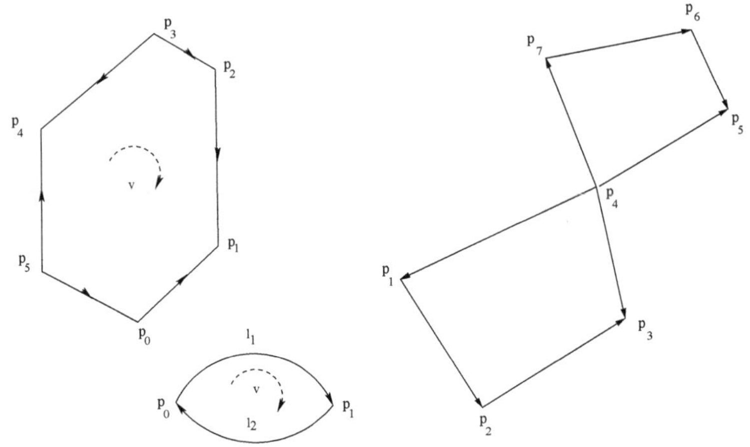

FIGURE 1. Examples of scripts

of $\operatorname{supp} \partial(c_j^k)$ does not determine the orientations $\pm\partial(c_j^k)$. The tightness condition however ensures that on $\operatorname{supp} \partial(c_j^k)$ there can only be two states of closed orientation given by $\pm\partial(c_j^k)$, so that the state of orientation on each cell c_j^k can be fully identified with the state of orientation on the boundary. The tightness condition also implies a number of interesting geometric properties for scripts, such as a line has two endpoints or a 2-cell is a polygon. In a forthcoming paper we prove that using tightness one can determine when a two-dimensional script corresponds to an oriented two-dimensional manifold.

Definition 3.7. A k-cell c is called a *k-simplex* if either c is a point (the case $k = 0$), or the boundary ∂c of c is a tight $(k-1)$-surface that is the sum (with coefficients ± 1) of $k+1$ different $(k-1)$-cells that are also $(k-1)$-simplexes. A *simplicial script* is a tight geoscript for which all cells are simplexes.

Definition 3.8. A *geomap* $G : \mathfrak{S} \to \mathfrak{S}'$ between two tight geoscripts $\mathfrak{S}$ and $\mathfrak{S}'$ is a collection of linear maps

$$g_k : \mathfrak{S}_k \to \mathfrak{S}'_k$$

with the following two properties:

(a) the image of every k-surface $C_k \in \mathfrak{S}_k$ is a k-surface $C'_k \in \mathfrak{S}'_k$, e.g., on a k-cell c_j^k we have:

$$g_k(c_j^k) = \sum \mu_j^{k,s} {c'}_s^k, \qquad \mu_j^{k,s} \in \{-1, 1\}.$$

(b) for each k, the natural extension of g_k to a set of k-chains fulfills the relation:

$$\partial g_k(C_k) = g_{k-1}(\partial C_k).$$

Moreover, g_k is called *tight* if it maps tight surfaces to tight surfaces.

The notion of geomap can be used to define when two geoscripts are isomorphic. Let $\mathfrak{S} = \{\mathfrak{S}_{-1}, \mathfrak{S}_0, \mathfrak{S}_1, \dots\}$ and $\mathfrak{T} = \{\mathfrak{T}_{-1}, \mathfrak{T}_0, \mathfrak{T}_1, \dots\}$ be two geoscripts and suppose we have a geomap given by $g_k : \mathfrak{S}_k \to \mathfrak{C}(\mathfrak{T})_k$, from $\mathfrak{S}_k$ to the chains of $\mathfrak{T}_k$, that is such that for every cell $c_j^k \in \mathfrak{S}_k$,

$$g_k(c_j^k) = \pm d_j^k,$$

where $d_j^k \in \mathfrak{T}_k$, i.e., suppose that g_k is a bijection up to the sign between $\mathfrak{S}_k$ and $\mathfrak{T}_k$. Then we say that script $\mathfrak{S}$ is *isomorphic* to script $\mathfrak{T}$. It means essentially that one can change the signs of the cells provided one makes the necessary adjustments for the boundary map ∂, and these adjustments are determined by the relations $\partial g_k = g_{k-1}\partial$.

Definition 3.9. A geoscript $\mathfrak{S}'$ is called a *refinement* of a given geoscript $\mathfrak{S}$ if there exists an injective geomap $G = \{g_k\}_k : \mathfrak{S} \to \mathfrak{S}'$. A refinement is called *tight* if every g_k is tight and if for each k, there exists only one surface C'_k inside the image $g_k(c_j^k)$ for which

$$\partial C'_k = \partial g_k(c_j^k).$$

Theorem 3.10. *Any tight geoscript $\mathfrak{S}$ admits a refinement to a simplicial script $\mathfrak{S}$.*

Proof. The proof is done by induction over k, and it is left as an exercise for the avid reader. □

We define the analog of the homology groups of a tight script $\mathfrak{S}$, due to the fact that the boundary operators $\partial : \mathfrak{C}_{k+1} \to \mathfrak{C}_k$ obey $\partial^2 = 0$ in all dimensions k. We define:

$$\mathcal{H}_k(\mathfrak{S}) =: Z_k(\mathfrak{S})/B_k(\mathfrak{S}),$$

where $Z_k(\mathfrak{S})$ is the group of (closed oriented) k-cycles, and $B_k(\mathfrak{S})$ is the group of boundaries of $(k+1)$-chains of $\mathfrak{S}$.

Definition 3.11. We define the *inner product* of k-chains by

$$\left\langle \sum_s \alpha_s c_s^k, \sum_s \beta_s c_s^k \right\rangle := \sum \alpha_s \beta_s.$$

Then the *exterior derivative* d on chains is defined by

$$dc_j^k := \sum_\ell \mu_j^{k,\ell} c_\ell^{k+1},$$

naturally extended to the module of chains, and subject to the condition

$$\langle dc_j^k, c_\ell^{k+1} \rangle = \langle c_j^k, \partial c_\ell^{k+1} \rangle.$$

Similarly for the differential operators d, one can define the corresponding cohomology groups $\mathcal{H}^k(\mathfrak{S})$ of a tight script:

$$\mathcal{H}^k(\mathfrak{S}) =: Z^k(\mathfrak{S})/B^k(\mathfrak{S}),$$

where $Z^k(\mathfrak{S})$ is the group of $(k+1)$-chains closed with respect to d, and $B^k(\mathfrak{S})$ is the group of coboundaries (in the image of d) of k-chains of $\mathfrak{S}$.

Note that in the case of a simplicial script $\mathfrak{S}$ being built by making use of usual (triangular) simplexes, it is similar to the usual notion of a simplicial complex. From dimension 3 and up a non-simplicial script does not uniquely determine the topology of the supporting space, though. Therefore, in a certain sense, scripts are a more loose concept than the traditional abstract simplexes.

4. The discrete Dirac and Laplace operators on scripts

Let f be a function defined on a tight script $\mathfrak{S}$ with integer, real, complex, or Clifford algebra values. For example, if $\mathfrak{S}$ is has dimension 2, f is defined by

$$f = f_0 + f_1 + f_2,$$
$$f_0 = \sum_{j \in \mathfrak{S}_0} f_{0j} p_j, \qquad f_1 = \sum_{j \in \mathfrak{S}_1} f_{1j} l_j, \qquad f_2 = \sum_{j \in \mathfrak{S}_2} f_{2j} v_j.$$

Definition 4.1. The discrete Hodge–Dirac operator for a tight script $\mathfrak{S}$ is defined as

$$\not{\partial} = \partial + d,$$

acting on the corresponding parts of a function f.

For example, in the case $n = 2$, we have:

$$\begin{aligned}\not{\partial} f &= \partial f_1 + (d f_0 + \partial f_2) + d f_1 \\ &= \sum_{j \in \mathfrak{S}_1} f_{1j} \partial l_j + \left(\sum_{j \in \mathfrak{S}_0} f_{0j} d p_j + \sum_{j \in \mathfrak{S}_2} f_{2j} \partial v_j \right) + \sum_{j \in \mathfrak{S}_1} f_{1j} d l_j.\end{aligned}$$

Definition 4.2. The discrete Laplace operator on a tight script $\mathfrak{S}$ is defined by:

$$\Delta = \frac{1}{2}(\partial d + d\partial) = \frac{1}{2}(\partial + d)^2 = \frac{1}{2}\not{\partial}^2. \tag{4.1}$$

For example, for f as above, we have:

$$\begin{aligned}2\Delta f &= \partial(d f_0) + (\partial(d f_1) + d(\partial f_1)) + d(\partial f_2) \\ &= \sum_{j \in \mathfrak{S}_0} f_{0j} \partial(d p_j) + \sum_{j \in \mathfrak{S}_1} f_{1j} (\partial(d l_j) + d(\partial l_j)) + \sum_{j \in \mathfrak{S}_2} f_{2j} d(\partial P_j).\end{aligned}$$

Note that the Laplace operator defined above acts on all of $\mathfrak{S}$, not only on vertices (points). Let us remark that the above definition can be seen as a concretization of the abstract Hodge–Laplace operator [27]. Normally, the abstract definition is given in terms of the exterior derivative d and its adjoint d^*, but in our context we do not need to formally introduce the operator d^*. Furthermore, one of the requirements for the discretization of the abstract Hodge–Dirac operator in [27] is that the exterior derivative commutes with bounded (or smoothed) projections. This is not a trivial study and restricts their approach to simplicial complexes. While this is natural in the context of looking at simplicial decomposition of domains our setting is more general.

5. Classic examples of scripts

We give below concrete descriptions and computations of scripts obtained from classical examples of topological spaces.

5.1. A Möbius strip

As a topological space, the Möbius strip is obtained from a rectangle, identifying one pair of opposite edges in reverse orientation. In order to make it a tight geo-script, denoted by $\mathfrak{S}_M$, we obtain the same result by gluing two rectangles along one edge, as described in Figure 2. The script containes four points, six lines and

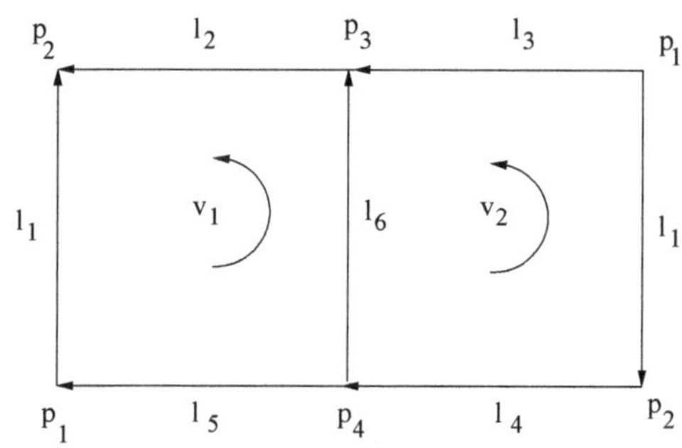

FIGURE 2. The Möbius script

two planes:

$$\mathfrak{S}_0 = \{p_1, p_2, p_3, p_4\}, \quad \mathfrak{S}_1 = \{l_1, l_2, l_3, l_4, l_5, l_6\}, \quad \mathfrak{S}_2 = \{v_1, v_2\}.$$

The boundary operator ∂ acts on the Möbius script as follows:

$$\begin{aligned} \partial l_1 &= p_2 - p_1, & \partial l_2 &= p_2 - p_3, \\ \partial l_3 &= p_3 - p_1, & \partial l_4 &= p_4 - p_2, \\ \partial l_5 &= p_1 - p_4, & \partial l_6 &= p_3 - p_4. \end{aligned}$$

Note that $\partial(l_2 + l_3 + l_4 + l_5) = 0$, so $l_2 + l_3 + l_4 + l_5$ is a tight closed curve. Next, we have:

$$\begin{aligned} \partial v_1 &= -l_1 + l_2 - l_5 + l_6, \\ \partial v_2 &= -l_1 + l_3 - l_4 - l_6. \end{aligned}$$

Note that all linear combinations of the boundaries above have coefficients ± 1. Also, one can easily check that the boundary operator squares to 0, as desired. For example:

$$\begin{aligned} \partial\left(\partial v_1\right) &= -\partial l_1 + \partial l_2 - \partial l_5 + \partial l_6 \\ &= -(p_2 - p_1) + (p_2 - p_3) - (p_1 - p_4) + (p_3 - p_4) = 0. \end{aligned}$$

Similarly $\partial(\partial v_2) = 0$. Therefore, the Möbius script is a tight geoscript. Topologically is equivalent to a CW-complex consisting of one 2-cell: $(v_1 + v_2)$, three 1-cells: $l_1, (l_2 + l_3), (l_4 + l_5)$, and two 0-cells: p_1, p_2.

The script homology of $\mathfrak{S}_M$ is obtained in a similar fashion as one computes the homology of a CW-complex. In more detail, consider the sequence of chains:

$$0 \xrightarrow{\partial} \mathfrak{C}_2 \xrightarrow{\partial} \mathfrak{C}_1 \xrightarrow{\partial} \mathfrak{C}_0 \xrightarrow{\partial} 0.$$

We note that

$$\partial(v_1 + v_2) = -2l_1 + (l_2 + l_3) - (l_4 + l_5),$$

so the image of the boundary of the sum of the two planes is non-empty. Its kernel is 0, so $\mathcal{H}_2(\mathfrak{S}_M) = 0$. Next,

$$\partial l_1 = p_2 - p_1, \quad \partial(l_2 + l_3) = p_2 - p_1, \quad \partial(l_4 + l_5) = p_1 - p_2,$$

therefore up to a sign, $(l_2+l_3)-l_1$ and $(l_4+l_5)-l_1$ are homologous cycles. The kernel is two-dimensional (three line generators and the image is one-dimensional), so isomorphic to $\mathbb{Z}^2$. It follows that $\mathcal{H}_1(\mathfrak{S}_M) = \mathbb{Z}$. Similarly, one obtains $\mathcal{H}_0(\mathfrak{S}_M) = \mathbb{Z}$.

The differential operator d acts on the script $\mathfrak{S}_M$ as follows:

$$\begin{aligned} dp_1 &= -l_1 - l_3 + l_5, \\ dp_2 &= l_1 + l_2 - l_4, \\ dp_3 &= -l_2 + l_3 + l_6, \\ dp_4 &= l_4 - l_5 - l_6, \end{aligned}$$

and

$$dl_1 = -v_1 - v_2, \quad dl_2 = v_1, \quad dl_3 = v_2, \quad dl_4 = -v_2,$$
$$dl_5 = -v_1, \quad dl_6 = v_1 - v_2.$$

For the cohomology of the script $\mathfrak{S}_M$, we study the sequence:

$$0 \xrightarrow{d} \mathfrak{C}_0 \xrightarrow{d} \mathfrak{C}_1 \xrightarrow{d} \mathfrak{C}_2 \xrightarrow{d} 0.$$

Note that $\sum_{j=1}^{4} dp_j = 0$, therefore $\mathcal{H}^0(\mathfrak{S}_M) = \mathbb{Z}$. Because

$$-dl_1 = d(l_2 + l_3) = -d(l_4 + l_5) = v_1 + v_2,$$

therefore $d(2l_1 + (l_2 + l_3) - (l_4 + l_5)) = 0$, so the kernel of d on lines is $3 - 2 = 1$-dimensional. Moreover, the image of d on lines and the kernel of d on planes are both generated by $v_1 + v_2$. Summarizing, the script cohomology of the Möbius strip is indeed, as expected:

$$\mathcal{H}^0(\mathfrak{S}_M) = \mathcal{H}^1(\mathfrak{S}_M) = \mathbb{Z}, \qquad \mathcal{H}^2(\mathfrak{S}_M) = 0.$$

Since the discrete Dirac operator is defined as $\partial\!\!\!/ = \partial + d$, using

$$f = f_0 + f_1 + f_2, \qquad f_0 = \sum_{j=1}^{4} f_{0j}p_j, \quad f_1 = \sum_{j=1}^{6} f_{1j}l_j, \quad f_2 = \sum_{j=1}^{2} f_{2j}v_j,$$

we have:

$$\partial\!\!\!/ f = \partial f_1 + (df_0 + \partial f_2) + df_1.$$

Computations yield to:

$$\partial f_1 = \begin{bmatrix} -1 & 0 & -1 & 0 & 1 & 0 \\ 1 & 1 & 0 & -1 & 0 & 0 \\ 0 & -1 & 1 & 0 & 0 & 1 \\ 0 & 0 & 0 & 1 & -1 & -1 \end{bmatrix} [f_{1j}]^t [p_j],$$

where $[f_{1j}]^t$ is the column vector of the corresponding 6 inputs, and $[p_j]$ is the row vector of the four points. Similarly we obtain:

$$df_0 = \begin{bmatrix} -1 & 1 & 0 & 0 \\ 0 & 1 & -1 & 0 \\ -1 & 0 & 1 & 0 \\ 0 & -1 & 0 & 1 \\ 1 & 0 & 0 & -1 \\ 0 & 0 & 1 & -1 \end{bmatrix} [f_{0j}]^t [l_j],$$

where we notice that the matrix above is the transpose of the previous one for ∂f_1, as it should. Next we get:

$$\partial f_2 = \begin{bmatrix} -1 & -1 \\ 1 & 0 \\ 0 & 1 \\ 0 & -1 \\ -1 & 0 \\ 1 & -1 \end{bmatrix} [f_{2j}]^t [l_j],$$

and

$$df_1 = \begin{bmatrix} -1 & 1 & 0 & 0 & -1 & 1 \\ -1 & 0 & 1 & -1 & 0 & -1 \end{bmatrix} [f_{1j}]^t [v_j],$$

Put together, the Dirac operator in matrix form is given as:

$$\partial\!\!\!/_M = \left(\begin{array}{cccc|cccccc|cc}
0 & 0 & 0 & 0 & -1 & 0 & -1 & 0 & 1 & 0 & 0 & 0 \\
0 & 0 & 0 & 0 & 1 & 1 & 0 & -1 & 0 & 0 & 0 & 0 \\
0 & 0 & 0 & 0 & 0 & -1 & 1 & 0 & 0 & 1 & 0 & 0 \\
0 & 0 & 0 & 0 & 0 & 0 & 0 & 1 & -1 & -1 & 0 & 0 \\
\hline
-1 & 1 & 0 & 0 & 0 & 0 & 0 & 0 & 0 & 0 & -1 & -1 \\
0 & 1 & -1 & 0 & 0 & 0 & 0 & 0 & 0 & 0 & 1 & 0 \\
-1 & 0 & 1 & 0 & 0 & 0 & 0 & 0 & 0 & 0 & 0 & 1 \\
0 & -1 & 0 & 1 & 0 & 0 & 0 & 0 & 0 & 0 & 0 & -1 \\
1 & 0 & 0 & -1 & 0 & 0 & 0 & 0 & 0 & 0 & -1 & 0 \\
0 & 0 & 1 & -1 & 0 & 0 & 0 & 0 & 0 & 0 & 1 & -1 \\
\hline
0 & 0 & 0 & 0 & -1 & 1 & 0 & 0 & -1 & 1 & 0 & 0 \\
0 & 0 & 0 & 0 & -1 & 0 & 1 & -1 & 0 & -1 & 0 & 0
\end{array}\right)$$

with eigenvalues $-2, 0, 2$ of multiplicities $5, 2, 5$, respectively.

The discrete Laplace operator is given by:

$$\begin{aligned}2\Delta f &= \partial(df_0) + (\partial(df_1) + d(\partial f_1)) + d(\partial f_2)\\ &= \sum_{j=0}^{3} f_{0j}\partial(dp_j) + \sum_{j=1}^{8} f_{1j}\left(\partial(dl_j) + d(\partial l_j)\right) + \sum_{j=1}^{4} f_{2j} d(\partial v_j).\end{aligned}$$

In matrix form we obtain:

$$\begin{aligned}2\Delta_M f = &\begin{bmatrix} 3 & -1 & -1 & -1\\ -1 & 3 & -1 & -1\\ -1 & -1 & 3 & -1\\ -1 & -1 & -1 & 3\end{bmatrix}[f_{0j}]^t[p_j]\\ &+\begin{bmatrix} 4 & 0 & 0 & 0 & 0 & 0\\ 0 & 3 & -1 & -1 & -1 & 0\\ 0 & -1 & 3 & -1 & -1 & 0\\ 0 & -1 & -1 & 3 & -1 & 0\\ 0 & -1 & -1 & -1 & 3 & 0\\ 0 & 0 & 0 & 0 & 0 & 4\end{bmatrix}[f_{1j}]^t[l_j]\\ &+\begin{bmatrix} 4 & 0\\ 0 & 4\end{bmatrix}[f_{2j}]^t[v_j]\,.\end{aligned}$$

In matrix form, the Laplacian of the Möbius script $\mathfrak{S}_M$ is given by the square of the Dirac matrix:

$$\Delta_M = \frac{1}{2}\left(\begin{array}{cccc|cccccc|cc} 3 & -1 & -1 & -1 & 0 & 0 & 0 & 0 & 0 & 0 & 0 & 0\\ -1 & 3 & -1 & -1 & 0 & 0 & 0 & 0 & 0 & 0 & 0 & 0\\ -1 & -1 & 3 & -1 & 0 & 0 & 0 & 0 & 0 & 0 & 0 & 0\\ -1 & -1 & -1 & 3 & 0 & 0 & 0 & 0 & 0 & 0 & 0 & 0\\ \hline 0 & 0 & 0 & 0 & 4 & 0 & 0 & 0 & 0 & 0 & 0 & 0\\ 0 & 0 & 0 & 0 & 0 & 3 & -1 & -1 & -1 & 0 & 0 & 0\\ 0 & 0 & 0 & 0 & 0 & -1 & 3 & -1 & -1 & 0 & 0 & 0\\ 0 & 0 & 0 & 0 & 0 & -1 & -1 & 3 & -1 & 0 & 0 & 0\\ 0 & 0 & 0 & 0 & 0 & -1 & -1 & -1 & 3 & 0 & 0 & 0\\ 0 & 0 & 0 & 0 & 0 & 0 & 0 & 0 & 0 & 4 & 0 & 0\\ \hline 0 & 0 & 0 & 0 & 0 & 0 & 0 & 0 & 0 & 0 & 4 & 0\\ 0 & 0 & 0 & 0 & 0 & 0 & 0 & 0 & 0 & 0 & 0 & 4\end{array}\right).$$

5.2. The torus

Consider the torus T equipped with the script defined as in Figure 3. Topologically it is obtained by identifying the opposite sides of a rectangle with the same orientation. The boundary operator acts as follows:

$$\begin{aligned}\partial l_1 &= p_1 - p_0, & \partial l_2 &= p_0 - p_1,\\ \partial l_3 &= p_0 - p_2, & \partial l_4 &= p_2 - p_0,\\ \partial l_5 &= p_3 - p_2, & \partial l_6 &= p_2 - p_3,\\ \partial l_7 &= p_1 - p_3, & \partial l_8 &= p_3 - p_1,\end{aligned}$$

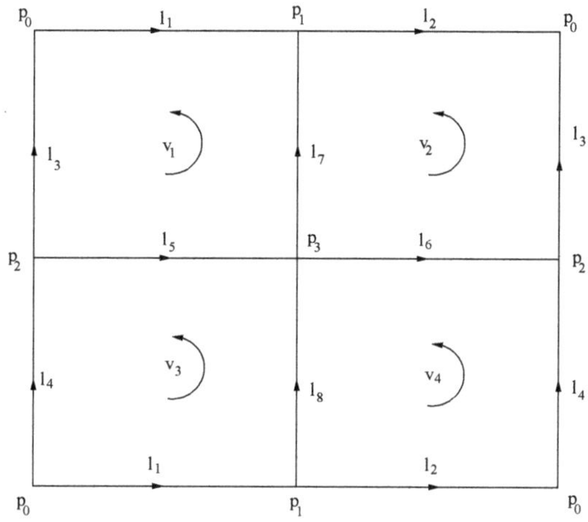

FIGURE 3. Torus script

and

$$\begin{aligned}\partial v_1 &= l_5 + l_7 - l_1 - l_3,\\ \partial v_2 &= l_6 + l_3 - l_2 - l_7,\\ \partial v_3 &= l_1 + l_8 - l_5 - l_4,\\ \partial v_4 &= l_2 + l_4 - l_6 - l_8.\end{aligned}$$

All 0, 1 and 2-cells are tight cells, so the script $\mathfrak{S}_T$ above is a tight geoscript.

For the script homology of the torus, we consider the sequence of chains:

$$0 \xrightarrow{\partial} \mathfrak{C}_2 \xrightarrow{\partial} \mathfrak{C}_1 \xrightarrow{\partial} \mathfrak{C}_0 \xrightarrow{\partial} 0,$$

and we note that:

$$\sum_{j=1}^{4} \partial l_j = 0, \qquad \sum_{j=1}^{4} \partial v_j = 0,$$

so, if we denote the line sums $l_{12} := (l_1 + l_2)$ and $l_{34} := (l_3 + l_4)$, and γ the sum of all v_j, we have:

$$\partial(l_{12}) = \partial(l_{34}) = 0, \qquad \partial\gamma = 0.$$

It turns out that l_{12} and l_{34} are a basis of $\mathcal{H}_1(\mathfrak{S}_T)$ and γ is the generator of $\mathcal{H}_2(\mathfrak{S}_T)$, i.e., we capture the script homology of the torus:

$$\mathcal{H}_0(\mathfrak{S}_T) = \mathbb{Z}, \qquad \mathcal{H}_1(\mathfrak{S}_T) = \mathbb{Z} \oplus \mathbb{Z}, \qquad \mathcal{H}_2(\mathfrak{S}_T) = \mathbb{Z}.$$

At the differential operator level, we obtain:

$$\begin{aligned} dp_0 &= -l_1 + l_2 + l_3 - l_4, \\ dp_1 &= l_1 - l_2 + l_7 - l_8, \\ dp_2 &= -l_3 + l_4 - l_5 + l_6, \\ dp_3 &= l_5 - l_6 - l_7 + l_8, \end{aligned}$$

and

$$\begin{aligned} dl_1 &= -v_1 + v_3, & dl_2 &= -v_2 + v_4, \\ dl_3 &= -v_1 + v_2, & dl_4 &= -v_3 + v_4, \\ dl_5 &= v_1 - v_3, & dl_6 &= -v_4 + v_2, \\ dl_7 &= v_1 - v_2, & dl_8 &= v_3 - v_4. \end{aligned}$$

Note that $\sum_{j=0}^{3} dp_j = 0$, $\sum_{j=1}^{8} l_j = 0$, and using the notation for the sum of two lines, $l_{ij} := l_i + l_j$, we have:

$$\begin{aligned} d(l_{15} - l_{26}) &= 0, & d(p_0 + p_2) &= -d(p_1 + p_3) = -(l_{15} - l_{26}), \\ d(l_{37} - l_{48}) &= 0, & d(p_0 + p_1) &= -d(p_2 + p_3) = (l_{37} - l_{48}). \end{aligned}$$

Therefore each $\mathcal{H}^0(\mathfrak{S}_T)$ and $\mathcal{H}^2(\mathfrak{S}_T)$ have one generator, and $\mathcal{H}^1(\mathfrak{S}_T)$ has two generators. Summarizing, we obtain the script cohomology groups of $\mathfrak{S}_T$ is given by:

$$\mathcal{H}^0(\mathfrak{S}_T) = \mathbb{Z}, \qquad \mathcal{H}^1(\mathfrak{S}_T) = \mathbb{Z} \oplus \mathbb{Z}, \qquad \mathcal{H}^2(\mathfrak{S}_T) = \mathbb{Z}.$$

Since the discrete Dirac operator is defined as $\not{\partial} = \partial + d$, using

$$f = f_0 + f_1 + f_2, \qquad f_0 = \sum_{j=0}^{3} f_{0j} p_j, \quad f_1 = \sum_{j=1}^{8} f_{1j} l_j, \quad f_2 = \sum_{j=1}^{4} f_{2j} v_j,$$

we have:

$$\not{\partial} f = \partial f_1 + (df_0 + \partial f_2) + df_1.$$

Computations yield to:

$$\partial f_1 = \begin{bmatrix} -1 & 1 & 1 & -1 & 0 & 0 & 0 & 0 \\ 1 & -1 & 0 & 0 & 0 & 0 & 1 & -1 \\ 0 & 0 & -1 & 1 & -1 & 1 & 0 & 0 \\ 0 & 0 & 0 & 0 & 1 & -1 & -1 & 1 \end{bmatrix} [f_{1j}]^t [p_j],$$

where $[f_{1j}]^t$ is the column vector the corresponding 8 inputs, and $[p_j]$ is the row vector of the four points. Similarly we obtain:

$$df_0 = \begin{bmatrix} -1 & 1 & 0 & 0 \\ 1 & -1 & 0 & 0 \\ 1 & 0 & -1 & 0 \\ -1 & 0 & 1 & 0 \\ 0 & 0 & -1 & 1 \\ 0 & 0 & 1 & -1 \\ 0 & 1 & 0 & -1 \\ 0 & -1 & 0 & 1 \end{bmatrix} [f_{0j}]^t [l_j],$$

where we notice that the matrix above is the transpose of the previous one for ∂f_1. Next we get:

$$\partial f_2 = \begin{bmatrix} -1 & 0 & 1 & 0 \\ 0 & -1 & 0 & 1 \\ -1 & 1 & 0 & 0 \\ 0 & 0 & -1 & 1 \\ 1 & 0 & -1 & 0 \\ 0 & 1 & 0 & -1 \\ 1 & -1 & 0 & 0 \\ 0 & 0 & 1 & -1 \end{bmatrix} [f_{2j}]^t [l_j],$$

and

$$df_1 = \begin{bmatrix} -1 & 0 & -1 & 0 & 1 & 0 & 1 & 0 \\ 0 & -1 & 1 & 0 & 0 & 1 & -1 & 0 \\ 1 & 0 & 0 & -1 & -1 & 0 & 0 & 1 \\ 0 & 1 & 0 & 1 & 0 & -1 & 0 & -1 \end{bmatrix} [f_{1j}]^t [v_j],$$

Put together, the Dirac operator on the torus script in matrix form is given as:

$$\left(\begin{array}{cccc|cccccccc|cccc}
0 & 0 & 0 & 0 & -1 & 1 & 1 & -1 & 0 & 0 & 0 & 0 & 0 & 0 & 0 & 0 \\
0 & 0 & 0 & 0 & 1 & -1 & 0 & 0 & 0 & 0 & 1 & -1 & 0 & 0 & 0 & 0 \\
0 & 0 & 0 & 0 & 0 & 0 & -1 & 1 & -1 & 1 & 0 & 0 & 0 & 0 & 0 & 0 \\
0 & 0 & 0 & 0 & 0 & 0 & 0 & 0 & 1 & -1 & -1 & 1 & 0 & 0 & 0 & 0 \\
\hline
-1 & 1 & 0 & 0 & 0 & 0 & 0 & 0 & 0 & 0 & 0 & 0 & -1 & 0 & 1 & 0 \\
1 & -1 & 0 & 0 & 0 & 0 & 0 & 0 & 0 & 0 & 0 & 0 & 0 & -1 & 0 & 1 \\
1 & 0 & -1 & 0 & 0 & 0 & 0 & 0 & 0 & 0 & 0 & 0 & -1 & 1 & 0 & 0 \\
-1 & 0 & 1 & 0 & 0 & 0 & 0 & 0 & 0 & 0 & 0 & 0 & 0 & 0 & -1 & 1 \\
0 & 0 & -1 & 1 & 0 & 0 & 0 & 0 & 0 & 0 & 0 & 0 & 1 & 0 & -1 & 0 \\
0 & 0 & 1 & -1 & 0 & 0 & 0 & 0 & 0 & 0 & 0 & 0 & 0 & 1 & 0 & -1 \\
0 & 1 & 0 & -1 & 0 & 0 & 0 & 0 & 0 & 0 & 0 & 0 & 1 & -1 & 0 & 0 \\
0 & -1 & 0 & 1 & 0 & 0 & 0 & 0 & 0 & 0 & 0 & 0 & 0 & 0 & 1 & -1 \\
\hline
0 & 0 & 0 & 0 & -1 & 0 & -1 & 0 & 1 & 0 & 1 & 0 & 0 & 0 & 0 & 0 \\
0 & 0 & 0 & 0 & 0 & -1 & 1 & 0 & 0 & 1 & -1 & 0 & 0 & 0 & 0 & 0 \\
0 & 0 & 0 & 0 & 1 & 0 & 0 & -1 & -1 & 0 & 0 & 1 & 0 & 0 & 0 & 0 \\
0 & 0 & 0 & 0 & 0 & 1 & 0 & 1 & 0 & -1 & 0 & -1 & 0 & 0 & 0 & 0
\end{array}\right)$$

with eigenvalues $-2\sqrt{2}, -2, 0, 2, 2\sqrt{2}$ of multiplicities $2, 4, 4, 4, 2$, respectively.

The discrete Laplace operator is given by:

$$\begin{aligned} 2\Delta_T f &= \partial(df_0) + (\partial(df_1) + d(\partial f_1)) + d(\partial f_2) \\ &= \sum_{j=0}^{3} f_{0j}\partial(dp_j) + \sum_{j=1}^{8} f_{1j}\left(\partial(dl_j) + d(\partial l_j)\right) + \sum_{j=1}^{4} f_{2j} d(\partial v_j). \end{aligned}$$

We obtain:

$$\begin{aligned} \Delta_T f = &\begin{bmatrix} 2 & -1 & -1 & 0 \\ -1 & 2 & 0 & -1 \\ -1 & 0 & 2 & -1 \\ 0 & -1 & -1 & 2 \end{bmatrix} [f_{0j}]^t [p_j] \\ &+ \begin{bmatrix} 2 & -1 & 0 & 0 & -1 & 0 & 0 & 0 \\ -1 & 2 & 0 & 0 & 0 & -1 & 0 & 0 \\ 0 & 0 & 2 & -1 & 0 & 0 & -1 & 0 \\ 0 & 0 & -1 & 2 & 0 & 0 & 0 & -1 \\ -1 & 0 & 0 & 0 & 2 & -1 & 0 & 0 \\ 0 & -1 & 0 & 0 & -1 & 2 & 0 & 0 \\ 0 & 0 & -1 & 0 & 0 & 0 & 2 & -1 \\ 0 & 0 & 0 & -1 & 0 & 0 & -1 & 2 \end{bmatrix} [f_{1j}]^t [l_j] \\ &+ \begin{bmatrix} 2 & -1 & -1 & 0 \\ -1 & 2 & 0 & -1 \\ -1 & 0 & 2 & -1 \\ 0 & -1 & -1 & 2 \end{bmatrix} [f_{2j}]^t [v_j] . \end{aligned}$$

In matrix form, the Laplacian of the script for the torus T is given by the square of the Dirac matrix:

$$\left(\begin{array}{cccc|cccccccc|cccc}
2 & -1 & -1 & 0 & 0 & 0 & 0 & 0 & 0 & 0 & 0 & 0 & 0 & 0 & 0 & 0 \\
-1 & 2 & 0 & -1 & 0 & 0 & 0 & 0 & 0 & 0 & 0 & 0 & 0 & 0 & 0 & 0 \\
-1 & 0 & 2 & -1 & 0 & 0 & 0 & 0 & 0 & 0 & 0 & 0 & 0 & 0 & 0 & 0 \\
0 & -1 & -1 & 2 & 0 & 0 & 0 & 0 & 0 & 0 & 0 & 0 & 0 & 0 & 0 & 0 \\
\hline
0 & 0 & 0 & 0 & 2 & -1 & 0 & 0 & -1 & 0 & 0 & 0 & 0 & 0 & 0 & 0 \\
0 & 0 & 0 & 0 & -1 & 2 & 0 & 0 & 0 & -1 & 0 & 0 & 0 & 0 & 0 & 0 \\
0 & 0 & 0 & 0 & 0 & 0 & 2 & -1 & 0 & 0 & -1 & 0 & 0 & 0 & 0 & 0 \\
0 & 0 & 0 & 0 & 0 & 0 & -1 & 2 & 0 & 0 & 0 & -1 & 0 & 0 & 0 & 0 \\
0 & 0 & 0 & 0 & -1 & 0 & 0 & 0 & 2 & -1 & 0 & 0 & 0 & 0 & 0 & 0 \\
0 & 0 & 0 & 0 & 0 & -1 & 0 & 0 & -1 & 2 & 0 & 0 & 0 & 0 & 0 & 0 \\
0 & 0 & 0 & 0 & 0 & 0 & -1 & 0 & 0 & 0 & 2 & -1 & 0 & 0 & 0 & 0 \\
0 & 0 & 0 & 0 & 0 & 0 & 0 & -1 & 0 & 0 & -1 & 2 & 0 & 0 & 0 & 0 \\
\hline
0 & 0 & 0 & 0 & 0 & 0 & 0 & 0 & 0 & 0 & 0 & 0 & 2 & -1 & -1 & 0 \\
0 & 0 & 0 & 0 & 0 & 0 & 0 & 0 & 0 & 0 & 0 & 0 & -1 & 2 & 0 & -1 \\
0 & 0 & 0 & 0 & 0 & 0 & 0 & 0 & 0 & 0 & 0 & 0 & -1 & 0 & 2 & -1 \\
0 & 0 & 0 & 0 & 0 & 0 & 0 & 0 & 0 & 0 & 0 & 0 & 0 & -1 & -1 & 2
\end{array}\right).$$

5.3. The Klein bottle

Consider the Klein bottle equipped with the script $\mathfrak{S}_K$ defined as in Figure 4. It is obtained from a rectangle identifying one pair of opposite sides with the same orientations, and the other pair is identified with opposite line orientations. We

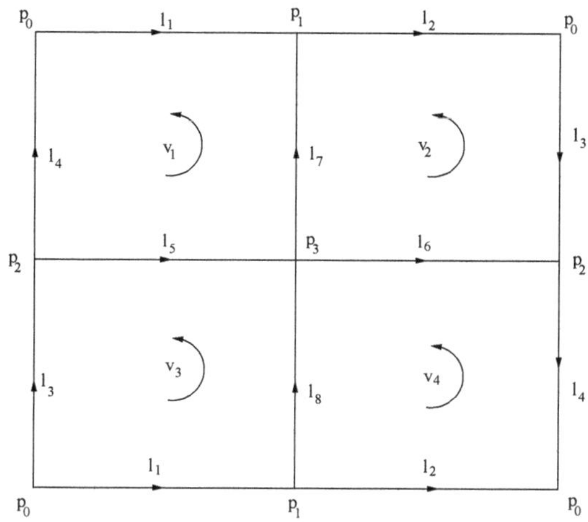

FIGURE 4. Klein script

obtain the following script geometry of the Klein script:

$$\begin{aligned} \partial l_1 &= p_1 - p_0, & \partial l_2 &= p_0 - p_1, \\ \partial l_3 &= p_2 - p_0, & \partial l_4 &= p_0 - p_2, \\ \partial l_5 &= p_3 - p_2, & \partial l_6 &= p_2 - p_3, \\ \partial l_7 &= p_1 - p_3, & \partial l_4 &= p_3 - p_1, \end{aligned}$$

and

$$\begin{aligned} \partial v_1 &= l_5 + l_7 - l_1 - l_4, \\ \partial v_2 &= l_6 - l_3 - l_2 - l_7, \\ \partial v_3 &= l_1 + l_8 - l_5 - l_3, \\ \partial v_4 &= l_2 - l_4 - l_6 - l_8. \end{aligned}$$

Note again that this is a tight script, and we have:

$$\sum_{j=1}^{4} \partial v_j = -2(l_3 + l_4)\,.$$

If we denote the line sum $l_{12} := (l_1 + l_2)$ and $l_{34} := (l_3 + l_4)$, and γ is the sum of all v_j, then l_{12} is a generator for $\mathcal{H}_1(\mathfrak{S}_K)$ modulo torsion, and l_{34} is a torsion element of $\mathcal{H}_1(\mathfrak{S}_K)$. But γ is not a cycle anymore, as $\partial\gamma = -2l_{34}$, i.e., we recapture the script homology of the Klein bottle:

$$\mathcal{H}_0(\mathfrak{S}_K) = \mathbb{Z}, \qquad \mathcal{H}_1(\mathfrak{S}_K) = \mathbb{Z} \oplus \mathbb{Z}_2, \qquad \mathcal{H}_2(\mathfrak{S}_K) = 0.$$

At the differential operator level, we obtain:

$$\begin{aligned} dp_0 &= -l_1 + l_2 - l_3 + l_4, \\ dp_1 &= l_1 - l_2 + l_7 - l_8, \\ dp_2 &= l_3 - l_4 - l_5 + l_6, \\ dp_3 &= l_5 - l_6 - l_7 + l_8, \end{aligned}$$

and

$$\begin{aligned} dl_1 &= -v_1 + v_3, & dl_2 &= -v_2 + v_4, \\ dl_3 &= -v_2 - v_3, & dl_4 &= -v_1 - v_4, \\ dl_5 &= v_1 - v_3, & dl_6 &= -v_4 + v_2, \\ dl_7 &= v_1 - v_2, & dl_8 &= v_3 - v_4. \end{aligned}$$

Note that $\sum_{j=0}^{3} dp_j = 0$ and $d((l_1 - l_2) + (l_3 - l_4)) = 0$. Therefore $\mathcal{H}^1(\mathfrak{S}_K)$ is generated by one element. Next, we have:

$$dl_{12} = -(v_1 + v_2) + (v_3 + v_4), \quad dl_{34} = -(v_1 + v_2) - (v_3 + v_4),$$

therefore, because $d(l_{12} + l_{34}) = -2(v_1 + v_2)$, we obtain that $\mathcal{H}^2(\mathfrak{S}_K) = \mathbb{Z}_2$. This yields the script cohomology groups of $\mathfrak{S}_K$:

$$\mathcal{H}^0(\mathfrak{S}_K) = \mathbb{Z}, \qquad \mathcal{H}^1(\mathfrak{S}_K) = \mathbb{Z}, \qquad \mathcal{H}^2(\mathfrak{S}_K) = \mathbb{Z}_2.$$

The discrete Dirac operator for the Klein script above is given by the formula:

$$\partial\!\!\!/ f = \partial f_1 + (df_0 + \partial f_2) + df_1,$$

where

$$f = f_0 + f_1 + f_2, \qquad f_0 = \sum_{j=0}^{3} f_{0j} p_j, \quad f_1 = \sum_{j=1}^{8} f_{1j} l_j, \quad f_2 = \sum_{j=1}^{4} f_{2j} v_j.$$

Computations yield the following results:

$$\partial f_1 = \begin{bmatrix} -1 & 1 & -1 & 1 & 0 & 0 & 0 & 0 \\ 1 & -1 & 0 & 0 & 0 & 0 & 1 & -1 \\ 0 & 0 & 1 & -1 & -1 & 1 & 0 & 0 \\ 0 & 0 & 0 & 0 & 1 & -1 & -1 & 1 \end{bmatrix} [f_{1j}]^t [p_j],$$

$$df_0 = \begin{bmatrix} -1 & 1 & 0 & 0 \\ 1 & -1 & 0 & 0 \\ -1 & 0 & 1 & 0 \\ 1 & 0 & -1 & 0 \\ 0 & 0 & -1 & 1 \\ 0 & 0 & 1 & -1 \\ 0 & 1 & 0 & -1 \\ 0 & -1 & 0 & 1 \end{bmatrix} [f_{0j}]^t [l_j].$$

Next we get:

$$\partial f_2 = \begin{bmatrix} -1 & 0 & 1 & 0 \\ 0 & -1 & 0 & 1 \\ 0 & -1 & -1 & 0 \\ -1 & 0 & 0 & -1 \\ 1 & 0 & -1 & 0 \\ 0 & 1 & 0 & -1 \\ 1 & -1 & 0 & 0 \\ 0 & 0 & 1 & -1 \end{bmatrix} [f_{2j}]^t [l_j],$$

and

$$df_1 = \begin{bmatrix} -1 & 0 & 0 & -1 & 1 & 0 & 1 & 0 \\ 0 & -1 & -1 & 0 & 0 & 1 & -1 & 0 \\ 1 & 0 & -1 & 0 & -1 & 0 & 0 & 1 \\ 0 & 1 & 0 & -1 & 0 & -1 & 0 & -1 \end{bmatrix} [f_{1j}]^t [v_j],$$

In matrix form, the Dirac operator for the Klein script is:

$$\left(\begin{array}{cccc|cccccccc|cccc}
0 & 0 & 0 & 0 & -1 & 1 & -1 & 1 & 0 & 0 & 0 & 0 & 0 & 0 & 0 & 0 \\
0 & 0 & 0 & 0 & 1 & -1 & 0 & 0 & 0 & 0 & 1 & -1 & 0 & 0 & 0 & 0 \\
0 & 0 & 0 & 0 & 0 & 0 & 1 & -1 & -1 & 1 & 0 & 0 & 0 & 0 & 0 & 0 \\
0 & 0 & 0 & 0 & 0 & 0 & 0 & 0 & 1 & -1 & -1 & 1 & 0 & 0 & 0 & 0 \\
\hline
-1 & 1 & 0 & 0 & 0 & 0 & 0 & 0 & 0 & 0 & 0 & 0 & -1 & 0 & 1 & 0 \\
1 & -1 & 0 & 0 & 0 & 0 & 0 & 0 & 0 & 0 & 0 & 0 & 0 & -1 & 0 & 1 \\
-1 & 0 & 1 & 0 & 0 & 0 & 0 & 0 & 0 & 0 & 0 & 0 & 0 & -1 & -1 & 0 \\
1 & 0 & -1 & 0 & 0 & 0 & 0 & 0 & 0 & 0 & 0 & 0 & -1 & 0 & 0 & -1 \\
0 & 0 & -1 & 1 & 0 & 0 & 0 & 0 & 0 & 0 & 0 & 0 & 1 & 0 & -1 & 0 \\
0 & 0 & 1 & -1 & 0 & 0 & 0 & 0 & 0 & 0 & 0 & 0 & 0 & 1 & 0 & -1 \\
0 & 1 & 0 & -1 & 0 & 0 & 0 & 0 & 0 & 0 & 0 & 0 & 1 & -1 & 0 & 0 \\
0 & -1 & 0 & 1 & 0 & 0 & 0 & 0 & 0 & 0 & 0 & 0 & 0 & 0 & 1 & -1 \\
\hline
0 & 0 & 0 & 0 & -1 & 0 & 0 & -1 & 1 & 0 & 1 & 0 & 0 & 0 & 0 & 0 \\
0 & 0 & 0 & 0 & 0 & -1 & -1 & 0 & 0 & 1 & -1 & 0 & 0 & 0 & 0 & 0 \\
0 & 0 & 0 & 0 & 1 & 0 & -1 & 0 & -1 & 0 & 0 & 1 & 0 & 0 & 0 & 0 \\
0 & 0 & 0 & 0 & 0 & 1 & 0 & -1 & 0 & -1 & 0 & -1 & 0 & 0 & 0 & 0
\end{array}\right)$$

with eigenvalues $-2\sqrt{2}, -2, -\sqrt{2}, 0, \sqrt{2}, 2, 2\sqrt{2}$ of multiplicities $2, 3, 2, 2, 2, 3, 2$, respectively.

The Laplace operator is given by:

$$\Delta_K f = \begin{bmatrix} 2 & -1 & -1 & 0 \\ -1 & 2 & 0 & -1 \\ -1 & 0 & 2 & -1 \\ 0 & -1 & -1 & 2 \end{bmatrix} [f_{0j}]^t [p_j] + \begin{bmatrix} 2 & -1 & 0 & 0 & -1 & 0 & 0 & 0 \\ -1 & 2 & 0 & 0 & 0 & -1 & 0 & 0 \\ 0 & 0 & 2 & -1 & 0 & 0 & 1 & -1 \\ 0 & 0 & -1 & 2 & 0 & 0 & -1 & 1 \\ -1 & 0 & 0 & 0 & 2 & -1 & 0 & 0 \\ 0 & -1 & 0 & 0 & -1 & 2 & 0 & 0 \\ 0 & 0 & 1 & -1 & 0 & 0 & 2 & -1 \\ 0 & 0 & -1 & 1 & 0 & 0 & -1 & 2 \end{bmatrix} [f_{1j}]^t [l_j]$$

$$+ \frac{1}{2} \begin{bmatrix} 4 & -1 & -2 & 1 \\ -1 & 4 & 1 & -2 \\ -2 & 1 & 4 & -1 \\ 1 & -2 & -1 & 4 \end{bmatrix} [f_{2j}]^t [v_j].$$

5.4. The real projective plane

We investigate several scripts for the projective plane $\mathbb{RP}^2$. The simplest one is given in Figure 5, but it is not a geoscript. Indeed, this script, denoted by $\mathfrak{S}_{\mathbb{RP}^2,1}$

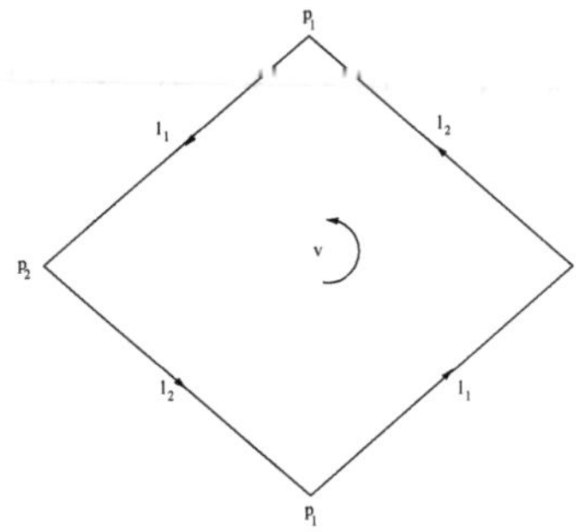

FIGURE 5. Simplest projective script

is characterized by

$$\mathfrak{S}_0 = \{p_1, p_2\}, \qquad \mathfrak{S}_1 = \{l_1, l_2\}, \qquad \mathfrak{S}_2 = \{v\},$$
$$\partial l_1 = p_2 - p_1, \qquad \partial l_2 = p_1 - p_2, \qquad \partial v = 2l_1 + 2l_2.$$

Therefore the boundary of the 2-chain v is not a geochain, as it contains coefficients different than ± 1.

Denoting the line sum $l_{12} := (l_1 + l_2)$ then l_{12} is representing the non-zero element of $\mathcal{H}_1(\mathfrak{S}_{\mathbb{RP}^2})$, and v is not a cycle, as $\partial v = 2l_{12}$, i.e., we obtain the script homology of the $\mathbb{RP}^2$:

$$\mathcal{H}_0(\mathfrak{S}_{\mathbb{RP}^2}) = \mathbb{Z}, \qquad \mathcal{H}_1(\mathfrak{S}_{\mathbb{RP}^2}) = \mathbb{Z}_2, \qquad \mathcal{H}_2(\mathfrak{S}_{\mathbb{RP}^2}) = 0.$$

For the differential operator d we get:

$$dp_1 = -(l_1 - l_2), \qquad dp_2 = l_1 - l_2,$$
$$dl_1 = dl_2 = v.$$

Note that the kernel of d on points is generated by $p_1 + p_2$, so $\mathcal{H}^0(\mathfrak{S}_{\mathbb{RP}^2}) = \mathbb{Z}$. Next, the kernel of d on lines and the image of d on points are both generated by $l_1 - l_2$, which yields to $\mathcal{H}^1(\mathfrak{S}_{\mathbb{RP}^2}) = 0$. The image of $d(l_1 + l_2) = 2v$, therefore $\mathcal{H}^2(\mathfrak{S}_{\mathbb{RP}^2}) = \mathbb{Z}_2$.

We compute the Dirac and Laplace operators on this script, yielding:

$$\partial\!\!\!/_{\mathbb{RP}^2} = \left(\begin{array}{cc|cc|c} 0 & 0 & -1 & 1 & 0 \\ 0 & 0 & 1 & -1 & 0 \\ \hline -1 & 1 & 0 & 0 & 2 \\ 1 & -1 & 0 & 0 & 2 \\ \hline 0 & 0 & 2 & 2 & 0 \end{array}\right)$$

with eigenvalues $-2\sqrt{2}, -2, 0, 2, 2\sqrt{2}$, all having multiplicities 1.

The Laplacian operator for this script depicting $\mathbb{RP}^2$ is:

$$\Delta_{\mathbb{RP}^2} = \frac{1}{2}\left(\begin{array}{cc|cc|c} 1 & -1 & 0 & 0 & 0 \\ -1 & 1 & 0 & 0 & 0 \\ \hline 0 & 0 & 3 & 1 & 0 \\ 0 & 0 & 1 & 3 & 0 \\ \hline 0 & 0 & 0 & 0 & 4 \end{array}\right).$$

In order to obtain a tight geoscript for the projective plane, we add one more point p_0 and four lines to the script above, as in Figure 6. We obtain:

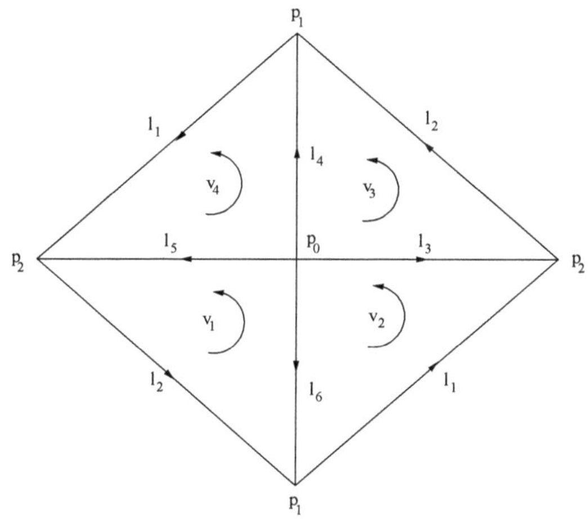

FIGURE 6. A tight geoscript for $\mathbb{RP}^2$

$$\begin{aligned} \partial l_1 &= p_2 - p_1, & \partial l_2 &= p_1 - p_2, \\ \partial l_3 &= p_2 - p_0, & \partial l_4 &= p_1 - p_0, \\ \partial l_5 &= p_2 - p_0, & \partial l_6 &= p_1 - p_0, \end{aligned}$$

and

$$\begin{aligned} \partial v_1 &= l_2 - l_6 + l_5, \\ \partial v_2 &= l_1 - l_3 + l_6, \\ \partial v_3 &= l_2 - l_4 + l_3, \\ \partial v_4 &= l_1 - l_5 + l_4. \end{aligned}$$

At the differential operator level, we obtain:

$$\begin{aligned} dp_0 &= -l_3 - l_4 - l_5 - l_6, \\ dp_1 &= -l_1 + l_2 + l_4 + l_6, \\ dp_2 &= l_1 - l_2 + l_3 + l_5, \end{aligned}$$

and

$$\begin{aligned} dl_1 &= v_2 + v_4, & dl_2 &= v_1 + v_3,\\ dl_3 &= -v_2 + v_3, & dl_4 &= -v_3 + v_4,\\ dl_5 &= v_1 - v_4, & dl_6 &= v_2 - v_1. \end{aligned}$$

The homology and cohomology of this real projective plane script is computed in a similar way as in the case of the first $\mathbb{RP}^2$ script, yielding, of course, the same result.

The discrete Dirac operator for the script above is given by:

$$\partial\!\!\!/ f = \partial f_1 + (df_0 + \partial f_2) + df_1,$$

where

$$f = f_0 + f_1 + f_2, \qquad f_0 = \sum_{j=0}^{2} f_{0j} p_j, \quad f_1 = \sum_{j=1}^{6} f_{1j} l_j, \quad f_2 = \sum_{j=1}^{4} f_{2j} v_j.$$

Computations yield the following results:

$$\partial f_1 = \begin{bmatrix} 0 & 0 & -1 & -1 & -1 & -1 \\ -1 & 1 & 0 & 1 & 0 & 1 \\ 1 & -1 & 1 & 0 & 1 & 0 \end{bmatrix} [f_{1j}]^t [p_j],$$

$$df_0 = \begin{bmatrix} 0 & -1 & 1 \\ 0 & 1 & -1 \\ -1 & 0 & 1 \\ -1 & 1 & 0 \\ -1 & 0 & 1 \\ -1 & 1 & 0 \end{bmatrix} [f_{0j}]^t [l_j],$$

$$\partial f_2 = \begin{bmatrix} 0 & 1 & 0 & 1 \\ 1 & 0 & 1 & 0 \\ 0 & -1 & 1 & 0 \\ 0 & 0 & -1 & 1 \\ 1 & 0 & 0 & -1 \\ -1 & 1 & 0 & 0 \end{bmatrix} [f_{2j}]^t [l_j],$$

and

$$df_1 = \begin{bmatrix} 0 & 1 & 0 & 0 & 1 & -1 \\ 1 & 0 & -1 & 0 & 0 & 1 \\ 0 & 1 & 1 & -1 & 0 & 0 \\ 1 & 0 & 0 & 1 & -1 & 0 \end{bmatrix} [f_{1j}]^t [v_j],$$

In full matrix form we obtain

$$\left(\begin{array}{ccc|cccccc|cccc}
0 & 0 & 0 & 0 & 0 & -1 & -1 & -1 & -1 & 0 & 0 & 0 & 0 \\
0 & 0 & 0 & -1 & 1 & 0 & 1 & 0 & 1 & 0 & 0 & 0 & 0 \\
0 & 0 & 0 & 1 & -1 & 1 & 0 & 1 & 0 & 0 & 0 & 0 & 0 \\
\hline
0 & -1 & 1 & 0 & 0 & 0 & 0 & 0 & 0 & 0 & 1 & 0 & 1 \\
0 & 1 & -1 & 0 & 0 & 0 & 0 & 0 & 0 & 1 & 0 & 1 & 0 \\
-1 & 0 & 1 & 0 & 0 & 0 & 0 & 0 & 0 & 0 & -1 & 1 & 0 \\
-1 & 1 & 0 & 0 & 0 & 0 & 0 & 0 & 0 & 0 & 0 & -1 & 1 \\
-1 & 0 & 1 & 0 & 0 & 0 & 0 & 0 & 0 & 1 & 0 & 0 & -1 \\
-1 & 1 & 0 & 0 & 0 & 0 & 0 & 0 & 0 & -1 & 1 & 0 & 0 \\
\hline
0 & 0 & 0 & 0 & 1 & 0 & 0 & 1 & -1 & 0 & 0 & 0 & 0 \\
0 & 0 & 0 & 1 & 0 & -1 & 0 & 0 & 1 & 0 & 0 & 0 & 0 \\
0 & 0 & 0 & 0 & 1 & 1 & -1 & 0 & 0 & 0 & 0 & 0 & 0 \\
0 & 0 & 0 & 1 & 0 & 0 & 1 & -1 & 0 & 0 & 0 & 0 & 0
\end{array}\right).$$

The eigenvalues are: $-\sqrt{6}, -\sqrt{2}, 0, \sqrt{2}, \sqrt{6}$ with multiplicities $3, 3, 1, 3, 3$, respectively.

In matrix form, the Laplacian is given by the square of the Dirac matrix:

$$\frac{1}{2}\left(\begin{array}{ccc|cccccc|cccc}
4 & -2 & -2 & 0 & 0 & 0 & 0 & 0 & 0 & 0 & 0 & 0 & 0 \\
-2 & 4 & -2 & 0 & 0 & 0 & 0 & 0 & 0 & 0 & 0 & 0 & 0 \\
-2 & -2 & 4 & 0 & 0 & 0 & 0 & 0 & 0 & 0 & 0 & 0 & 0 \\
\hline
0 & 0 & 0 & 4 & -2 & 0 & 0 & 0 & 0 & 0 & 0 & 0 & 0 \\
0 & 0 & 0 & -2 & 4 & 0 & 0 & 0 & 0 & 0 & 0 & 0 & 0 \\
0 & 0 & 0 & 0 & 0 & 4 & 0 & 2 & 0 & 0 & 0 & 0 & 0 \\
0 & 0 & 0 & 0 & 0 & 0 & 4 & 0 & 2 & 0 & 0 & 0 & 0 \\
0 & 0 & 0 & 0 & 0 & 2 & 0 & 4 & 0 & 0 & 0 & 0 & 0 \\
0 & 0 & 0 & 0 & 0 & 0 & 2 & 0 & 4 & 0 & 0 & 0 & 0 \\
\hline
0 & 0 & 0 & 0 & 0 & 0 & 0 & 0 & 0 & 3 & -1 & 1 & -1 \\
0 & 0 & 0 & 0 & 0 & 0 & 0 & 0 & 0 & -1 & 3 & -1 & 1 \\
0 & 0 & 0 & 0 & 0 & 0 & 0 & 0 & 0 & 1 & -1 & 3 & -1 \\
0 & 0 & 0 & 0 & 0 & 0 & 0 & 0 & 0 & -1 & 1 & -1 & 3
\end{array}\right).$$

A third script for $\mathbb{RP}^2$ is in the spirit of the torus and the Klein bottle, as given in Figure 7.

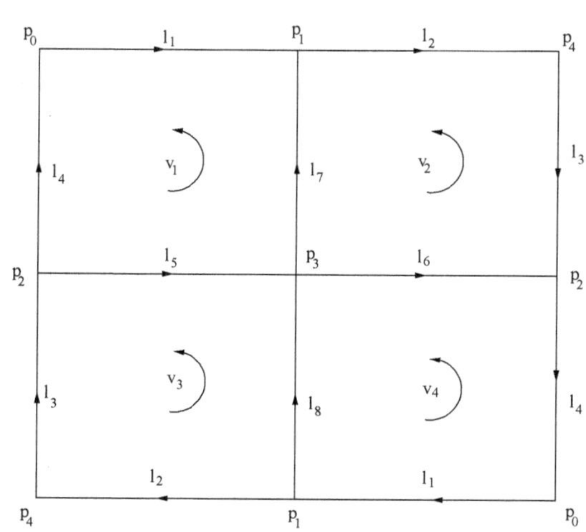

FIGURE 7. A third projective script

We obtain:

$$\begin{aligned} \partial l_1 &= p_1 - p_0, & \partial l_2 &= p_4 - p_1, \\ \partial l_3 &= p_2 - p_4, & \partial l_4 &= p_0 - p_2, \\ \partial l_5 &= p_3 - p_2, & \partial l_6 &= p_2 - p_3, \\ \partial l_7 &= p_1 - p_3, & \partial l_4 &= p_3 - p_1, \end{aligned}$$

and

$$\begin{aligned} \partial v_1 &= l_5 + l_7 - l_1 - l_4, \\ \partial v_2 &= l_6 - l_3 - l_2 - l_7, \\ \partial v_3 &= -l_2 + l_8 - l_5 - l_3, \\ \partial v_4 &= -l_1 - l_4 - l_6 - l_8. \end{aligned}$$

We obtain

$$\sum_{j=1}^{4} \partial v_j = -2(l_1 + l_2 + l_3 + l_4).$$

Denoting the line sum $l_{1234} := (l_1 + l_2 + l_3 + l_4)$ and γ the sum of all v_j, then l_{1234} is a representing the non-zero element of $\mathcal{H}_1(\mathfrak{S}_{\mathbb{RP}^2})$, and γ is not a cycle, as $\partial\gamma = -2l_{1234}$, i.e., we get again:

$$\mathcal{H}_1(\mathfrak{S}_{\mathbb{RP}^2}) \simeq \mathbb{Z}_2, \qquad \mathcal{H}_2(\mathfrak{S}_{\mathbb{RP}^2}) = 0.$$

The cohomology is calculated in a similar way.

Similar to the computations above for the Klein bottle, we obtain the following matrix form for the discrete Dirac operator on $\mathbb{RP}^2$:

$$\left(\begin{array}{ccccc|cccccccc|cccc}
0 & 0 & 0 & 0 & 0 & -1 & 0 & 0 & 1 & 0 & 0 & 0 & 0 & 0 & 0 & 0 & 0 \\
0 & 0 & 0 & 0 & 0 & 1 & -1 & 0 & 0 & 0 & 0 & 1 & -1 & 0 & 0 & 0 & 0 \\
0 & 0 & 0 & 0 & 0 & 0 & 0 & 1 & -1 & -1 & 1 & 0 & 0 & 0 & 0 & 0 & 0 \\
0 & 0 & 0 & 0 & 0 & 0 & 0 & 0 & 0 & 1 & -1 & -1 & 1 & 0 & 0 & 0 & 0 \\
0 & 0 & 0 & 0 & 0 & 0 & 1 & -1 & 0 & 0 & 0 & 0 & 0 & 0 & 0 & 0 & 0 \\
\hline
-1 & 1 & 0 & 0 & 0 & 0 & 0 & 0 & 0 & 0 & 0 & 0 & 0 & -1 & 0 & 0 & -1 \\
0 & -1 & 0 & 0 & 1 & 0 & 0 & 0 & 0 & 0 & 0 & 0 & 0 & 0 & -1 & -1 & 0 \\
0 & 0 & 1 & 0 & -1 & 0 & 0 & 0 & 0 & 0 & 0 & 0 & 0 & 0 & -1 & -1 & 0 \\
1 & 0 & -1 & 0 & 0 & 0 & 0 & 0 & 0 & 0 & 0 & 0 & 0 & -1 & 0 & 0 & -1 \\
0 & 0 & -1 & 1 & 0 & 0 & 0 & 0 & 0 & 0 & 0 & 0 & 0 & 1 & 0 & -1 & 0 \\
0 & 0 & 1 & -1 & 0 & 0 & 0 & 0 & 0 & 0 & 0 & 0 & 0 & 0 & 1 & 0 & -1 \\
0 & 1 & 0 & -1 & 0 & 0 & 0 & 0 & 0 & 0 & 0 & 0 & 0 & 1 & -1 & 0 & 0 \\
0 & -1 & 0 & 1 & 0 & 0 & 0 & 0 & 0 & 0 & 0 & 0 & 0 & 0 & 0 & 1 & -1 \\
\hline
0 & 0 & 0 & 0 & 0 & -1 & 0 & 0 & -1 & 1 & 0 & 1 & 0 & 0 & 0 & 0 & 0 \\
0 & 0 & 0 & 0 & 0 & 0 & -1 & -1 & 0 & 0 & 1 & -1 & 0 & 0 & 0 & 0 & 0 \\
0 & 0 & 0 & 0 & 0 & 0 & -1 & -1 & 0 & -1 & 0 & 0 & 1 & 0 & 0 & 0 & 0 \\
0 & 0 & 0 & 0 & 0 & -1 & 0 & 0 & -1 & 0 & -1 & 0 & -1 & 0 & 0 & 0 & 0
\end{array}\right)$$

with eigenvalues $-2\sqrt{2}$, $-\sqrt{5+\sqrt{5}}$, $-\sqrt{5-\sqrt{5}}$, -2, $-\sqrt{2}$, 0, $\sqrt{2}$, 2, $\sqrt{5-\sqrt{5}}$, $\sqrt{5+\sqrt{5}}$, $2\sqrt{2}$ with multiplicities 1, 2, 2, 3, 1, 3, 2, 2, 1, respectively.

The Laplacian for this script depicting $\mathbb{RP}^2$ is given by:

$$\left(\begin{array}{ccccc|cccccccc|cccc}
2 & -1 & -1 & 0 & 0 & 0 & 0 & 0 & 0 & 0 & 0 & 0 & 0 & 0 & 0 & 0 & 0 \\
-1 & 4 & 0 & -2 & 1 & 0 & 0 & 0 & 0 & 0 & 0 & 0 & 0 & 0 & 0 & 0 & 0 \\
-1 & 0 & 4 & -2 & -1 & 0 & 0 & 0 & 0 & 0 & 0 & 0 & 0 & 0 & 0 & 0 & 0 \\
0 & -2 & -2 & 4 & 0 & 0 & 0 & 0 & 0 & 0 & 0 & 0 & 0 & 0 & 0 & 0 & 0 \\
0 & -1 & -1 & 0 & 2 & 0 & 0 & 0 & 0 & 0 & 0 & 0 & 0 & 0 & 0 & 0 & 0 \\
\hline
0 & 0 & 0 & 0 & 0 & 4 & -1 & 0 & 1 & -1 & 1 & 0 & 0 & 0 & 0 & 0 & 0 \\
0 & 0 & 0 & 0 & 0 & -1 & 4 & 1 & 0 & 1 & -1 & 0 & 0 & 0 & 0 & 0 & 0 \\
0 & 0 & 0 & 0 & 0 & 0 & 1 & 4 & -1 & 0 & 0 & 1 & -1 & 0 & 0 & 0 & 0 \\
0 & 0 & 0 & 0 & 0 & 1 & 0 & -1 & 4 & 0 & 0 & -1 & 1 & 0 & 0 & 0 & 0 \\
0 & 0 & 0 & 0 & 0 & -1 & 1 & 0 & 0 & 4 & -2 & 0 & 0 & 0 & 0 & 0 & 0 \\
0 & 0 & 0 & 0 & 0 & 1 & -1 & 0 & 0 & -2 & 4 & 0 & 0 & 0 & 0 & 0 & 0 \\
0 & 0 & 0 & 0 & 0 & 0 & 0 & 1 & -1 & 0 & 0 & 4 & -2 & 0 & 0 & 0 & 0 \\
0 & 0 & 0 & 0 & 0 & 0 & 0 & -1 & 1 & 0 & 0 & -2 & 4 & 0 & 0 & 0 & 0 \\
\hline
0 & 0 & 0 & 0 & 0 & 0 & 0 & 0 & 0 & 0 & 0 & 0 & 0 & 4 & -1 & -1 & 2 \\
0 & 0 & 0 & 0 & 0 & 0 & 0 & 0 & 0 & 0 & 0 & 0 & 0 & -1 & 4 & 2 & -1 \\
0 & 0 & 0 & 0 & 0 & 0 & 0 & 0 & 0 & 0 & 0 & 0 & 0 & -1 & 2 & 4 & -1 \\
0 & 0 & 0 & 0 & 0 & 0 & 0 & 0 & 0 & 0 & 0 & 0 & 0 & 2 & -1 & -1 & 4
\end{array}\right).$$

Finally, we can obtain an $\mathbb{RP}^2$ by adding a rectangle v_3 to a Möbius strip M (see Subsection 5.1) with boundary

$$\partial v_3 = l_2 + l_3 + l_4 + l_5 ,$$

as in Figure 8. It follows that

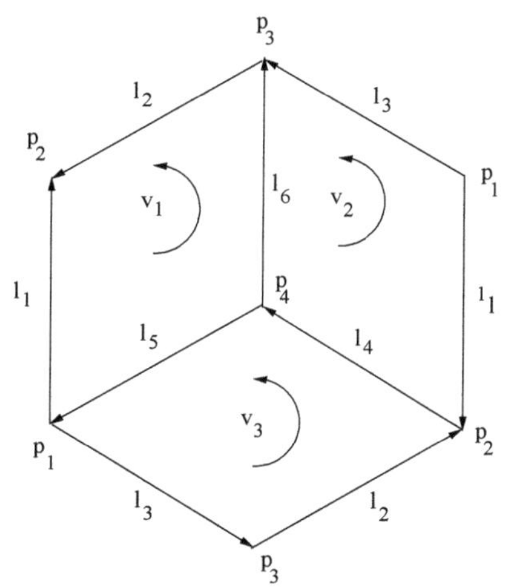

FIGURE 8. Projective plane obtained from a Möbius strip

$$\partial(v_1 + v_2 + v_3) = 2(l_2 + l_3 - l_1),$$

which is the generator of $\mathcal{H}_1(\mathfrak{S}_{\mathbb{RP}^2}) = \mathbb{Z}_2$.

The differential operator d acts on this script as follows:

$$dp_1 = -l_1 - l_3 + l_5,$$
$$dp_2 = l_1 + l_2 - l_4,$$
$$dp_3 = -l_2 + l_3 + l_6,$$
$$dp_4 = l_4 - l_5 - l_6,$$

and

$$dl_1 = -v_1 - v_2, \quad dl_2 = v_1 + v_3, \quad dl_3 = v_2 + v_3, \quad dl_4 = -v_2 + v_3,$$
$$dl_5 = -v_1 + v_3, \quad dl_6 = v_1 - v_2.$$

The resulting Dirac operator matrix for this script is:

$$\left(\begin{array}{cccc|cccccc|ccc}
0 & 0 & 0 & 0 & -1 & 0 & -1 & 0 & 1 & 0 & 0 & 0 & 0 \\
0 & 0 & 0 & 0 & 1 & 1 & 0 & -1 & 0 & 0 & 0 & 0 & 0 \\
0 & 0 & 0 & 0 & 0 & -1 & 1 & 0 & 0 & 1 & 0 & 0 & 0 \\
0 & 0 & 0 & 0 & 0 & 0 & 0 & 1 & -1 & -1 & 0 & 0 & 0 \\
\hline
-1 & 1 & 0 & 0 & 0 & 0 & 0 & 0 & 0 & 0 & -1 & -1 & 0 \\
0 & 1 & -1 & 0 & 0 & 0 & 0 & 0 & 0 & 0 & 1 & 0 & 1 \\
-1 & 0 & 1 & 0 & 0 & 0 & 0 & 0 & 0 & 0 & 0 & 1 & 1 \\
0 & -1 & 0 & 1 & 0 & 0 & 0 & 0 & 0 & 0 & 0 & -1 & 1 \\
1 & 0 & 0 & -1 & 0 & 0 & 0 & 0 & 0 & 0 & -1 & 0 & 1 \\
0 & 0 & 1 & -1 & 0 & 0 & 0 & 0 & 0 & 0 & 1 & -1 & 0 \\
\hline
0 & 0 & 0 & 0 & -1 & 1 & 0 & 0 & -1 & 1 & 0 & 0 & 0 \\
0 & 0 & 0 & 0 & -1 & 0 & 1 & -1 & 0 & -1 & 0 & 0 & 0 \\
0 & 0 & 0 & 0 & 0 & 1 & 1 & 1 & 1 & 0 & 0 & 0 & 0
\end{array}\right).$$

with eigenvalues $-2, 0, 2$ of multiplicities $6, 1, 6$, respectively. The Laplacian of this script is given by:

$$\frac{1}{2}\left(\begin{array}{cccc|cccccc|ccc}
3 & -1 & -1 & -1 & 0 & 0 & 0 & 0 & 0 & 0 & 0 & 0 & 0 \\
-1 & 3 & -1 & -1 & 0 & 0 & 0 & 0 & 0 & 0 & 0 & 0 & 0 \\
-1 & -1 & 3 & -1 & 0 & 0 & 0 & 0 & 0 & 0 & 0 & 0 & 0 \\
-1 & -1 & -1 & 3 & 0 & 0 & 0 & 0 & 0 & 0 & 0 & 0 & 0 \\
\hline
0 & 0 & 0 & 0 & 4 & 0 & 0 & 0 & 0 & 0 & 0 & 0 & 0 \\
0 & 0 & 0 & 0 & 0 & 4 & 0 & 0 & 0 & 0 & 0 & 0 & 0 \\
0 & 0 & 0 & 0 & 0 & 0 & 4 & 0 & 0 & 0 & 0 & 0 & 0 \\
0 & 0 & 0 & 0 & 0 & 0 & 0 & 4 & 0 & 0 & 0 & 0 & 0 \\
0 & 0 & 0 & 0 & 0 & 0 & 0 & 0 & 4 & 0 & 0 & 0 & 0 \\
0 & 0 & 0 & 0 & 0 & 0 & 0 & 0 & 0 & 4 & 0 & 0 & 0 \\
\hline
0 & 0 & 0 & 0 & 0 & 0 & 0 & 0 & 0 & 0 & 4 & 0 & 0 \\
0 & 0 & 0 & 0 & 0 & 0 & 0 & 0 & 0 & 0 & 0 & 4 & 0 \\
0 & 0 & 0 & 0 & 0 & 0 & 0 & 0 & 0 & 0 & 0 & 0 & 4
\end{array}\right).$$

5.5. Connected sum of two projective planes

First, we construct the following tight script $\mathfrak{S}_{2\mathbb{RP}^2}$ for a connected sum of two $\mathbb{RP}^2$, as in Figure 9. Topologically it is obtained from two circles, then one cuts a third "small" circle from each initial ones and glue them along their boundary (with the same orientation) to form the connected sum. In Figure 9 the two initial circles are: first $(l_1 + l_2)$ from upper left corner identified with $l_1 + l_2$ from bottom left corner – that gives a circle around the point p_1; the second circle is $l_4 + l_3$ from the upper right corner identified with $l_4 + l_3$ from lower right corner – a second circle around p_1. Finally, the gluing circle is $(l_6 + l_5)$, also around p_1. It is well known that topologically $\mathbb{RP}^2 \sharp \mathbb{RP}^2$ is equivalent to a Klein bottle. We obtain the following script geometry:

$$\partial l_1 = p_2 - p_1, \quad \partial l_2 = p_1 - p_2,$$
$$\partial l_3 = p_1 - p_3, \quad \partial l_4 = p_3 - p_1,$$
$$\partial l_5 = p_1 - p_0, \quad \partial l_6 = p_0 - p_1,$$
$$\partial l_7 = p_0 - p_1, \quad \partial l_8 = p_1 - p_0,$$

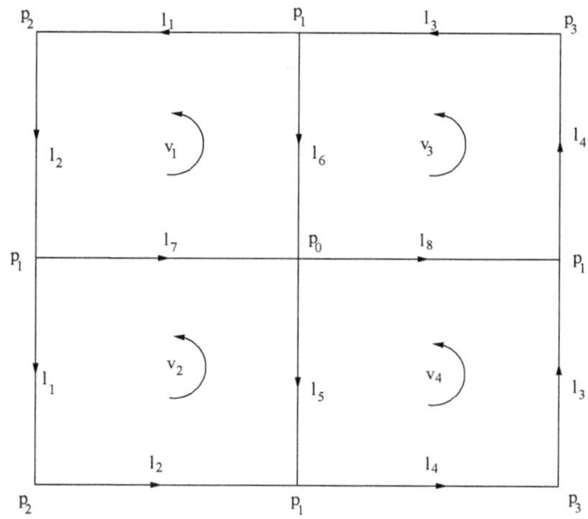

FIGURE 9. Connected sum of two projective planes

and

$$\begin{aligned}\partial v_1 &= l_1 + l_2 - l_6 + l_7,\\ \partial v_2 &= l_1 + l_2 - l_5 - l_7,\\ \partial v_3 &= l_3 + l_4 + l_6 + l_8,\\ \partial v_4 &= l_3 + l_4 + l_5 - l_8\,.\end{aligned}$$

Therefore,

$$\sum_{j=1}^{2} \partial v_j = 2(l_1 + l_2 + l_3 + l_4),$$

which lead to $\mathcal{H}_1(\mathfrak{S}_{2\mathbb{RP}^2}) \simeq \mathbb{Z} \oplus \mathbb{Z}_2$ and $\mathcal{H}_2(\mathfrak{S}_{2\mathbb{RP}^2}) = 0$, same script holomology of a Klein bottle.

At the differential operator level, we obtain:

$$\begin{aligned}dp_0 &= -l_5 + l_6 + l_7 - l_8,\\ dp_1 &= -l_1 + l_2 + l_3 - l_4 + l_5 - l_6 - l_7 + l_8,\\ dp_2 &= l_1 - l_2, \qquad dp_3 = l_4 - l_3,\end{aligned}$$

and

$$\begin{aligned}dl_1 &= v_1 + v_2, & dl_2 &= v_1 + v_2,\\ dl_3 &= v_3 + v_4, & dl_4 &= v_3 + v_4,\\ dl_5 &= -v_2 + v_4, & dl_6 &= -v_1 + v_3,\\ dl_7 &= v_1 - v_2, & dl_8 &= v_3 - v_4.\end{aligned}$$

In matrix form, the Dirac operator for the $\mathbb{RP}^2\sharp\mathbb{RP}^2$ script is:

$$\left(\begin{array}{cccc|cccccccc|cccc}
0 & 0 & 0 & 0 & 0 & 0 & 0 & 0 & -1 & 1 & 1 & -1 & 0 & 0 & 0 & 0\\
0 & 0 & 0 & 0 & -1 & 1 & 1 & -1 & 1 & -1 & -1 & 1 & 0 & 0 & 0 & 0\\
0 & 0 & 0 & 0 & 1 & -1 & 0 & 0 & 0 & 0 & 0 & 0 & 0 & 0 & 0 & 0\\
0 & 0 & 0 & 0 & 0 & 0 & -1 & 1 & 0 & 0 & 0 & 0 & 0 & 0 & 0 & 0\\
\hline
0 & -1 & 1 & 0 & 0 & 0 & 0 & 0 & 0 & 0 & 0 & 0 & 1 & 1 & 0 & 0\\
0 & 1 & -1 & 0 & 0 & 0 & 0 & 0 & 0 & 0 & 0 & 0 & 1 & 1 & 0 & 0\\
0 & 1 & 0 & -1 & 0 & 0 & 0 & 0 & 0 & 0 & 0 & 0 & 0 & 0 & 1 & 1\\
0 & -1 & 0 & 1 & 0 & 0 & 0 & 0 & 0 & 0 & 0 & 0 & 0 & 0 & 1 & 1\\
-1 & 1 & 0 & 0 & 0 & 0 & 0 & 0 & 0 & 0 & 0 & 0 & 0 & -1 & 0 & 1\\
1 & -1 & 0 & 0 & 0 & 0 & 0 & 0 & 0 & 0 & 0 & 0 & -1 & 0 & 1 & 0\\
1 & -1 & 0 & 0 & 0 & 0 & 0 & 0 & 0 & 0 & 0 & 0 & 1 & -1 & 0 & 0\\
-1 & 1 & 0 & 0 & 0 & 0 & 0 & 0 & 0 & 0 & 0 & 0 & 0 & 0 & 1 & -1\\
\hline
0 & 0 & 0 & 0 & 1 & 1 & 0 & 0 & 0 & -1 & 1 & 0 & 0 & 0 & 0 & 0\\
0 & 0 & 0 & 0 & 1 & 1 & 0 & 0 & -1 & 0 & -1 & 0 & 0 & 0 & 0 & 0\\
0 & 0 & 0 & 0 & 0 & 0 & 1 & 1 & 0 & 1 & 0 & 1 & 0 & 0 & 0 & 0\\
0 & 0 & 0 & 0 & 0 & 0 & 1 & 1 & 1 & 0 & 0 & -1 & 0 & 0 & 0 & 0
\end{array}\right)$$

with eigenvalues

$$-\sqrt{7+\sqrt{17}}, -\sqrt{6}, -\sqrt{7-\sqrt{17}}, -2, -\sqrt{2}, 0, \sqrt{2}, 2, \sqrt{7-\sqrt{17}}, \sqrt{6}, \sqrt{7+\sqrt{17}}$$

of multiplicities $1, 1, 1, 2, 2, 2, 2, 2, 1, 1, 1$, respectively.

The Laplace operator is given by:

$$2\Delta_{2\mathbb{RP}^2} f = \begin{bmatrix} 4 & -4 & 0 & 0\\ -4 & 8 & -2 & -2\\ 0 & -2 & 2 & 0\\ 0 & -2 & 0 & 2 \end{bmatrix} [f_{0j}]^t [p_j]$$

$$+ \begin{bmatrix} 4 & 0 & -1 & 1 & -2 & 0 & 1 & -1\\ 0 & 4 & 1 & -1 & 0 & -2 & -1 & 1\\ -1 & 1 & 4 & 0 & 2 & 0 & -1 & 1\\ 1 & -1 & 0 & 4 & 0 & 2 & 1 & -1\\ -2 & 0 & 2 & 0 & 4 & -2 & -1 & 1\\ 0 & 2 & 0 & 2 & -2 & 4 & 1 & -1\\ 1 & -1 & -1 & 1 & -1 & 1 & 4 & -2\\ -1 & 1 & 1 & -1 & 1 & -1 & -2 & 4 \end{bmatrix} [f_{1j}]^t [l_j]$$

$$+ \frac{1}{2} \begin{bmatrix} 4 & 1 & -1 & 0\\ 1 & 4 & 0 & -1\\ -1 & 0 & 4 & 1\\ 0 & -1 & 1 & 4 \end{bmatrix} [f_{2j}]^t [v_j] .$$

A different method of obtaining the connected sum $\mathbb{RP}^2\sharp\mathbb{RP}^2$ is done by attaching two Möbius bands on the same boundary, as the projective plane $\mathbb{RP}^2$ minus a disk is topological equivalent to a Möbius strip. Looking back at Figure 2, we attach to it a second Möbius band as in Figure 10. The two strips glued together on the same four points p_1, p_2, p_3, p_4, containing two more lines l_7, l_8 and two more planes v_3, v_4 as in Figure 10. To the computations of Subsection 5.1 we add the following boundaries:

$$\partial l_7 = p_1 - p_2, \qquad \partial l_8 = p_4 - p_3,$$
$$\partial v_3 = -l_2 + l_5 - l_7 + l_8, \quad \partial v_4 = -l_3 + l_4 - l_7 - l_8.$$

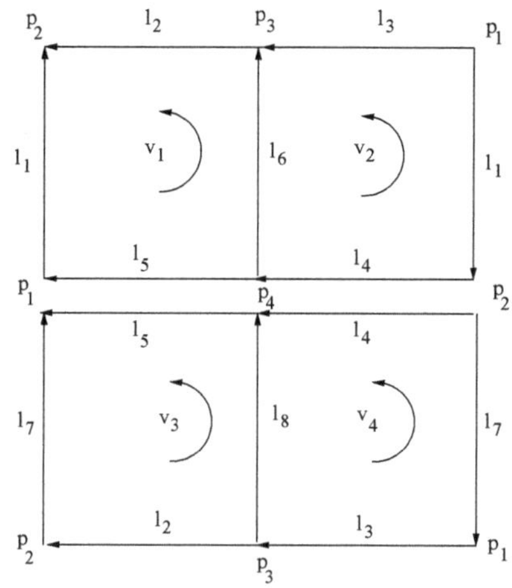

FIGURE 10. Connecting two Möbius strips

Here we note that

$$\partial(v_1 + v_2 - v_3 - v_4) = 2(l_7 - l_1),$$

is the generator of the $\mathcal{H}_1(\mathfrak{S}_{\mathbb{RP}^2}) = \mathbb{Z}_2$ homology. Note that we obtain the same script (different labelling) as the Klein bottle (see Figure 4), for which we performed all the necessary computations in subsection 5.3.

6. Outlook

In [13, 15] the authors present an approach to discrete differential modeling, which includes notions of discrete differential forms on simplexes and discrete manifolds, discrete boundary and co-boundary operators, discrete Hodge decomposition, and a discrete version of the Poincaré lemma. The same can and will be studied in the case of Script Geometry although some of necessary tools need to be developed since Script Geometry is a more loose concept than working with simplicial complexes.

In [14] the authors describe their approach to the theory of discrete exterior calculus (DEC). They introduce notions of discrete vector fields and operators acting on them, e.g., discrete divergence and curl, which has applications such as a discrete Hodge decomposition of 3D vector fields on irregular grids. A closely related work is discrete mechanics, where the main idea is to discretize the variational principle itself rather than the Euler–Lagrange equations. The discretization is not intended on time only, DEC methods are used in spatially extended mechanics, i.e., classical field theory. Furthermore, this theory is also widely applied in discrete electromagnetism which is another field were we see applications of Script Geometry in the future. This is also closely linked with a principal question in finite element exterior calculus. Up to now the commutativity of the exterior derivative with bounded projections to sub-meshes was only being shown for simplicial decompositions of domains. From our point of view this is still a major drawback for applying this calculus to more general type of meshes. Here, Script Geometry could be the basis for a more general approach.

In [10, 11, 12] a complete function theory has been established for a Dirac type operator on the grid $\mathbb{Z}^n$, including Taylor series, Fueter polynomials, and a discrete Cauchy–Kovalevskaya theorem. We look forward to relate this work to script geometry.

Acknowledgement

The first and second authors were supported by Portuguese funds through the CIDMA – Center for Research and Development in Mathematics and Applications, and the Portuguese Foundation for Science and Technology ("FCT – Fundação para a Ciência e a Tecnologia"), within project UID/MAT/ 0416/2013.

References

[1] D.N. Arnold, R.S. Falk, R. Winther, "Finite element exterior Calculus, homological techniques, and applications", Acta Numerica 2006, 1–155.

[2] F. Baaske, S. Bernstein, H. De Ridder, F. Sommen, "On solutions of a discretized heat equation in discrete Clifford analysis", J. Diff. Eq. Appl. 20 (2), pp. 271–295.

[3] M. Belkin, J. Sun, Y. Wang, "Discrete Laplace operator on meshed surfaces", SCG08, College Park, Maryland, 2008

[4] F. Brackx, D. De Schepper, F. Sommen, L. Van de Voorde, "Discrete Clifford Analysis: A Germ of Function Theory", in: I. Sabadini, M. Shapiro, F. Sommen (eds.), Hypercomplex Analysis, Birkhäuser, 37–53, 2009.

[5] P. Cerejeiras, N. Faustino, N. Vieira, "Numerical Clifford analysis for nonlinear Schrödinger problem", Numerical Methods for Partial Differential Equations, 24(4) (2008) 1181–1202, 2008.

[6] P. Cerejeiras, U. Kähler, M. Ku, "Discrete Hilbert boundary value problems on half lattices", Journal of Difference Equations and Applications, 21 (12) (2015), 1277–1304.

[7] P. Cerejeiras, U. Kähler, M. Ku, F. Sommen, "Discrete Hardy spaces", Journal of Fourier Analysis and Applications, 20 (2014), 715–750.

[8] H. De Schepper, F. Sommen, L. Van de Voorde, "A Basic Framework for Discrete Clifford Analysis", Exp. Math. 18:4, 2009.

[9] H. De Ridder, "Discrete Clifford analysis: function theory and integral transforms", PhD-thesis, Ghent, 2013.

[10] H. De Ridder, H. De Schepper, F. Sommen, "Fueter polynomials in discrete Clifford analysis", Math. Zeit. 272 (1-2) (2012), pp. 253–268.

[11] H. De Ridder, H. De Schepper, F. Sommen. "Taylor series expansion in discrete Clifford analysis", Compl. Anal. Oper. Th. 8 (2), (2013), pp. 485–511.

[12] H. De Ridder, H. De Schepper, F. Sommen. “The Cauchy–Kovalevskaya Extension Theorem in Discrete Clifford Analysis”, Comm. Pure Appl. Math. 10 (4) (2011), 1097–1109.

[13] M. Desbrun, E. Kanso, Y. Tong, “Discrete Differential Forms for Computational Modeling”, in: Discrete Differential Geometry, Volume 38 of the series Oberwolfach Seminars, 287–324, 2008.

[14] M. Desbrun, A.N. Hirani, J.E. Marsden, “Discrete Exterior Calculus for Variational Problems in Computer Vision and Graphics”, in 42nd IEEE Conference on Decision and Control Proceedings (CDC ’03), 42 (2003), 4902–4907.

[15] M. Desbrun, M. Leok, J.E. Marsden, “Discrete Poincaré lemma”, Appl. Num. Math. 53 (2005), 231–248.

[16] A. Dimakis, F. Müller-Hoissen, “Discrete Differential Calculus, Graphs, Topologies and Gauge Theory”, J. Math. Phys. **35**, 6703–6735, 1994.

[17] N. Faustino, *Discrete Clifford Analysis*, PhD Thesis, University of Aveiro, 2008.

[18] N. Faustino, U. Kähler, “Fischer decomposition for difference Dirac operators” Adv. Appl. Clifford Algebras 17-1 (2007).

[19] N. Faustino, U. Kähler, F.Sommen, “Discrete Dirac operators in Clifford Analysis”, Adv. Appl. Clifford Algebras 17:3 (2007).

[20] N. Faustino, G. Ren, “(Discrete) Almansi Type Decompositions: An umbral calculus framework based on $\mathfrak{osp}(1|2)$ symmetries”, Math. Meth. Appl. Sci. 11 (2011), 305–318.

[21] E. Forgy, U. Schreiber, “Discrete differential geometry on causal graphs”, arXiv: math-ph/0407005v1, 2004.

[22] K. Gürlebeck, A. Hommel, “On finite difference potentials and their fundamental solutions”, Adv. Appl.Clifford Algebras 11:2 (2001).

[23] K. Gürlebeck, A. Hommel, “On finite difference potentials and their applications in a discrete function theory”, Math. Methods Appl. Sci. 25:16–18 (2002)

[24] P. Herholz, “General discrete Laplace operators on polygonal meshes”, Diplomarbeit, Humboldt Univ. Berlin, 2012

[25] R. P. Isaacs, “A finite difference function theory”, Univ. Nac. Tucumán, Revista A, *2*, 177–201, 1941.

[26] I. Kanamori, N. Kawamoto, “Dirac–Kähler Fermion from Clifford Product with Noncommutative Differential Form on a Lattice”, Int. J. Mod. Phys. A19 (2004), 695–736.

[27] P. Leopardi, A. Stern, The abstract Hodge–Dirac operator and its stable discretization, arXiv:1401.1576, 2014.

[28] L. Lovasz, “Discrete Analytic Functions: An Exposition”, Surveys in differential geometry, Vol. IX, 1–44, 2004.

[29] S. Smirnov, “Discrete Complex Analysis and Probability”, Proceedings of the ICM 2010, Hyderabad, 2010, 1–27.

[30] A. Stern, “Discrete Differential Forms for Numerical Relativity”, PCGM 2008.

[31] M. Wardetzky, S. Mathur, F. Kälberer, E. Grinspun, “Discrete Laplace operators: no free lunch”, Eurographics Symposium on Geometry Processing, 2007.

[32] J. Vaz, "Clifford-like calculus over lattices", Adv. Appl. Cliff. Alg., 7(1) (1997), 37–70.

Paula Cerejeiras and Uwe Kähler
Department of Mathematics
Universidade de Aveiro
P-3810-159 Aveiro, Portugal
e-mail: pceres@ua.pt
ukaehler@ua.pt

Franciscus Sommen
Clifford Research Group
Department of Mathematical Analysis
Ghent University, Galglaan 2
B-9000 Ghent, Belgium
e-mail: fs@cage.ugent.be

Adrian Vajiac
Chapman University
One University Drive
Orange CA 92866
e-mail: avajiac@chapman.edu

Modern Trends in Hypercomplex Analysis
Trends in Mathematics, 111–142

A Panorama on Quaternionic Spectral Theory and Related Functional Calculi

Fabrizio Colombo and David P. Kimsey

Abstract. In this paper we offer an overview of the state of the art of quaternionic spectral theory. Precisely we review some functional calculi and the quaternionic spectral theorem based on the S-spectrum. We start with the S-functional calculus which is the Riesz–Dunford functional calculus for quaternionic operators which suggested the existence of the S-spectrum of quaternionic operators, then we introduce the Spectral Theorem based on the S-spectrum which is of fundamental importance for the formulation of quaternionic quantum mechanics. Moreover we discuss the quaternionic H^∞-functional calculus that is the quaternionic analogue of the H^∞-functional calculus for sectorial operators introduced by A. McIntosh. In the case a quaternionic linear operator is the infinitesimal generator quaternionic group of linear operators by the Laplace–Stieltjes transform we extend the Philips functional calculus in this setting. The W-functional calculus and the F-functional calculus are monogenic functional calculi, in the spirit of the monogenic functional calculus introduced by A. McIntosh, but both calculi are based on slice hyperholomorphic functions and on manipulations of their Cauchy formulas.

Mathematics Subject Classification (2010). Primary 47A10, Secondary 47A60.

Keywords. Quaternionic spectral theorem based on the S-spectrum, S-functional calculus, H^∞-functional calculus, Philips functional calculus, W-functional calculus, F-functional calculus.

1. Introduction

The spectral theorem for normal linear operators on a complex Hilbert space, see [25], is a crucial result to define functions of operators. Moreover, the spectral theorem has an incredibly large number of applications. E.g., the spectral theorem plays an important role in stating the axioms of quantum mechanics, because it give the structure of the solution of the Schrödinger equation. The interest in spectral theory for quaternionic operators is motivated by the paper [11] of Birkhoff and von Neumann, on the logic of quantum mechanics, who showed that Schrödinger equation can be written basically in the complex or quaternionic setting.

Important contributions to the development of the quaternionic version of quantum mechanics can be found in [1, 26, 28, 34], but the correct notion of spectrum for quaternionic operators was still missing until the introduction of the S-spectrum, see the book [21] and the references therein.

In classical functional analysis the most important tool to define functions of a linear operator A acting on a complex Banach space X is the theory of holomorphic functions. Replacing the Cauchy kernel by the resolvent operator of A in the Cauchy formula for holomorphic functions we obtain the so-called Riesz–Dunford functional calculus, see [24]. Under certain conditions, this calculus can be extended to unbounded linear operators acting on a Banach space.

In the classical case the Riesz–Dunford functional calculus and the spectral theorem are both based on the same notion of spectrum, for a bounded operator A, its spectrum $\sigma(A)$ is defined as

$$\sigma(A) = \{\lambda \in \mathbb{C} \ : \ \lambda I - A \text{ is not invertible in } B(X)\}$$

where $B(X)$ denotes the Banach space of all bounded linear operators acting on X endowed with the natural norm. The resolvent set is defined as

$$\rho(A) := \mathbb{C} \setminus \sigma(A)$$

and the resolvent operator is defined by

$$R(\lambda, A) := (\lambda I - A)^{-1}, \qquad \lambda \in \rho(A).$$

Several important properties of operators follows from that fact that $R(\lambda, A) : \rho(A) \to B(X)$ is a holomorphic function operator-valued.

In this introduction we will concentrate on the notions of spectra of a quaternionic linear operators to understand the difficulties behind the quaternionic spectral theorem and the hyperholomorphic functional calculi. In the sections that follow we will discuss in detail the spectral theorem and the functional calculi based on the S-spectrum.

An element s in the set of quaternions $\mathbb{H}$, is denoted by $s = s_0 + s_1e_1 + s_2e_2 + s_3e_3$, its conjugate is $s = s_0 - s_1e_1 - s_2e_2 - s_3e_3$ where e_1, e_2 and e_3 are the imaginary units of the quaternion s, $\text{Re}(s) = s_0$ is the real part and the norm $|s|$ is such that $|s|^2 = s_0^2 + s_1^2 + s_2^2 + s_3^2$. We denote by $\mathcal{B}(V)$ the left Banach space of all bounded right linear operators acting on the two-sided quaternionic Banach space V. Observe that when we consider a right linear quaternionic operator (the case of left linear operators is similar) we have two possibilities. The left spectrum $\sigma_L(T)$ of $T \in \mathcal{B}(V)$ is defined by

$$\sigma_L(T) = \{s \in \mathbb{H} \quad : \quad s\mathcal{I} - T \quad \text{is not invertible in } \mathcal{B}(V)\},$$

where the notation $s\mathcal{I}$ in $\mathcal{B}(V)$ means that $(s\mathcal{I})(v) = sv$.

The right spectrum $\sigma_R(T)$ of T is associated with the right eigenvalue problem, i.e., the search for those $s \in \mathbb{H}$ such that there exists a nonzero vector v satisfying

$$T(v) = vs.$$

It is important to note that if s is a right eigenvalue, then all quaternions belonging to the sphere $r^{-1}sr$, $r \in \mathbb{H} \setminus \{0\}$, are also eigenvalues. But observe that the operator $\mathcal{I}s - T$ associated to the right eigenvalue problem is not linear, so it is not clear what is the resolvent operator to be considered. The left resolvent operator $\mathcal{R}_L(s,T)$ is defined by

$$\mathcal{R}_L(s,T) := (s\mathcal{I} - T)^{-1}, \quad s \notin \sigma_L(T),$$

but it is not known what notion of hyperholomorphicity it satisfies.

Because of this ambiguity in the definition of the quaternionic spectrum one may be tempted to consider Fueter regular functions and see if their Cauchy kernel suggests what kind of resolvent operator and spectrum one should consider.

Precisely, the Cauchy kernel $\mathcal{G}$ of Fueter regular functions is defined by

$$\mathcal{G}(s,q) = \frac{\bar{s} - \bar{q}}{|s-q|^4} = \frac{(s-q)^{-1}}{|s-q|^2} = (s-q)^{-2}(\overline{s} - \overline{q})^{-1}$$

and is both left and right Fueter regular on $\mathbb{H}\setminus\{0\}$. It admits the series expansion

$$\sum_{n\geq 0} \sum_{\nu \in \sigma_n} P_\nu(q)\mathcal{G}_\nu(s), \quad \text{for} \quad |s| < |q| \tag{1.1}$$

where σ_n is a set of indices of permutations and

$$\mathcal{G}_\nu(s) := \frac{\partial^n}{\partial x_1^{n_1} \partial x_2^{n_2} \partial x_3^{n_3}} \mathcal{G}(s,0).$$

Here $P_\nu(q)$ denotes the homogeneous polynomials

$$P_\nu(q) = \frac{1}{k!} \sum_{\ell_1,\dots,\ell_k} z_{\ell_1} \dots z_{\ell_k},$$

where $z_j = x_j e_0 - x_0 e_j$, for $j = 1,2,3$, and the sum is taken over all different permutations of $\ell_1, \dots, \ell_k$. This polynomials $P_\nu(q)$ replace the powers q^n, $n \in \mathbb{N} \cup \{0\}$, that are not Fueter regular, in the Taylor series for this class of functions. When we replaces q by operator T and we suppose that $T = T_0 + e_1T_1 + e_2T_2 + e_3T_3$, where the bounded operators T_ℓ, $\ell = 0,1,2,3$ commute among themselves, then sum of the series (1.1) converges, for $\|T\| < |q|$, to

$$(s\mathcal{I} - T)^{-2}(\overline{s}\mathcal{I} - \overline{T})^{-1},$$

where $\overline{T} = T_0 - e_1T_1 - e_2T_2 - e_3T_3$; but in the case T_ℓ, $\ell = 0,1,2,3$ do not commute among themselves the sum is not known. It does not seem that it can reman as in the case of commuting components because of the structure of the Cauchy–Fueter kernel $\frac{s^{-1}}{|s|^2} = s^{-2}\overline{s}^{-1}$ we must be able to write $|s|^2 = s\overline{s}$ also for operators and this can be done if $T\overline{T} = T_0^2 + T_1^2 + T_2^2 + T_3^2$ that is when T_ℓ, $\ell = 0, 1, 2, 3$ commute among themselves.

We do not know what kind of hyperholomorphicity there is associated to the left resolvent operator $\mathcal{R}_L(s, T)$, but for the commuting case we can write

$$\mathcal{G}(s, T) = \mathcal{R}_L(s, T)^2 \mathcal{R}_L(\overline{s}, \overline{T})$$

when $s \notin \sigma_L(T)$ and $\overline{s} \notin \sigma_L(\overline{T})$ and $\mathcal{G}(s, T)$ is Cauchy–Fueter regular operator-valued.

Now let us come to the spectral theorem and to what physicists use in quaternionic quantum mechanics. In Adler's book [1], the right spectrum of quaternionic linear operators with eigenvalues is used. However, there is no notion of holomorphicity. Consequently, only a partial spectral description of linear operators appears in [1].

In the case of slice hyperholomorphicity the Cauchy kernel is given by the sum of the series

$$\sum_{n\geq 0} q^n s^{-1-n} = -(q^2 - 2q\mathrm{Re}(s) + |s|^2)^{-1}(q - \overline{s}), \quad for \quad |q| < |s|$$

and it does not depend on the commutativity of the components of q so that when one replaces q by an operator T with noncommuting components the sum of the above series remains the same. This crucial fact leads to the natural definition of the S-spectrum. The S-spectrum, see [21], is defined as

$$\sigma_S(T) = \{s \in \mathbb{H} \ : \ T^2 - 2\mathrm{Re}(s)T + |s|^2\mathcal{I} \quad \text{is not invertible in } \mathcal{B}(\mathrm{V})\},$$

while the S-resolvent set is

$$\rho_S(T) := \mathbb{H} \setminus \sigma_S(T).$$

Due to the noncommutativity of the quaternions and form the definition of slice hyperholomorphicity, there are two different Cauchy kernels and so there are two resolvent operators associated with a quaternionic linear operator: setting

$$Q_s(T) := (T^2 - 2\mathrm{Re}(s)T + |s|^2\mathcal{I})^{-1}, \quad s \in \rho_S(T)$$

the left and the right S-resolvent operators are defined as

$$S_L^{-1}(s, T) := -Q_s(T)(T - \overline{s}\mathcal{I}), \qquad S_R^{-1}(s, T) := -(T - \overline{s}\mathcal{I})Q_s(T), \quad s \in \rho_S(T), \tag{1.2}$$

respectively. These two S-resolvent operators are slice hyperholomorphic functions with values in $\mathcal{B}(V)$. Recently it was possible to prove the quaternionic spectral theorem based on the S-spectrum. This fact restores the analogy with the complex case in which the Riesz–Dunford functional calculus and the spectral theorem are based on the same notion of spectrum of a linear operator. To replace the complex spectral theory with the quaternionic spectral theory we have to replace the classical spectrum with the S-spectrum. We conclude by saying that the S-spectrum can be defined also for n-tuples of not necessarily commuting operators and the Riesz–Dunford functional calculus can be extended to this case using the notion of slice hyperholomorphic functions Clifford algebra-valued, see [21].

The plan of the paper is as follows. Section 1 is devoted to the problems that led to the definition of the S-spectrum. Section 2 contains the Cauchy formulas of slice hyperholomorphic functions. Section 3 contains the spectral theorem for bounded normal operators. Section 4 is devoted to relating the spectral representation in Section 3 with known results in the finite-dimensional case. Sections 5 is devoted to a treatment of the S-functional calculus for both bounded and unbounded operators. Section 6 is devoted to the quaternionic version of the H^∞-functional calculus. Section 7 treats the Philips functional calculus which is based on the bilateral quaternionic Laplace–Stieltjes transform and it applies to infinitesimal generators of quaternionic groups. In Section 8 we offer the quaternionic version of the W-functional calculus. In Section 9 we recall the Fueter mapping theorem in integral form and the related F-functional calculus.

2. Slice hyperholomorphic functions

In this section we recall some results on slice hyperholomorphic functions, more details can be found in the books [4, 21, 30]. We denote by $\mathbb{S}$ the 2-sphere of purely imaginary quaternions of modulus 1:

$$\mathbb{S} = \{q = x_1e_1 + x_2e_2 + x_3e_3 \in \mathbb{H} \mid q^2 = -1\}$$

and we recall that for any $i \in \mathbb{S}$ we can define a complex plane $\mathbb{C}_i$ whose elements are of the form $q = u + iv$ for $u,\ v \in \mathbb{R}$. Any quaternion q belongs to a suitable complex plane: if we set

$$i_q := \begin{cases} \frac{q}{|q|}, & \text{if } q \neq 0 \\ \text{any } i \in \mathbb{S}, & \text{if } q = 0, \end{cases}$$

then $q = u + i_q v$ with $u = \mathrm{Re}(q)$ and $v = |q|$, so, it follows that, the skew field of quaternions $\mathbb{H}$ can be seen as

$$\mathbb{H} = \bigcup_{i \in \mathbb{S}} \mathbb{C}_i.$$

For any $q = u + i_q v \in \mathbb{H}$ we define the set

$$[q] := \{u + iv \mid i \in \mathbb{S}\}.$$

A possible way to define slice hyperholomorphic functions is the following.

Definition 2.1 (Slice hyperholomorphic function). Let $U \subset \mathbb{H}$ be open and let $f : U \to \mathbb{H}$ be a real differentiable function. For any $i \in \mathbb{S}$, let

$$f_i := f|_{U \cap \mathbb{C}_i}$$

denote the restriction of f to the plane $\mathbb{C}_i$. The function f is called left slice hyperholomorphic if, for any $i \in \mathbb{S}$,

$$\frac{1}{2}\left(\frac{\partial}{\partial u}f_i(q) + i\frac{\partial}{\partial v}f_i(q)\right) = 0 \quad \text{for all } q = u + iv \in U \cap \mathbb{C}_i \tag{2.1}$$

and right slice hyperholomorphic if, for any $i \in \mathbb{S}$,

$$\frac{1}{2}\left(\frac{\partial}{\partial u}f_i(q) + \frac{\partial}{\partial v}f_i(q)i\right) = 0 \quad \text{for all } q = u + iv \in U \cap \mathbb{C}_i. \tag{2.2}$$

A left (or right) slice hyperholomorphic function that satisfies $f(U \cap \mathbb{C}_i) \subset \mathbb{C}_i$ for every $i \in \mathbb{S}$ is called intrinsic.

We denote the set of all left slice hyperholomorphic functions on U by $\mathcal{SH}^L(U)$, the set of all right slice hyperholomorphic functions on U by $\mathcal{SH}^R(U)$ and the set of all intrinsic functions by $\mathcal{N}(U)$.

Intrinsic functions are important because the multiplication and composition operations preserve slice hyperholomorphicity. This is not true for arbitrary slice hyperholomorphic functions.

Definition 2.2. The left slice hyperholomorphic Cauchy kernel is

$$S_L^{-1}(s,q) = -(q^2 - 2\mathrm{Re}(s)q + |s|^2)^{-1}(q - \overline{s}) \quad \text{for } q \notin [s]$$

and the right slice hyperholomorphic Cauchy kernel is

$$S_R^{-1}(s,q) = -(q - \overline{s})(q^2 - 2\mathrm{Re}(s)q + |s|^2)^{-1} \quad \text{for } q \notin [s].$$

Definition 2.3. Let $U \subseteq \mathbb{H}$. We say that U is axially symmetric if, for all $u+iv \in U$, the whole 2-sphere $[u + iv]$ is contained in U.

Definition 2.4. Let $U \subseteq \mathbb{H}$ be a domain in $\mathbb{H}$. We say that U is a slice domain (s-domain for short) if $U \cap \mathbb{R}$ is nonempty and if $U \cap \mathbb{C}_i$ is a domain in $\mathbb{C}_i$ for all $i \in \mathbb{S}$.

So we can state the Cauchy formulas:

Theorem 2.5. *Let $U \subset \mathbb{H}$ be an axially symmetric slice domain such that its boundary $\partial(U \cap \mathbb{C}_i)$ in $\mathbb{C}_i$ consists of a finite number of continuously differentiable Jordan curves. Let $i \in \mathbb{S}$ and set $ds_i = -i\, ds$. If f is left slice hyperholomorphic on an open set that contains $\overline{U}$, then*

$$f(q) = \frac{1}{2\pi}\int_{\partial(U\cap\mathbb{C}_i)} S_L^{-1}(s,q)\, ds_i\, f(s) \quad \text{for all } q \in U.$$

If f is right slice hyperholomorphic on an open set that contains $\overline{U}$, then

$$f(q) = \frac{1}{2\pi}\int_{\partial(U\cap\mathbb{C}_i)} f(s)\, ds_i\, S_R^{-1}(s,q) \quad \text{for all } q \in U.$$

The above integrals depend neither on the open set U nor on the complex plane $\mathbb{C}_i$ for $i \in \mathbb{S}$.

In the sequel we will need the following definitions.

Definition 2.6 (Argument function). Let $s \in \mathbb{H} \setminus \{0\}$. We define $\arg(s)$ as the unique number $\theta \in [0, \pi]$ such that $s = |s|e^{\theta i_s}$.

Observe that $\theta = \arg(s)$ does not depend on the choice of i_s if $s \in \mathbb{R} \setminus \{0\}$ since $p = |p|e^{0i}$ for any $i \in \mathbb{S}$ if $p > 0$ and $p = |p|e^{\pi i}$ for any $i \in \mathbb{S}$ if $p < 0$.

Let $\vartheta \in [0, \pi]$ we define the sets

$$\mathcal{S}_\vartheta = \{s \in \mathbb{H} \mid \ |\arg(p)| \leq \vartheta \text{ or } s = 0\},$$

$$\mathcal{S}_\vartheta^0 = \{s \in \mathbb{H} \mid \ |\arg(p)| < \vartheta\}. \tag{2.3}$$

We now introduce the following subsets of the set of slice hyperholomorphic functions that consist of bounded slice hyperholomorphic functions.

Definition 2.7. Let $\mu \in (0, \pi]$. We set

$$\mathcal{SH}_L^\infty(\mathcal{S}_\mu^0) = \{f \in \mathcal{SH}_L(\mathcal{S}_\mu^0) \quad \text{such that } \|f\|_\infty := \sup_{s \in \mathcal{S}_\mu^0} |f(s)| < \infty\},$$

$$\mathcal{SH}_R^\infty(\mathcal{S}_\mu^0) = \{f \in \mathcal{SH}_R(\mathcal{S}_\mu^0) \quad \text{such that } \|f\|_\infty := \sup_{s \in \mathcal{S}_\mu^0} |f(s)| < \infty\},$$

$$\mathcal{N}^\infty(\mathcal{S}_\mu^0) := \{f \in \mathcal{N}(\mathcal{S}_\mu^0) \quad \text{such that } \|f\|_\infty := \sup_{s \in \mathcal{S}_\mu^0} |f(s)| < \infty\}.$$

With these spaces of bounded functions with suitable decay conditions and with the subclass of closed operators whose S-resolvent operators satisfy appropriate conditions we can extend the H^∞-functional calculus in the quaternionic setting.

3. The spectral theorem based on the S-spectrum

The spectral theorem for quaternionic normal matrices based on the right spectrum is originally due to [38] and, also independently, to [12] (see also the survey papers [46] and [27]). Moreover, in the literature there are some papers on the quaternionic spectral theorem see [28, 43, 45], where the notion of spectrum is not made clear. In the paper [10], using the quaternionic version of Herglotz theorem (see [9]) it is proved the spectral theorem for quaternionic unitary operator based on the S-spectrum. The spectral theorem for bounded and unbounded normal operators based on the S-spectrum was proved in [8]. In [32], the spectral theorem based on S-spectrum is proved for compact normal operators on a quaternionic Hilbert space. In this section we show the structure of the spectral theorem for normal bounded quaternionic operators on a Hilbert space taken form [8].

Definition 3.1. Let $\mathcal{H}$ be a right linear quaternionic Hilbert space, endowed with an $\mathbb{H}$-valued inner product $\langle \cdot, \cdot \rangle$ which satisfies, for every $\alpha, \beta \in \mathbb{H}$, and $x, y, z \in \mathcal{H}$, the relations:

$$\begin{aligned}\langle x,y\rangle &= \overline{\langle y,x\rangle}.\\ \langle x,x\rangle &\geq 0 \quad \text{and} \quad \|x\|^2 := \langle x,x\rangle = 0 \Longleftrightarrow x=0.\\ \langle x\alpha+y\beta,z\rangle &= \langle x,z\rangle\alpha+\langle y,z\rangle\beta.\\ \langle x,y\alpha+z\beta\rangle &= \bar{\alpha}\langle x,y\rangle+\bar{\beta}\langle x,z\rangle.\end{aligned}$$

Theorem 3.2 ([31]). *Let $T\in\mathcal{L}(\mathcal{H})$. The following statements hold:*

(i) *If T is positive, then $\sigma_S(T)\subseteq[0,\infty)$. If, in particular, $T\in\mathcal{B}(\mathcal{H})$ is positive, then*

$$\sigma_S(T)\subseteq[0,\|T\|].$$

(ii) *If T is self-adjoint, then $\sigma_S(T)\subseteq\mathbb{R}$. If, in particular, $T\in\mathcal{B}(\mathcal{H})$ is self-adjoint, then*

$$\sigma_S(T)\subseteq[-\|T\|,\|T\|].$$

(iii) *If T is anti self-adjoint, then $\sigma_S(T)\subseteq\{p\in\mathbb{H}\ :\ \mathrm{Re}(p)=0\}$. If, in particular, $T\in\mathcal{B}(\mathcal{H})$ is anti self-adjoint, then*

$$\sigma_S(T)\subseteq\{p\in\mathbb{H}\ :\ \mathrm{Re}(p)=0 \text{ and } |p|\leq\|T\|\}.$$

(iv) *If T is unitary, then $\sigma_S(T)\subseteq\mathbb{S}$.*

Throughout this paper, we will only be considering right quaternionic Hilbert spaces. Consequently, we will use *quaternionic Hilbert space* in place of *right quaternionic Hilbert space.*

The following result can be found in Proposition 2.6 of [31] with the caveat that the inner product in [31] is antilinear in the first variable and linear in the second variable.

Theorem 3.3. *Let $\mathcal{N}$ be an orthonormal basis of a quaternionic Hilbert space $\mathcal{H}$. Then every $x\in\mathcal{H}$ can be decomposed uniquely via*

$$x=\sum_{z\in\mathcal{N}} z\langle x,z\rangle, \tag{3.1}$$

where

$$\sum_{z\in\mathcal{N}} z\langle x,z\rangle := \sup\Big\{\sum_{z\in\mathcal{N}_f} z\langle x,z\rangle : \mathcal{N}_f \text{ is a non-empty finite subset of } \mathcal{N}\Big\}.$$

Definition 3.4. Let $J\in\mathcal{B}(\mathcal{H})$ be anti self-adjoint and unitary and $j\in\mathbb{S}$. Let $\mathcal{H}^j_\pm$ denote the closed complex (with respect to the complex plane $\mathbb{C}_j$) subspaces given by

$$\mathcal{H}^j_\pm=\{x\in\mathcal{H}:Jx=\pm xj\}. \tag{3.2}$$

We will now formulate some useful results from [31] in the following lemma.

Lemma 3.5. *If J is an anti self-adjoint and unitary operator and $j\in\mathbb{S}$, then:*

(i) $\mathcal{H}^j_\pm\neq\{0\}$.

(ii) *As a $\mathbb{C}_j$-Hilbert space, $\mathcal{H}$ admits the following direct sum decomposition*

$$\mathcal{H} = \mathcal{H}_+^j \oplus \mathcal{H}_-^j. \tag{3.3}$$

Definition 3.6. Fix an orthonormal basis $\mathcal{N}$ of a quaternionic Hilbert space $\mathcal{H}$. The *left scalar multiplication* L_p of $\mathcal{H}$ induced by $\mathcal{N}$ is the map

$$(p,x) \in \mathbb{H} \times \mathcal{H} \mapsto px \in \mathcal{H}$$

given by

$$px := \sum_{y \in \mathcal{N}} yp\langle x, y\rangle.$$

Lemma 3.7 (Statement (a) of Proposition 3.8 in [31]). *Let $\mathcal{H}$ be a quaternionic Hilbert space. If $J \in \mathcal{B}(\mathcal{H})$ is an anti self-adjoint and unitary operator, then corresponding to any fixed $j \in \mathbb{S}$, there exists a left-scalar multiplication L_p so that*

$$J = L_j.$$

In the following theorem, we will make use of the operator $|T| := (T^*T)^{1/2}$ for $T \in \mathcal{B}(\mathcal{H})$. See Section 2.4 of [31] for a definition of the square root of a positive operator which relies on the functional calculus therein.

Theorem 3.8. *Let $T \in \mathcal{B}(\mathcal{H})$ be normal. Then there exist uniquely determined operators $A := (1/2)(T + T^*)$ and $B := (1/2)|T - T^*|$ which both belong to $\mathcal{B}(\mathcal{H})$ and an operator $J \in \mathcal{B}(\mathcal{H})$ which is uniquely determined on $\{\mathrm{Ker}(T - T^*)\}^{\perp}$ so that the following properties hold:*

(i) *$T = A + J\,B$.*
(ii) *A is self-adjoint and B is positive.*
(iii) *J is anti self-adjoint and unitary.*
(iv) *A, B and J mutually commute.*
(v) *For any fixed $j \in \mathbb{S}$, there exists an orthonormal basis $\mathcal{N}_j$ of $\mathcal{H}$ with the property that $J = L_j$.*

Definition 3.9. Let $\Omega \subseteq \mathbb{H}$. We call Ω *axially symmetric* if $p_0 + ip_1 \in \Omega$ with $p_0, p_1 \in \mathbb{R}$ and $i \in \mathbb{S}$, then $p_0 + jp_1 \in \Omega \cap \mathbb{C}_j$ for all $j \in \mathbb{S}$.

Definition 3.10. Let $\Omega \subseteq \mathbb{H}$ be an axially symmetric set and let $D \subseteq \mathbb{R}^2$ be such that

$$D = \{(u,v) \in \mathbb{R}^2 : u + jv \in \Omega \text{ for some } j \in \mathbb{S}\}.$$

Let $\mathcal{S}(\Omega, \mathbb{H})$ denote the quaternionic linear space of slice continuous functions, i.e., $\mathcal{S}(\Omega, \mathbb{H})$ consists of functions $f : \Omega \to \mathbb{H}$ of the form

$$f(u + jv) = f_0(u,v) + jf_1(u,v) \quad \text{for} \quad (u,v) \in D \quad \text{and for} \quad j \in \mathbb{S},$$

where f_0 and f_1 are continuous $\mathbb{H}$-valued functions on D so that

$$f_0(u,v) = f_0(u,-v) \quad \text{and} \quad f_1(u,v) = -f_1(u,-v).$$

If f_0 and f_1 are real-valued, then we say that the continuous slice function f is *intrinsic*. The subspace of intrinsic continuous slice functions is denoted by $\mathcal{S}_{\mathbb{R}}(\Omega, \mathbb{H})$.

Remark 3.11. We observe that if $f \in \mathcal{S}_{\mathbb{R}}(\Omega, \mathbb{H})$ and we consider the restriction of f to $\Omega_j := \Omega \cap \mathbb{C}_j$, where $j \in \mathbb{S}$, then f has values in $\mathbb{C}_j$. This fact makes clear the notation $\mathcal{S}_{\mathbb{R}}(\Omega_j, \mathbb{C}_j)$.

With those tools at hand we can state the main steps that lead to the spectral theorem based on the S-spectrum. The following result relates continuous functions and slice continuous functions.

Lemma 3.12 (See Lemma 4.1 in [8]). *Let $C(\Omega_j^+, \mathbb{R})$ denote the set of real-valued continuous functions on $\Omega_j^+ = \sigma_S(T) \cap \mathbb{C}_j^+$ and $\mathcal{S}_{\mathbb{R}}(\Omega_j, \mathbb{R})$ denote the set of real-valued functions in $\mathcal{S}_{\mathbb{R}}(\Omega_j, \mathbb{H})$, where $\Omega_j = \sigma_S(T) \cap \mathbb{C}_j$. Let $C_0(\Omega_j^+, \mathbb{R})$ denote the subset of functions $f \in C(\Omega_j^+, \mathbb{R})$ such that $f|_{\Omega_j \cap \mathbb{R}} = 0$. The following statements hold:*

(i) *There exists a bijection between $C(\Omega_j^+, \mathbb{R})$ and $\mathcal{S}_{\mathbb{R}}(\Omega_j, \mathbb{R})$.*
(ii) *If $\Omega_j^+ \cap \mathbb{R} \neq \emptyset$, then there exists a bijection between $C_0(\Omega_j^+, \mathbb{R})$ and purely imaginary functions in $\mathcal{S}_{\mathbb{R}}(\Omega_j, \mathbb{H})$.*
(iii) *If $\Omega_j^+ \cap \mathbb{R} = \emptyset$, then there exists a bijection between $C(\Omega_j^+, \mathbb{R})$ and purely imaginary functions in $\mathcal{S}_{\mathbb{R}}(\Omega_j, \mathbb{H})$.*

Now we construct the spectral measure associated to the normal operator $T = A + JB \in \mathcal{B}(\mathcal{H})$ be normal, fix $x \in \mathcal{H}$ and let

$$\ell_x(g) = \langle g(T)x, x \rangle, \quad g \in C(\Omega_j^+, \mathbb{R}),$$

where $g(T) = f_0(B, A)$ is constructed by the continuous functional calculus. The operator ℓ_x is a real-valued bounded linear functional on $C(\Omega_j^+, \mathbb{R})$. Moreover, ℓ_x is a positive functional.

The Riesz representation theorem for continuous functions yields the existence of a uniquely determined positive-valued measure μ_x (for a fixed $j \in \mathbb{S}$) so that

$$\ell_x(g) = \int_{\Omega_j^+} g(p) d\mu_x(p), \quad g \in C(\Omega_j^+, \mathbb{R}). \tag{3.4}$$

By polarization we get

$$\langle g(T)x, y \rangle = \int_{\Omega_j^+} g(p) d\mu_{x,y}(p), \quad g \in C(\Omega_j^+, \mathbb{R}), \tag{3.5}$$

where

$$\begin{aligned} 4\mu_{x,y} = \; & \mu_{x+y} - \mu_{x-y} + e_1 \mu_{x+ye_1} - e_1 \mu_{x-ye_1} \\ & + e_1 \mu_{x-ye_2} e_3 - e_1 \mu_{x+ye_2} e_3 + \mu_{x+ye_3} e_3 - \mu_{x-ye_3} e_3. \end{aligned} \tag{3.6}$$

The measure $\mu_{x,y}$ has the following properties (see Lemma 4.3 in [8]):

(i) $\mu_{x\alpha+y\beta,z} = \mu_{x,z}\alpha + \mu_{y,z}\beta, \quad \alpha, \beta \in \mathbb{H}.$
(ii) $\mu_{x,y\alpha+z\beta} = \bar{\alpha}\mu_{x,y} + \bar{\beta}\mu_{x,z}, \quad \alpha, \beta \in \mathbb{H}.$
(iii) $|\mu_{x,y}(\sigma_S(T) \cap \mathbb{C}_j^+)| \leq \|x\| \|y\|.$
(iv) $\bar{\mu}_{x,y} = \mu_{y,x},$

for all $x, y, z \in \mathcal{H}$. So thanks to the above properties from the Riesz representation theorem for quaternionic Hilbert spaces it follows that there exists a bounded linear operator E such that

$$\mu_{x,y}(\sigma) = \langle x, E(\sigma)y\rangle, \quad \sigma \in \mathfrak{B}(\sigma_S(T) \cap \mathbb{C}_j^+), \tag{3.7}$$

where $\mathfrak{B}(\sigma_S(T) \cap \mathbb{C}_j^+)$ denotes the Borel set in $\sigma_S(T) \cap \mathbb{C}_j^+$.

Theorem 3.13 (See Theorem 4.5 in [8]). *The $\mathcal{B}(\mathcal{H})$-valued countably additive measure E_j (that we denote by E), given by* (3.7), *for all $\sigma, \tau \in \mathfrak{B}(\sigma_S(T) \cap \mathbb{C}_j^+)$, enjoys the following properties:*

(i) $E(\sigma) = E(\sigma)^*$.
(ii) $\|E(\sigma)\| \leq 1$.
(iii) $E(\emptyset) = 0$ *and* $E(\sigma_S(T) \cap \mathbb{C}_j^+) = I_{\mathcal{H}}$.
(iv) $E(\sigma \cap \tau) = E(\sigma)E(\tau)$.
(v) $E(\sigma)^2 = E(\sigma)$.
(vi) *If* $\sigma_S(T) \cap \mathbb{R} = \emptyset$ *(respectively,* $\sigma_S(T) \cap \mathbb{R} \neq \emptyset$*), then* $E(\sigma)$ *commutes with* $f(T)$ *for all* $f \in C(\sigma_S(T) \cap \mathbb{C}_j^+, \mathbb{C}_j)$ *(respectively,* $f \in C_0(\sigma_S(T) \cap \mathbb{C}_j^+, \mathbb{C}_j)$*).*
(vii) $E(\sigma)$ *and* $E(\tau)$ *commute.*

One of the crucial facts in the proof of the spectral theorem is to glue together the components $f_0(T)$, $f_1(T)$, of the operators $f(T)$, and the operator J. It turns out that, if $T = A + JB$, and $f = f_0 + jf_1$ is an intrinsic slice continuous function then by the continuous functional calculus for intrinsic functions in [31] we have

$$f(T) = f_0(A, B) + Jf_1(A, B)$$

that we write in a more compact way as

$$f(T) = f_0(T) + Jf_1(T)$$

where $f_0(T)$, $f_1(T)$ and J commute among themselves so, see the proof of Lemma 4.4 in [8], we have

$$\begin{aligned}\langle f(T)x, y\rangle &= \langle f_0(T)x, y\rangle + \langle f_1(T)Jx, y\rangle \\ &= \int_{\Omega_j^+} f_0(p)d\langle E(p)x, y\rangle + \int_{\Omega_j^+} f_1(p)d\langle E(p)Jx, y\rangle, \quad x, y \in \mathcal{H}.\end{aligned}$$

Using the properties of J and the decomposition of Theorem 3.8 we have the two equivalent version of the spectral theorem.

Theorem 3.14 (Theorem 4.7 in [8]). *Suppose $T \in \mathcal{B}(\mathcal{H})$ is normal, let $J \in \mathcal{B}(\mathcal{H})$ be as in the decomposition of Theorem* 3.8 *and fix $j \in \mathbb{S}$. Let $\Omega_j^+ = \sigma_S(T) \cap \mathbb{C}_j^+$ and $\Pi_\pm^j$ denote the orthogonal projection onto $\mathcal{H}_\pm^j$ given in* (3.3)*, respectively. If $\sigma_S(T) \cap \mathbb{R} = \emptyset$ (respectively $\sigma_S(T) \cap \mathbb{R} \neq \emptyset$), then there exists a unique spectral measure E_j on Ω_j^+ so that*

$$\langle f(T)x, y\rangle = \int_{\Omega_j^+} f_0(p)\, d\langle E_j(p)x, y\rangle + \int_{\Omega_j^+} f_1(p)\, d\langle JE_j(p)x, y\rangle, \quad x, y \in \mathcal{H}, \tag{3.8}$$

or, equivalently,

$$\langle f(T)x,y\rangle = \int_{\Omega_j^+} f(p)\, d\langle \Pi_+^j E_j(p)x,y\rangle + \int_{\Omega_j^+} \overline{f(p)}\, d\langle \Pi_-^j E_j(p)x,y\rangle, \quad x,y \in \mathcal{H}, \tag{3.9}$$

for $f = f_0 + f_1 j \in C(\Omega_j^+, \mathbb{C}_j)$ *(respectively,* $C_0(\Omega_j^+, \mathbb{C}_j)$*), where* f_0 *and* f_1 *are real-valued. Moreover, upon identifying the complex planes* $\mathbb{C}_k$ *with* $\mathbb{C}_j$ *in the natural way by the mapping* φ_{jk}*, we have* $E_j(\varphi_{jk}(\sigma)) = E_k(\sigma), \sigma \in \mathfrak{B}(\Omega_k^+)$ *for all* $j,k \in \mathbb{S}$.

Remark 3.15. It should be noted that a *spectral mapping principle* for $f(T)$, as in Theorem 3.14, holds in the following sense:

$$f(\sigma_S(T) \cap \mathbb{C}_j^+) = \sigma_S(f(T)) \cap \mathbb{C}_j^+, \quad j \in \mathbb{S}. \tag{3.10}$$

Formula (3.10) has been shown in Theorem 7.8 in [31]. One can also check that formula (3.10) holds by using Theorem 3.14.

We recall that in the paper [8] a very general functional calculus is constructed (see Section 5 in [8]), so that the Spectral Theorem 3.14 holds not only for continuous functions but for a larger class. Moreover, the functional calculus in Section 5 in [8] and the spectral theorem for bounded operators allows to extend the spectral theorem to unbounded normal operators (see Theorem 6.2 in [8]) and to define, for these operators, a functional calculus see Corollary 6.6 in [8]).

4. The spectral theorem in the finite-dimensional case

We will begin the section with a version of the spectral theorem for a normal matrix $T \in \mathbb{H}^{n\times n}$, where $\mathbb{H}^{n\times n}$ denotes the set of all $n \times n$ matrices with entries in $\mathbb{H}$, originally due to Lee [38] and, also independently, to Brenner [12] (see also the survey papers [46] and [27]).

Theorem 4.1 ([38], [12]). *Let* $T \in \mathbb{H}^{n\times n}$ *be normal. Corresponding to any* $j \in \mathbb{S}$*, there exist a unitary matrix* U_j *and* $\lambda_1, \dots, \lambda_n \in \mathbb{C}_j^+$*, where* $\mathbb{C}_j^+ = \{x + jy : x \in \mathbb{R}$ *and* $y \geq 0\}$*, such that*

$$T = U \mathrm{diag}(\lambda_1, \dots, \lambda_n) U^*. \tag{4.1}$$

The goal of this section is to explain how (4.1) and (3.8) are related when $f(p) = p$. An important ingredient in the proof of (3.8) is the decomposition

$$T = A + BJ, \tag{4.2}$$

where A, B and J are mutually commuting operators belonging to $\mathcal{B}(\mathcal{H})$ with the following properties: $A = A^*$, B is positive and $J^* = J^{-1} = -J$. Note that A and B are uniquely determined by T, however J is only unique on a certain subspace of $\mathcal{H}$. The decomposition (4.2) was established in [31] using basic operator factorization results. Moreover, (4.2) was proven in [31] without any dependence on a spectral theorem.

We first note that if $f(p)=p$ and $\mathcal{H}=\mathbb{H}^n$ in (3.8), then

$$\langle Tx,y\rangle=\int_{\sigma_S(T)\cap\mathbb{C}_j^+}\mathrm{Re}(p)d\langle E(p)x,y\rangle+\int_{\sigma_S(T)\cap\mathbb{C}_j^+}\mathrm{Im}(p)d\langle JE(p)x,y\rangle,$$

where $x,y\in\mathbb{H}^n$. Since $T\in\mathbb{H}^{n\times n}$, we have $\sigma_S(T)=\sigma_R(T)$, and hence

$$\begin{aligned}\langle Tx,y\rangle&=\sum_{m=1}^{n}\{\mathrm{Re}(\lambda_m)\langle E(\lambda_m)x,y\rangle+\mathrm{Im}(\lambda_m)\langle JE(\lambda_m)x,y\rangle\}\\&=\sum_{m=1}^{n}\langle\{\mathrm{Re}(\lambda_m)+J\mathrm{Im}(\lambda_m)\}E(\lambda_m)x,y\rangle.\end{aligned}\tag{4.3}$$

On the other hand, from (4.1), we have

$$T=UD_1U^*+UD_2jU^*,$$

where

$$D_1=\mathrm{diag}(\mathrm{Re}(\lambda_1),\dots,\mathrm{Re}(\lambda_n))\quad\text{and}\quad D_2=\mathrm{diag}(\mathrm{Im}(\lambda_1),\dots,\mathrm{Im}(\lambda_n)).$$

Moreover, if we let $J=UjU^*$, then we have $A=UD_1U^*$ and $B=UD_1U^*$ such that $A=A^*$, B is positive and $J^*=J^{-1}=-J$ and A, B and J mutually commute. Thus,

$$U\mathrm{diag}(\mathrm{Re}(\lambda_1),\dots,\mathrm{Re}(\lambda_n))U^*=\sum_{m=1}^{n}\mathrm{Re}(\lambda_m)E(\lambda_m)\tag{4.4}$$

and

$$U\mathrm{diag}(\mathrm{Im}(\lambda_1),\dots,\mathrm{Im}(\lambda_n))U^*=\sum_{m=1}^{n}\mathrm{Im}(\lambda_m)E(\lambda_m).\tag{4.5}$$

5. The S-functional calculus

The discovery of the Cauchy formula of slice hyperholomorphic functions [18] allowed the full understanding of the quaternionic functional calculus, see [7, 15, 16]. In this section we recall the basic facts and for more details we suggest the original papers [7, 15, 16] and the book [21]. Let V be a right vector space on $\mathbb{H}$. A map $T:V\to V$ is said to be a right linear operator if

$$T(u+v)=T(u)+T(v),\qquad T(us)=T(u)s,\quad\text{for all } s\in\mathbb{H},\ u,v\in V.$$

By $\mathrm{End}^R(V)$ we denote the set of right linear operators acting on V. In the sequel, we will consider only two-sided vector spaces V, otherwise the set $\mathrm{End}^R(V)$ is neither a left nor a right vector space over $\mathbb{H}$. With this assumption, $\mathrm{End}^R(V)$ becomes both a left and a right vector space on $\mathbb{H}$ with respect to the operations

$$(sT)(v):=sT(v),\qquad(Ts)(v):=T(sv),\quad\text{for all } s\in\mathbb{H},\ v\in V.\tag{5.1}$$

In particular (5.1) gives $(s\mathcal{I})(v)=(\mathcal{I}s)(v)=sv$. Similar considerations can be done when we consider V as a left vector space on $\mathbb{H}$ and a map $T:V\to V$ is a left linear operator. In the sequel, if we do not specify, we will always assume that V is a two-sided quaternionic Banach space with norm $\|\cdot\|$. The two-sided vector space $\mathcal{B}(V)$ of all right linear bounded operators on V is a Banach spaces if endowed with the natural norms:

$$\|T\| := \sup_{v\in V} \frac{\|T(v)\|}{\|v\|}.$$

We will also denote by $\mathcal{L}(V)$ the set of (right) linear operators. In this paper we will mainly consider right linear operators for simplicity but all the results that follows can be applied to left linear operators with suitable modifications of the statements. The following theorem makes precise the heuristic discussion in the introduction.

Theorem 5.1. *Let $T \in \mathcal{B}(V)$ and let $s \in \mathbb{H}$. Then, for $\|T\| < |s|$:*

(1) *the operator $(T - \overline{s}\mathcal{I})^{-1}\, s(T - \overline{s}\mathcal{I}) - T$ is the inverse of $\sum_{n\geq 0} T^n s^{-1-n}$ and*

$$\sum_{n\geq 0} T^n s^{-1-n} = -(T^2 - 2\mathrm{Re}(s)T + |s|^2\mathcal{I})^{-1}(T - \overline{s}\mathcal{I}), \tag{5.2}$$

(2) *the operator $(T - \overline{s}\mathcal{I})\, s(T - \overline{s}\mathcal{I})^{-1} - T$ is the inverse of $\sum_{n\geq 0} s^{-1-n}T^n$ and*

$$\sum_{n\geq 0} s^{-1-n}T^n = -(T - \overline{s}\mathcal{I})(T^2 - 2\mathrm{Re}(s)T + |s|^2\mathcal{I})^{-1}. \tag{5.3}$$

The above result involves just the powers T^n of the operator T and in the case T is written as $T = T_0 + iT_1 + jT_2 + kT_3$, where T_ℓ, $\ell = 0, 1, 2, 3$ are the bounded operators we do not have to require that T_ℓ mutually commute. This is the first crucial point. It is from the above theorem that the notion of S-spectrum naturally shows up.

Definition 5.2 (The S-spectrum and the S-resolvent set). Let $T \in \mathcal{B}(V)$. We define the S-spectrum $\sigma_S(T)$ of T as:

$$\sigma_S(T) = \{s \in \mathbb{H} : \ T^2 - 2\ \mathrm{Re}(s)T + |s|^2\mathcal{I} \quad \text{is not invertible in } \mathcal{B}(\mathrm{V})\}.$$

The S-resolvent set $\rho_S(T)$ is defined by

$$\rho_S(T) = \mathbb{H} \setminus \sigma_S(T).$$

The notion of S-spectrum of a linear quaternionic operator T is suggested by the definition of S-resolvent operator that is the kernel useful for the quaternionic functional calculus. From the definition of the S-spectrum directly follows its spherical nature.

Theorem 5.3 (Structure of the S-spectrum). *Let $T \in \mathcal{L}(V)$ and let $p = p_0 + p_1 i \in \sigma_S(T)$. Then all the elements of the sphere $[p_0 + ip_1]$ belong to $\sigma_S(T)$.*

Definition 5.4 (The S-resolvent operators). Let V be a two-sided quaternionic Banach space, $T \in \mathcal{B}(V)$ and $s \in \rho_S(T)$. We define the left S-resolvent operator as

$$S_L^{-1}(s, T) := -(T^2 - 2\mathrm{Re}(s)T + |s|^2\mathcal{I})^{-1}(T - \overline{s}\mathcal{I}), \tag{5.4}$$

and the right S-resolvent operator as

$$S_R^{-1}(s,T) := -(T - \overline{s}\mathcal{I})(T^2 - 2\mathrm{Re}(s)T + |s|^2\mathcal{I})^{-1}. \tag{5.5}$$

In this setting there is a major difference with respect to the complex case. Indeed, the resolvent equation involves both the S-resolvent operators. This means that both formulations (using the two Cauchy formulas) of the S-functional calculus are necessary.

Theorem 5.5 (see [7]). *Let $T \in \mathcal{B}(V)$. For $s, p \in \rho_S(T)$ with $s \notin [p]$, it is*

$$\begin{aligned} S_R^{-1}(s,T)S_L^{-1}(p,T)v = & \big[[S_R^{-1}(s,T) - S_L^{-1}(p,T)]p \\ & - \overline{s}[S_R^{-1}(s,T) - S_L^{-1}(p,T)]\big](p^2 - 2s_0p + |s|^2)^{-1}v, \quad v \in V. \end{aligned} \tag{5.6}$$

We conclude with the fact that the spectrum of a bounded linear operator is a nonempty and compact set also in the quaternionic setting.

Theorem 5.6 (Compactness of *S*-spectrum). *Let $T \in \mathcal{B}(V)$. Then the S-spectrum $\sigma_S(T)$ is a compact nonempty subset of $\mathbb{H}$.*

Definition 5.7. Let V be a two-sided quaternionic Banach space, $T \in \mathcal{B}(V)$ and let $U \subset \mathbb{H}$ be an axially symmetric s-domain that contains the S-spectrum $\sigma_S(T)$ be such that $\partial(U \cap \mathbb{C}_j)$ is union of a finite number of continuously differentiable Jordan curves for every $j \in \mathbb{S}$. We say that U is a T-admissible open set.

Definition 5.8. Let V be a two-sided quaternionic Banach space, $T \in \mathcal{B}(V)$ and let W be an open set in $\mathbb{H}$.

(i) A function $f \in \mathcal{SH}^L(W)$ is said to be locally left slice hyperholomorphic on $\sigma_S(T)$ if there exists a T-admissible domain U contained in $\mathbb{H}$ and such that $\overline{U} \subset W$. We denote by $\mathcal{SH}^L_{\sigma_S(T)}$ the set of locally slice hyperholomorphic functions on $\sigma_S(T)$.

(ii) A function $f \in \mathcal{SH}^R(W)$ is said to be locally right slice hyperholomorphic on $\sigma_S(T)$ if there exists a T-admissible domain U contained in $\mathbb{H}$ and such that $\overline{U} \subset W$. We denote by $\mathcal{SH}^R_{\sigma_S(T)}$ the set of locally slice hyperholomorphic functions on $\sigma_S(T)$.

Definition 5.9 (The (quaternionic) *S*-functional calculus). Let V be a two-sided quaternionic Banach space and $T \in \mathcal{B}(V)$. Let $U \subset \mathbb{H}$ be a T-admissible domain and set $ds_i = -dsi$. We define

$$f(T) = \frac{1}{2\pi}\int_{\partial(U\cap\mathbb{C}_i)} S_L^{-1}(s,T)\, ds_i\, f(s), \quad \text{for } f \in \mathcal{SH}^L_{\sigma_S(T)}, \tag{5.7}$$

and

$$f(T) = \frac{1}{2\pi}\int_{\partial(U\cap\mathbb{C}_i)} f(s)\, ds_i\, S_R^{-1}(s,T), \quad \text{for } f \in \mathcal{SH}^R_{\sigma_S(T)}. \tag{5.8}$$

Remark 5.10. The S-functional calculus is well defined because the two integrals depend neither on the open set U, that contains the S-spectrum of T, nor on the imaginary unit $\imath \in \mathbb{S}$.

Associated to the quaternionic functional calculus there are the Riesz projectors. We suppose that $T \in \mathcal{B}(V)$ and $\sigma_S(T) = \sigma_{1S}(T) \cup \sigma_{2S}(T)$, with

$$\mathrm{dist}\,(\sigma_{1S}(T), \sigma_{2S}(T)) > 0.$$

Let U_1 and U_2 be two axially symmetric slice domains such that $\sigma_{1S}(T) \subset U_1$ and $\sigma_{2S}(T) \subset U_2$, with $\overline{U}_1 \cap \overline{U}_2 = \emptyset$. We define that operators

$$P_\ell := \frac{1}{2\pi} \int_{\partial(U_\ell \cap \mathbb{C}_i)} S_L^{-1}(s,T)\, ds_i, \qquad \ell = 1, 2. \tag{5.9}$$

Using the S-resolvent equation it follows that P_ℓ are projectors, that is $P_\ell^2 = P_\ell$, and $TP_\ell = P_\ell T$ for $\ell = 1, 2$.

The most important properties of the Riesz–Dunford functional calculus are also shared also by the S-functional calculus. In fact, the S-functional calculus agrees with that natural functional calculus for polynomials, thanks to the linearity and the fact that

$$T^m = \frac{1}{2\pi} \int_{\partial(U \cap \mathbb{C}_i)} S_L^{-1}(s,T)\; ds_i\; s^m, \qquad m \in \mathbb{N} \cup \{0\}$$

with obvious meaning of the symbols. In general it is not true that the product of two slice hyperholomorphic functions is still slice hyperholomorphic, but when we consider the multiplication by an intrinsic function, on the correct side, we preserve the slice hyperholomorphicity (left or right). In the case $(fg)(T)$ is defined we have the product rule $(fg)(T) = f(T)g(T)$ of the S-functional calculus. For intrinsic functions we also have the spectral mapping theorem and the spectral radius theorem, see [7, 15, 16]. The Taylor Formula has been proved in [22].

5.1. The case of unbounded operators

The S-functional calculus can be extended to unbounded operators but there is a strong condition to assume. As in the classical setting the function f has to be holomorphic at infinity. In the quaternionic setting we say that f is a slice hyperholomorphic function at ∞ if $f(q)$ is slice hyperholomorphic function in a set $D'(\infty, r) = \{q \in \mathbb{H} \,:\, |q| > r\}$, for some $r > 0$, and $\lim_{q\to\infty} f(q)$ exists and it is finite. We set $f(\infty)$ to be the value of this limit. We will see how this conditions can be removed in the H^∞-functional calculus or with the Phillips functional calculus, where stronger conditions are assumed on the operator.

Definition 5.11. Let $T \in \mathcal{L}(V)$ be densely defined and let $\mathcal{R}_s(T) : \mathcal{D}(T^2) \to V$ be given by

$$\mathcal{R}_s(T)x = \{T^2 - 2\mathrm{Re}(s)T + |s|^2\mathcal{I}\}x, \quad x \in \mathcal{D}(T^2).$$

The S-resolvent set of T is defined as follows

$$\begin{aligned}\rho_S(T) =\{s \in V \,:\, &\mathrm{Ker}(\mathcal{R}_s(T)) = \{0\},\ \mathrm{Ran}(\mathcal{R}_s(T)) \text{ is dense}\\ &\text{in } V \text{ and } \mathcal{R}_s(T)^{-1} \in \mathcal{B}(V)\}.\end{aligned}$$

The S-spectrum is defined as

$$\sigma_S(T) = \mathbb{H} \setminus \rho_S(T).$$

We denote by $\mathcal{K}(V)$ the set of right linear closed operators $T : \mathcal{D}(T) \subset V \to V$, such that $\mathcal{D}(T)$ is dense in V. In the case of unbounded operators the S-spectrum in not necessarily bounded and we have to take into account the point at infinity. So we define the extended S-spectrum of T as

$$\overline{\sigma}_S(T) := \sigma_S(T) \cup \{\infty\}.$$

For operators $T \in \mathcal{K}(V)$ the definition of the set of locally slice hyperholomorphic functions on $\sigma_S(T)$ has to take into account the point at infinity so with consider $\mathcal{SH}^L_{\overline{\sigma}_S(T)}$, and also, analogously for $\mathcal{SH}^R_{\overline{\sigma}_S(T)}$.

The open set U related to $f \in \mathcal{SH}^L_{\overline{\sigma}_S(T)}$ (resp. $\mathcal{SH}^R_{\overline{\sigma}_S(T)}$) need not to be connected. Moreover, as in the classical functional calculus, U in general depends on f and can be unbounded. In the case of unbounded operators we will always require that the S-resolvent set is not empty.

Definition 5.12 (The S-functional calculus for unbounded operators). Consider $k \in \mathbb{R}$ and the function

$$\Phi : \overline{\mathbb{H}} \to \overline{\mathbb{H}}$$

defined by $p = \Phi(s) = (s-k)^{-1}$, $\Phi(\infty) = 0$, $\Phi(k) = \infty$. Let $T \in \mathcal{K}(V)$ with $\rho_S(T) \cap \mathbb{R} \neq \emptyset$ and suppose that $f \in \mathcal{SH}^L_{\overline{\sigma}_S(T)}$ (resp. $\mathcal{SH}^R_{\overline{\sigma}_S(T)}$). Let us consider

$$\phi(p) := f(\Phi^{-1}(p))$$

and the bounded linear operator defined by

$$A := (T - k\mathcal{I})^{-1}, \quad \textit{for some} \quad k \in \rho_S(T) \cap \mathbb{R}.$$

We define, in both cases, the operator $f(T)$ as

$$f(T) = \phi(A). \tag{5.10}$$

Since we assume that $\rho_S(T) \neq \emptyset$, we can define the operator

$$Q_s(T) := (T^2 - 2\mathrm{Re}(s)T + |s|^2\mathcal{I})^{-1}, \quad s \in \rho_S(T),$$

which is called the pseudo-resolvent (operator) of T. For right linear operators the S-resolvent operators that are defined on the whole space V are given by:

Definition 5.13. Let $T \in \mathcal{K}(V)$. The left S-resolvent operator is defined as

$$S_L^{-1}(s,T) := Q_s(T)\overline{s} - TQ_s(T), \qquad s \in \rho_S(T), \tag{5.11}$$

and the right S-resolvent operator is defined as

$$S_R^{-1}(s,T) := -(T - \mathcal{I}\overline{s})Q_s(T), \qquad s \in \rho_S(T). \tag{5.12}$$

From the above definitions of S-resolvent operators we get an integral representation of the S-functional calculus. The following theorem also shows that the S-functional calculus for unbounded operators is well defined because it does not depend on the point $k \in \mathbb{R}$ that we choose to define $f(T) = \phi(A)$.

Theorem 5.14. *Let V be a two-sided quaternionic Banach space and let W be a T-admissible open set. Let $T \in \mathcal{K}(V)$ with $\rho_S(T) \cap \mathbb{R} \neq \emptyset$. Then the operator $f(T)$ defined in* (5.10) *is independent of $k \in \rho_S(T) \cap \mathbb{R}$, and, for $f \in \mathcal{R}^L_{\overline{\sigma}_S(T)}$ and $v \in V$, we have*

$$f(T)v = f(\infty)\mathcal{I}v + \frac{1}{2\pi} \int_{\partial(W \cap \mathbb{C}_i)} S_L^{-1}(s,T)\, ds_i\, f(s)v, \tag{5.13}$$

and for $f \in \mathcal{R}^R_{\overline{\sigma}_S(T)}$ and $v \in V$, we have

$$f(T)v = f(\infty)\mathcal{I}v + \frac{1}{2\pi} \int_{\partial(W \cap \mathbb{C}_i)} f(s)\, ds_i\, S_R^{-1}(s,T)v. \tag{5.14}$$

For left linear operators a similar result holds. The proof of Theorem 5.14 is based on the following non trivial relations between the S-resolvent of T and the S-resolvent of A:

$$S_L^{-1}(s,T)v = p\mathcal{I}v - S_L^{-1}(p,A)p^2 v, \quad v \in V, \tag{5.15}$$

and

$$S_R^{-1}(s,T)v = p\mathcal{I}v - p^2 S_R^{-1}(p,A)v, \quad v \in V. \tag{5.16}$$

where $T \in \mathcal{K}(V)$, Φ, ϕ are as above and on the facts that $\Phi(\overline{\sigma}_S(T)) = \sigma_S(A)$ and $\phi(p) = f(\Phi^{-1}(p))$ determines a one-to-one correspondence between $f \in \mathcal{SH}_{\overline{\sigma}_S(T)}$ and $\phi \in \mathcal{SH}_{\overline{\sigma}_S(A)}$. The fractional powers have been studied in [23].

6. The H^∞-functional calculus

The H^∞-functional calculus has been introduced by A. McIntosh in [40] (see also [2]) and can be considered an extension of the Riesz–Dunford functional calculus to a class of unbounded operators. This calculus is connected with pseudo-differential operators, with Kato's square root problem, and with the study of evolution equations and, in particular, the characterization of maximal regularity and of the fractional powers of differential operators. In the recent paper [3] the H^∞-functional calculus has been extended to quaternionic operators and to n-tuples of not necessarily commuting operators. In this section we recall some of the main results related to this calculus in the quaternionic setting.

Definition 6.1 (Operator of type ω). Let $\omega \in [0,\pi)$ we say the linear operator $T : D(T) \subseteq V \to V$ is of type ω if

(i) T is closed and densely defined
(ii) $\sigma_S(T) \subset \mathcal{S}_\vartheta \cup \{\infty\}$
(iii) for every $\vartheta \in (\omega,\pi]$ there exists a positive constant C_ϑ such that

$$\|S_L^{-1}(s,T)\| \leq \frac{C_\vartheta}{|s|} \quad \text{for all non zero } s \in \mathcal{S}^0_\vartheta,$$

$$\|S_R^{-1}(s,T)\| \leq \frac{C_\vartheta}{|s|} \quad \text{for all non zero } s \in \mathcal{S}^0_\vartheta.$$

In order to define bounded functions of operators of type ω, we need a suitable subclasses of bounded slice hyperholomorphic functions that we have introduced in Section 1.

Definition 6.2. We define the spaces

$$\Psi_L(\mathcal{S}^0_\mu) = \left\{ f \in \mathcal{SH}^\infty_L(\mathcal{S}^0_\mu) \ : \ \exists\, \alpha > 0,\ c > 0\ |f(s)| \leq \frac{c|s|^\alpha}{1+|s|^{2\alpha}},\ \text{ for all } s \in \mathcal{S}^0_\mu \right\},$$

$$\Psi_R(\mathcal{S}^0_\mu) = \left\{ f \in \mathcal{SH}^\infty_R(\mathcal{S}^0_\mu) \ : \ \exists\, \alpha > 0,\ c > 0\ |f(s)| \leq \frac{c|s|^\alpha}{1+|s|^{2\alpha}},\ \text{ for all } s \in \mathcal{S}^0_\mu \right\},$$

$$\Psi(\mathcal{S}^0_\mu) = \left\{ f \in \mathcal{N}^\infty(\mathcal{S}^0_\mu) \ : \ \exists\, \alpha > 0,\ c > 0\ |f(s)| \leq \frac{c|s|^\alpha}{1+|s|^{2\alpha}},\ \text{ for all } s \in \mathcal{S}^0_\mu \right\}.$$

For operators of type ω and for slice hyperholomorphic functions as in Definition 6.2 we can define the following functional calculus.

Definition 6.3 (The S-functional calculus for operators of type ω). Let T be an operator of type ω. Let $i \in \mathbb{S}$, and let $\mathcal{S}^0_\mu$ be the sector defined above. Choose a piecewise smooth path Γ in $\mathcal{S}^0_\mu \cap \mathbb{C}_i$ that goes from $\infty e^{i\theta}$ to $\infty e^{-i\theta}$, for $\omega < \theta < \mu$, then

$$\psi(T) := \frac{1}{2\pi} \int_\Gamma S_L^{-1}(s,T)\, ds_i\, \psi(s), \qquad \text{for all} \quad \psi \in \Psi_L(\mathcal{S}^0_\mu), \tag{6.1}$$

$$\psi(T) := \frac{1}{2\pi} \int_\Gamma \psi(s)\, ds_i\, S_R^{-1}(s,T), \qquad \text{for all} \quad \psi \in \Psi_R(\mathcal{S}^0_\mu). \tag{6.2}$$

The S-functional calculus for operators of type ω is well defined because the integrals (6.1) and (6.2) depend neither on Γ nor on $i \in \mathbb{S}$, and they define bounded operators, see Theorem 4.9 in [3]. Moreover the linearity of the calculus is obvious and the product rule holds, see Theorem 4.12 in [3]. To define the H^∞-functional calculus we need the following set of operators:

Definition 6.4 (The set Ω). Let ω be a real number such that $0 \leq \omega \leq \pi$. We denote by Ω the set of linear operators T acting on a two-sided quaternionic Banach space such that:

(i) T is a linear operator of type ω;
(ii) T is one-to-one and with dense range.

Then we define the following function spaces according to the set of operators defined above:

Definition 6.5. Let ω and μ be real numbers such that $0 \leq \omega < \mu \leq \pi$, we set

$$\mathcal{F}_L(\mathcal{S}^0_\mu) = \{ f \in \mathcal{SH}_L(\mathcal{S}^0_\mu) : \ |f(s)| \leq C(|s|^k + |s|^{-k}) \ \text{ for some } \ k > 0 \text{ and } C > 0 \},$$

$$\mathcal{F}_R(\mathcal{S}^0_\mu) = \{ f \in \mathcal{SH}_R(\mathcal{S}^0_\mu) : \ |f(s)| \leq C(|s|^k + |s|^{-k}) \ \text{ for some } \ k > 0 \text{ and } C > 0 \},$$

$$\mathcal{F}(\mathcal{S}^0_\mu) = \{ f \in \mathcal{N}(\mathcal{S}^0_\mu) \quad : \ |f(s)| \leq C(|s|^k + |s|^{-k}) \ \text{ for some } \ k > 0 \text{ and } C > 0 \}.$$

To extend the S-functional calculus for operators of type ω we consider a quaternionic two-sided Banach space V, the operators in the class Ω, and

I) The non commutative algebra $\mathcal{F}_L(\mathcal{S}_\mu^0)$ (resp. $\mathcal{F}_R(\mathcal{S}_\mu^0)$).

II) The S-functional calculus Φ for operators of type ω

$$\Phi : \Psi_L(\mathcal{S}_\mu^0) \text{ (resp. } \Psi_R(\mathcal{S}_\mu^0)) \to \mathcal{B}(V), \quad \Phi : \phi \to \phi(T).$$

III) The commutative subalgebra of $\mathcal{F}_L(\mathcal{S}_\mu^0)$ consisting of intrinsic rational functions.

IV) The functions in $\mathcal{F}_L(\mathcal{S}_\mu^0)$ have at most polynomial growth. So taken an intrinsic rational functions ψ the operator $\psi(T)$ can be defined by the rational functional calculus, see Definition 3.7 in [3].

V) We assume that $\psi(T)$ is injective.

Definition 6.6 (The quaternionic H^∞ functional calculus). Let V be a two-sided quaternionic Banach space and let T belong to the set Ω. For $k \in \mathbb{N}$ consider the function

$$\psi(s) := \left(\frac{s}{1+s^2}\right)^{k+1}.$$

For $f \in \mathcal{F}_L(\mathcal{S}_\mu^0)$, and T right linear we define the extended functional calculus as

$$f(T) := (\psi(T))^{-1}(\psi f)(T). \tag{6.3}$$

For $f \in \mathcal{F}_R(\mathcal{S}_\mu^0)$, and T left linear we define the extended functional calculus as

$$f(T) := (f\psi)(T)(\psi(T))^{-1}. \tag{6.4}$$

We say that ψ regularizes f.

In the previous definition the operator $(\psi f)(T)$ (resp. $(f\psi)(T)$) is defined using the S-functional calculus Φ for operators of type ω, and $\psi(T)$ is defined by the rational functional calculus (see Definition 3.7 in [3]).

The following results is of crucial importance because it says that the quaternionic H^∞ functional calculus is well defined because it does not depend on the choice of ψ.

Theorem 6.7 (See Theorem 5.5 in [3]). *The definition of the functional calculus in* (6.3) *and in* (6.4) *does not depend on the choice of the intrinsic rational slice hyperholomorphic function* ψ.

This calculus can be applied to several operators in quaternionic analysis, for example the Cauchy–Fueter operator, quaternionic operators appearing in quaternionic quantum mechanics, and the global operator that annihilates slice hyperholomorphic functions:

$$|\underline{q}|^2 \frac{\partial}{\partial x_0} + \underline{q} \sum_{j=1}^{3} x_j \frac{\partial}{\partial x_j}, \quad \text{where} \quad \underline{q} = x_1 e_1 + x_2 e_2 + x_3 e_3.$$

7. The Phillips functional calculus for quaternionic groups

In the papers [5, 17, 33] the problem of the generation of quaternionic groups and semigroups is treated using the S-spectrum.

In this section we recall some important fact on the Phillips functional calculus for quaternionic operators developed in the paper [6].

A family of bounded right-linear operators $(\mathcal{U}(t))_{t\geq 0}$ on V is called a strongly continuous quaternionic semigroup if $\mathcal{U}(0) = \mathcal{I}$ and $\mathcal{U}(t_1 + t_2) = \mathcal{U}(t_1)\mathcal{U}(t_2)$ for $t_1, t_2 \geq 0$ and if $t \mapsto \mathcal{U}(t)v$ is a continuous function on $[0, \infty)$ for any $v \in V$.

Definition 7.1. Let $(\mathcal{U}(t))_{t\geq 0}$ be a strongly continuous quaternionic semigroup. Set

$$\mathcal{D}(T) = \left\{ v \in V : \lim_{h\to 0^+} \frac{1}{h}(\mathcal{U}(h)v - v) \text{ exists} \right\}$$

and

$$Tv = \lim_{h\to 0^+} \frac{1}{h}(\mathcal{U}(h)v - v), \quad v \in \mathcal{D}(T).$$

The operator T is called the quaternionic infinitesimal generator of the semigroup $(\mathcal{U}(t))_{t\geq 0}$.

We indicate that T is the infinitesimal generator of the semigroup $(\mathcal{U}(t))_{t\geq 0}$ by writing $\mathcal{U}_T(t)$ instead to $\mathcal{U}(t)$.

The set $\mathcal{D}(T)$ is a right subspace that is dense in V and $T : \mathcal{D}(T) \to V$ is a right linear closed quaternionic operator.

Theorem 7.2. *Let $(\mathcal{U}_T(t))_{t\geq 0}$ be a strongly continuous quaternionic semigroup and let T be its quaternionic infinitesimal generator. Then*

$$\omega_0 := \lim_{t\to\infty} \frac{1}{t} \ln \|\mathcal{U}_T(t)\| < \infty.$$

If $s \in \mathbb{H}$ with $\mathrm{Re}(s) > \omega_0$ then s belongs $\rho_S(T)$ and

$$S_R^{-1}(s, T) = \int_0^\infty e^{-ts}\mathcal{U}(t)\, dt.$$

The question whether a closed linear operator is the infinitesimal generator of a strongly continuous semigroup is answered by the Hille–Yosida–Phillips Theorem.

Theorem 7.3. *Let T be a closed linear operator with dense domain. Then T is the infinitesimal generator of a strongly continuous semigroup if and only if there exist constants $\omega \in \mathbb{R}$ and $M > 0$ such that $\sigma_S(T) \subset \{s \in \mathbb{H} : \mathrm{Re}(s) \leq \omega\}$ and such that for any $s_0 \in \mathbb{R}$ with $s_0 > \omega$*

$$\|(S_R^{-1}(s_0, T))^n\| \leq \frac{M}{(s_0 - \omega)^n} \qquad \text{for } n \in \mathbb{N}$$

We consider the problem to characterize when a strongly continuous semigroup of operators $(\mathcal{U}_T(t))_{t\geq 0}$ can be extended to a group $(\mathcal{Z}_T(t))_{t\in\mathbb{R}}$ of operators. This extension is unique if it exists and if the family $\mathcal{U}_-(t) = \mathcal{Z}_T(-t)$, $t \geq 0$, is a strongly continuous semigroup. Consider the identity

$$\frac{1}{h}[\mathcal{U}_-(h)v - v] = \frac{1}{-h}[-\mathcal{Z}_T(-2)[\mathcal{Z}_T(2-h)v - \mathcal{Z}_T(2)v]], \quad \text{for } h \in (0,1).$$

By taking the limit for $h \to 0$ we have that the infinitesimal generator of $\mathcal{U}_-(t)$ is $-T$ and $\mathcal{D}(-T) = \mathcal{D}(T)$. In this case T is called the quaternionic infinitesimal generator of the group $(\mathcal{Z}_T(t))_{t\in\mathbb{R}}$. The next theorem gives a necessary and sufficient condition such that a semigroup can be extended to a group, see Theorem 5.1 in [17].

Theorem 7.4. *An operator $T \in \mathcal{K}(V)$ is the quaternionic infinitesimal generator of a strongly continuous group of bounded quaternionic linear operators if and only if there exist real numbers $M > 0$ and $\omega \geq 0$ such that*

$$\|(S_R^{-1}(s_0, T))^n\| \leq \frac{M}{(|s_0| - \omega)^n}, \quad \text{for } \omega < |s_0|. \tag{7.1}$$

If T generates the group $(\mathcal{Z}_T(t))_{t\in\mathbb{R}}$, then $\|\mathcal{Z}_T(t)\| \leq Me^{\omega|t|}$.

Definition 7.5. We denote by $\mathbf{S}(T)$ the family of all quaternionic measures μ on $\mathbf{B}(\mathbb{R})$ such that

$$\int_{\mathbb{R}} d|\mu|(t)\, e^{(\omega+\varepsilon)|t|} < \infty$$

for some $\varepsilon = \varepsilon(\mu) > 0$. The function

$$\mathcal{L}(\mu)(s) = \int_{\mathbb{R}} d\mu(t)\, e^{-st}, \qquad -(\omega+\varepsilon) < \mathrm{Re}(s) < (\omega+\varepsilon)$$

is called the *quaternionic bilateral (right) Laplace–Stieltjes transform* of μ.

Definition 7.6. We denote by $\mathbf{V}(T)$ the set of *quaternionic bilateral Laplace–Stieltjes transforms of measures* in $\mathbf{S}(T)$.

Definition 7.7 (Functions of the quaternionic infinitesimal generator). Let T be the quaternionic infinitesimal generator of the strongly continuous group $(\mathcal{Z}_T(t))_{t\in\mathbb{R}}$ on a quaternionic Banach space V. For $f \in \mathbf{V}(T)$ with

$$f(s) = \int_{\mathbb{R}} d\mu(t)\, e^{-st} \quad \text{for } -(\omega+\varepsilon) < \mathrm{Re}(s) < \omega+\varepsilon,$$

and $\mu \in \mathbf{S}(T)$, we define the right linear operator $f(T)$ on V by

$$f(T)v = \int_{\mathbb{R}} d\mu(t)\, \mathcal{Z}_T(-t)v \qquad \text{for } v \in V. \tag{7.2}$$

Theorem 7.8. *For any $f \in \mathbf{V}(T)$, the operator $f(T)$ is bounded.*

It is important to note the in the case f is right slice hyperholomorphic at infinity then the functional calculus defined by quaternionic bilateral Laplace–Stieltjes transforms agrees with the S-functional calculus.

Theorem 7.9. *Let $f \in \mathbf{V}(T)$ and suppose that f is right slice hyperholomorphic at infinity. Then the operator $f(T)$ defined using the Laplace transform equals the operator $f[T]$ obtained from the S-functional calculus.*

The above theorem is more complicated to prove with respect to the classical case. It is based on an integral representation of $\mathcal{Z}_T(t)$, proved in Proposition 4.3 in [6]. Precisely, let α and c be real numbers such that

$$\omega < c < |\alpha|.$$

Then for any $u \in \mathcal{D}(T^2)$ we have

$$\mathcal{Z}_T(t)u = \frac{1}{2\pi} \int_{\partial(W_c \cap \mathbb{C}_i)} e^{ts}(\alpha - s)^{-2}\, ds_i\, S_R^{-1}(s,T)(\alpha\mathcal{I} - T)^2 u, \tag{7.3}$$

where W_c is the strip $W_c = \{s \in \mathbb{H} : -c < \mathrm{Re}(s) < c\}$ for $c > 0$ and we introduce the set $\partial(W_c \cap \mathbb{C}_i)$ for $i \in \mathbb{S}$. It consists of the two lines $s = c + i\tau$ and $s = -c - i\tau$, $\tau \in \mathbb{R}$, and their orientation is such that on $\mathbb{C}_i$ the orientation of $\partial(W_c \cap \mathbb{C}_i)$ is positive.

8. The W-functional calculus

This calculus has been introduced in [14] for monogenic functions here we reformulate it for the quaternionic setting. Using the Cauchy formula for slice hyperholomorphic functions it is possible to define an integral transforms that associate to a slice hyperholomorphic function a Fueter regular function, that are defined as:

Definition 8.1 (Fueter regular functions). Let U be an open set in $\mathbb{H}$. A real differentiable function $f : U \to \mathbb{H}$ is left Cauchy–Fueter (for brevity just Fueter) regular if

$$\frac{\partial}{\partial x_0} f(q) + e_1 \frac{\partial}{\partial x_1} f(q) + e_2 \frac{\partial}{\partial x_2} f(q) + e_3 \frac{\partial}{\partial x_3} f(q) = 0, \quad q \in U.$$

It is right Fueter regular if

$$\frac{\partial}{\partial x_0} f(q) + \frac{\partial}{\partial x_1} f(q) e_1 + \frac{\partial}{\partial x_2} f(q) e_2 + \frac{\partial}{\partial x_3} f(q) e_3 = 0, \quad q \in U.$$

Precisely, we use [44] to introduce an integral transform that associates to a slice hyperholomorphic functions a Fueter regular function of plane wave type. The following result is immediate, see [44] Section 1.1.

Proposition 8.2. *Suppose that the differentiable functions $(g_1, -g_2)$ satisfy the Cauchy–Riemann system in an open set of the complex plane identified with the set D of the pairs (u, p):*

$$\partial_u g_1(u,p) = -\partial_p g_2(u,p), \quad \partial_p g_1(u,p) = \partial_u g_2(u,p). \tag{8.1}$$

Let

$$U_D = \{x \in \mathbb{H} \ : \ x = u + \underline{\omega} p,\ (u,p) \in D,\ \underline{\omega} \in \mathbb{S}\}$$

and define the function $\tilde{G} : U_D \subseteq \mathbb{H} \to \mathbb{H}$

$$\tilde{G}(x) := g_1(u,p) - \underline{\omega}\, g_2(u,p). \tag{8.2}$$

Then $\tilde{G}(x)$ *is slice hyperholomorphic in* U_D.

When necessary, we will identify $\mathbb{H}$ with $\mathbb{R}^2 \times \mathbb{S}$ by setting $x \mapsto (x_0, p, \underline{\omega})$ and instead of $\tilde{G}(x)$ we will write $\tilde{G}(x_0, p, \underline{\omega})$ (keeping the symbol $\tilde{G}$ for the function).

Starting from the slice hyperholomorphic function $\tilde{G}(u, p, \underline{\omega})$ in (8.2) we can construct a Fueter regular function of plane wave type by replacing:

$$u = \langle \underline{x}, \underline{\omega} \rangle, \qquad p = x_0.$$

Suppose that the functions $(g_1, -g_2)$ satisfy the Cauchy–Riemann system and let us define the function

$$G(x_0, \langle \underline{x}, \underline{\omega} \rangle, \underline{\omega}) := g_1(\langle \underline{x}, \underline{\omega} \rangle, x_0) + \underline{\omega}\, g_2(\langle \underline{x}, \underline{\omega} \rangle, x_0), \quad \text{for } \underline{\omega} \in \mathbb{S}. \tag{8.3}$$

We recall a simple result stated in [44]:

Proposition 8.3. *The function G defined in* (8.3) *is left Fueter regular in the variable* $x = x_0 + \underline{x}$.

Definition 8.4. A function of the form (8.3) is called Fueter plane wave.

Definition 8.5 (The *W*-kernels). Let $S_L^{-1}(s,x)$, $S_R^{-1}(s,x)$ be the Cauchy kernels of left and of right slice hyperholomorphic functions, respectively, and let $\underline{\omega} \in \mathbb{S}$. For $\langle \underline{x}, \underline{\omega} \rangle - x_0 \underline{\omega} \notin [s]$ we define

$$\begin{aligned} W_{\underline{\omega}}^L(s,x) &:= S_L^{-1}(s, \langle \underline{x}, \underline{\omega} \rangle - x_0 \underline{\omega}) \\ &= -[(\langle \underline{x}, \underline{\omega} \rangle - x_0 \underline{\omega})^2 - 2s_0(\langle \underline{x}, \underline{\omega} \rangle - x_0 \underline{\omega}) + |s|^2]^{-1}(\langle \underline{x}, \underline{\omega} \rangle - x_0 \underline{\omega} - \overline{s}), \end{aligned}$$

and

$$\begin{aligned} W_{\underline{\omega}}^R(s,x) &:= S_R^{-1}(s, \langle \underline{x}, \underline{\omega} \rangle - x_0 \underline{\omega}) \\ &= -(\langle \underline{x}, \underline{\omega} \rangle - x_0 \underline{\omega} - \overline{s})[(\langle \underline{x}, \underline{\omega} \rangle - x_0 \underline{\omega})^2 - 2\mathrm{Re}(s)(\langle \underline{x}, \underline{\omega} \rangle - x_0 \underline{\omega}) + |s|^2]^{-1}, \end{aligned}$$

where $\underline{\omega} \in \mathbb{S}$ is considered as a parameter.

Observe that $W_{\underline{\omega}}^L$ and $W_{\underline{\omega}}^R$ are obtained by the change of variable $x \to \langle \underline{x}, \underline{\omega} \rangle - x_0 \underline{\omega}$ in the Cauchy kernels of slice hyperholomorphic functions and $\langle \underline{x}, \underline{\omega} \rangle - x_0 \underline{\omega}$ is still a paravector. We have the theorem:

Theorem 8.6. *Let $\underline{\omega} \in \mathbb{S}$ be a parameter and let $U \subset \mathbb{H}$ be an axially symmetric slice domain. Suppose that $\partial(U \cap \mathbb{C}_i)$ is a finite union of continuously differentiable Jordan curves for every $i \in \mathbb{S}$. Set $ds_i = -dsi$ for $I \in \mathbb{S}$. Assume that $\langle \underline{x}, \underline{\omega} \rangle - x_0 \underline{\omega} \in U$ and that O is an open set containing $\overline{U}$. Then the integrals*

$$\frac{1}{2\pi} \int_{\partial(U \cap \mathbb{C}_i)} W_{\underline{\omega}}^L(s,x)\, ds_i\, f(s), \quad \textit{for} \quad f \in \mathcal{SH}^L(O), \tag{8.4}$$

$$\frac{1}{2\pi} \int_{\partial(U \cap \mathbb{C}_i)} f(s)\, ds_i\, W_{\underline{\omega}}^R(s,x), \quad \textit{for} \quad f \in \mathcal{SH}^R(O), \tag{8.5}$$

depend neither on the open set U nor on the imaginary unit $i \in \mathbb{S}$.

Observe that thanks to the Cauchy formula for slice hyperholomorphic functions the integrals in (8.4) and (8.5) depend neither on the open set U nor on $i \in \mathbb{S}$.

Definition 8.7 (The W-transforms). Let $U \subset \mathbb{H}$ be an axially symmetric slice domain. Suppose that $\partial(U \cap \mathbb{C}_i)$ is a finite union of continuously differentiable Jordan curves for every $i \in \mathbb{S}$. Set $ds_i = -dsi$ for $i \in \mathbb{S}$. Assume that $\langle \underline{x}, \underline{\omega} \rangle - x_0 \underline{\omega} \in U$. If f is a (left) slice monogenic function on a set that contains $\overline{U}$ then we define the left W^L-transform as

$$\breve{f}_{\underline{\omega}}(x) = \frac{1}{2\pi} \int_{\partial(U\cap\mathbb{C}_i)} W^L_{\underline{\omega}}(s,x) ds_i f(s), \quad \textit{for} \quad f \in \mathcal{SH}^L(O). \tag{8.6}$$

If f is a right slice monogenic function then we define the right W^R-transform as

$$\breve{f}_{\underline{\omega}}(x) = \frac{1}{2\pi} \int_{\partial(U\cap\mathbb{C}_i)} f(s) ds_i W^R_{\underline{\omega}}(s,x), \quad \textit{for} \quad f \in \mathcal{SH}^R(O). \tag{8.7}$$

Here $\underline{\omega} \in \mathbb{S}$ is a parameter.

We observe that:

I) For every $\underline{\omega} \in \mathbb{S}$ the function $W^L_{\underline{\omega}}(s,x)$ is right slice hyperholomorphic in s and left Fueter regular in x for every x, s such that $(\langle \underline{x}, \underline{\omega} \rangle - x_0 \underline{\omega}) \notin [s]$. Moreover, the W^L-transform maps left slice hyperholomorphic functions f into left Fueter regular plane wave functions $\breve{f}_{\underline{\omega}}$.

II) For every $\underline{\omega} \in \mathbb{S}$ the function $W^R_{\underline{\omega}}(s,x)$ is left slice hyperholomorphic in s and right Fueter regular in x for every x, s such that $(\langle \underline{x}, \underline{\omega} \rangle - x_0 \underline{\omega}) \notin [s]$. Moreover, the W^R-transform maps right slice hyperholomorphic functions f into right Fueter regular plane wave functions $\breve{f}_{\underline{\omega}}$.

We now make a refinement of the above result in order to prepare the definition of a variation of the Fueter functional calculus, that is the functional calculus based of the Cauchy–Fueter formula, namely the W-functional calculus.

Theorem 8.8. *Let $T = T_0 + e_1T_1 + e_2T_2 + e_3T_3 \in \mathcal{B}(V)$ be such that for $\|T\| < |s|$ where $s \in \mathbb{H}$. Assume that $\underline{\omega} \in \mathbb{S}$ and define the operator*

$$A_{\underline{\omega}} := \sum_{j=1}^{3} T_j \omega_j - T_0 \underline{\omega}.$$

Then we have:

I) *$A_{\underline{\omega}}$ belongs to $\mathcal{B}(V)$.*

II) *The operator $A^2_{\underline{\omega}} - 2\mathrm{Re}(s) A_{\underline{\omega}} + |s|^2 \mathcal{I}$ is invertible for $\|T\| < |s|$ for all $\underline{\omega} \in \mathbb{S}$.*

III) *For $\|T\| < |s|$ for all $\underline{\omega} \in \mathbb{S}$ we have*

$$\sum_{m\geq 0} A^m_{\underline{\omega}} s^{-1-m} = -(A^2_{\underline{\omega}} - 2\mathrm{Re}(s) A_{\underline{\omega}} + |s|^2 \mathcal{I})^{-1} (A_{\underline{\omega}} - \overline{s}\mathcal{I}), \tag{8.8}$$

$$\sum_{m\geq 0} s^{-1-m} A^m_{\underline{\omega}} = -(A_{\underline{\omega}} - \overline{s}\mathcal{I})(A^2_{\underline{\omega}} - 2Re(s) A_{\underline{\omega}} + |s|^2 \mathcal{I})^{-1}. \tag{8.9}$$

The above theorem motivates the notion of W-spectrum.

Definition 8.9 (The W-spectrum and the W-resolvent set). Let $T \in \mathcal{B}(V)$ and let $\underline{\omega} \in \mathbb{S}$, we define the operators

$$A_{\underline{\omega}} = \sum_{j=1}^{3} T_j \omega_j - T_0 \underline{\omega}$$

and

$$Q_{\underline{\omega}}(T, s) := A_{\underline{\omega}}^2 - 2s_0 A_{\underline{\omega}} + |s|^2 \mathcal{I}.$$

We define the W-spectrum $\sigma_W(T)$ of T as:

$$\sigma_W(T, \underline{\omega}) = \{s \in \mathbb{R}^{n+1} \ : \ Q_{\underline{\omega}}(T, s) \quad \text{is not invertible in } \mathcal{B}(V)\}.$$

The W-resolvent set $\rho_W(T)$ is defined by

$$\rho_W(T, \underline{\omega}) = \mathbb{H} \setminus \sigma_W(T, \underline{\omega}).$$

The proofs of the following two results, which will be useful in the sequel, follow as in the case of the S-spectrum.

Theorem 8.10 (Structure of the W-spectrum). *Let $T \in \mathcal{B}(V)$, $\underline{\omega} \in \mathbb{S}$, and let $p = p_0 + p_1 j \in [p_0 + p_1 j] \subset \mathbb{H} \setminus \mathbb{R}$, such that $p \in \sigma_W(T, \underline{\omega})$. Then all the elements of the 2-sphere $[p_0 + p_1 j]$ belong to $\sigma_W(T, \underline{\omega})$. Thus the W-spectrum consists of real points and/or 2-spheres.*

Theorem 8.11 (Compactness of the W-spectrum). *Let $T \in \mathcal{B}(V)$, $\underline{\omega} \in \mathbb{S}$. Then the W-spectrum $\sigma_W(T, \underline{\omega})$ is a compact nonempty set.*

Definition 8.12 (W-resolvent operators). Let $T \in \mathcal{B}(V)$, $\underline{\omega} \in \mathbb{S}$ and $A_{\underline{\omega}} := \sum_{j=1}^{3} T_j \omega_j - T_0 \underline{\omega}$. For $s \in \rho_W(T)$ we define the left W-resolvent operator by

$$W_{\underline{\omega}}^L(s, T) = -(A_{\underline{\omega}}^2 - 2\mathrm{Re}(s) A_{\underline{\omega}} + |s|^2 \mathcal{I})^{-1} (A_{\underline{\omega}} - \overline{s} \mathcal{I}), \tag{8.10}$$

and the right W-resolvent operator by

$$W_{\underline{\omega}}^R(s, T) = -(A_{\underline{\omega}} - \overline{s} \mathcal{I})(A_{\underline{\omega}}^2 - 2Re(s) A_{\underline{\omega}} + |s|^2 \mathcal{I})^{-1}. \tag{8.11}$$

Definition 8.13. Let $T \in \mathcal{B}(V)$, $\underline{\omega} \in \mathbb{S}$ and let $U \subset \mathbb{H}$ be an axially symmetric slice domain.

- We say that U is admissible (for T) if it contains the W-spectrum $\sigma_W(T, \underline{\omega})$, and if $\partial(U \cap \mathbb{C}_i)$ is union of a finite number of rectifiable Jordan curves for every $i \in \mathbb{S}$.
- Let O be an open set in $\mathbb{H}$. A function $f \in \mathcal{SH}^L(O)$ (resp. right $f \in \mathcal{SH}^R(O)$) is said to be locally left (resp. right) slice hyperholomorphic on $\sigma_W(T, \underline{\omega})$ if there exists an admissible domain $U \subset \mathbb{H}$ such that $\overline{U} \subset O$.
- We will denote by $\mathcal{SH}^L_{\sigma_W(T,\underline{\omega})}$ (resp. $\mathcal{SH}^R_{\sigma_W(T,\underline{\omega})}$) the set of locally left (resp. right) slice hyperholomorphic functions on $\sigma_W(T, \underline{\omega})$.

The proof of the next result follows the proof of Theorem 8.6 and so it is based on the corresponding proof in the case of the S-functional calculus.

Theorem 8.14. *Let $T \in \mathcal{B}(V)$ and let $\underline{\omega} \in \mathbb{S}$. Let U be an admissible set for T and set $ds_i = ds/i$. Then the integrals*

$$\frac{1}{2\pi}\int_{\partial(U\cap\mathbb{C}_i)} W_{\underline{\omega}}^L(s,T)\, ds_i\, f(s), \qquad \textit{for} \qquad f \in \mathcal{SH}^L_{\sigma_W(T,\underline{\omega})} \tag{8.12}$$

and

$$\frac{1}{2\pi}\int_{\partial(U\cap\mathbb{C}_i)} f(s)\, ds_i\, W_{\underline{\omega}}^R(s,T), \qquad \textit{for} \qquad f \in \mathcal{SH}^R_{\sigma_W(T,\underline{\omega})} \tag{8.13}$$

depend neither on the open set U nor on the imaginary unit $i \in \mathbb{S}$.

The above theorem is important because it shows that the following definition of the W-functional calculus is well posed.

Definition 8.15 (The W-functional calculus for bounded operators). Let $T \in \mathcal{B}(V)$ and let $\underline{\omega} \in \mathbb{S}$. Let U be an admissible set for T and set $ds_i = ds/i$. We define the W-functional calculus for bounded operators as

$$\breve{f}_{\underline{\omega}}(T) = \frac{1}{2\pi}\int_{\partial(U\cap\mathbb{C}_i)} W_{\underline{\omega}}^L(s,T)\, ds_i\, f(s), \qquad \textit{for} \qquad f \in \mathcal{SH}^L_{\sigma_W(T,\underline{\omega})}, \tag{8.14}$$

and

$$\breve{f}_{\underline{\omega}}(T) = \frac{1}{2\pi}\int_{\partial(U\cap\mathbb{C}_i)} f(s)\, ds_i\, W_{\underline{\omega}}^R(s,T), \qquad \textit{for} \qquad f \in \mathcal{SH}^R_{\sigma_W(T,\underline{\omega})}. \tag{8.15}$$

The W-functional calculus is a monogenic functional calculus that is based on slice hyperholomorphic functions but it produces operators $\breve{f}_{\underline{\omega}}(T)$ where $\breve{f}_{\underline{\omega}}(s)$ is a Fueter regular function. The W-functional calculus, and the F-functional calculus treated in the next section, are monogenic, or better to say Fueter functional calculi when we deal with the quaternionic setting, because they are in the spirit of the monogenic functional calculus (based on the Cauchy formula of monogenic functions) introduced and studied by A. McIntosh and is collaborators in a series of papers [36, 37, 39] and the book [35].

9. The F-functional calculus

The Fueter mapping theorem is one of the deepest results in hypercomplex analysis, see [29]. It gives a procedure to generate Cauchy–Fueter regular functions starting from holomorphic functions of a complex variable. In the case of Clifford algebra-valued functions the proof of the analogue of the Fueter mapping theorem is due to Sce [42] for n odd and to Qian [41] for the general case.

Using the Cauchy formula for slice hyperholomorphic functions it is possible to give the Fueter mapping theorem an integral representation. This has been done in [20], and in the recent paper [13] the authors introduce the formulations of the F-functional calculus and the F-resolvent equation which are important to study the Riesz projectors associated to this calculus. The extension to unbounded operators has been done in [19]. Here we recall some facts for the case of bounded operators.

Definition 9.1 (The F-kernel). Let q, $s \in \mathbb{H}$. We define, for $s \notin [q]$, the F^L-kernel as

$$F^L(s,q) := -4(s-\bar{q})(s^2 - 2\mathrm{Re}(q)s + |q|^2)^{-2},$$

and the F^R-kernel as

$$F^R(s,q) := -4(s^2 - 2\mathrm{Re}(q)s + |q|^2)^{-2}(s-\bar{q}).$$

With the above notation the Fueter mapping theorem in integral form becomes:

Theorem 9.2 (The Fueter mapping theorem in integral form). *Set $ds_i = ds/i$ and let $W \subset \mathbb{H}$ be an open set. Let U be a bounded axially symmetric slice domain such that $\overline{U} \subset W$. Suppose that the boundary of $U \cap \mathbb{C}_i$ consists of a finite number of rectifiable Jordan curves for any $i \in \mathbb{S}$.*

(a) *If $q \in U$ and $f \in \mathcal{SH}^L(W)$ then $\breve{f}(q) = \Delta f(q)$ is left Fueter regular and it admits the integral representation*

$$\breve{f}(q) = \frac{1}{2\pi}\int_{\partial(U\cap\mathbb{C}_i)} F^L(s,q)ds_i f(s), \tag{9.1}$$

(b) *If $q \in U$ and $f \in \mathcal{SH}^R(W)$ then $\breve{f}(q) = \Delta f(q)$ is right Fueter regular and it admits the integral representation*

$$\breve{f}(q) = \frac{1}{2\pi}\int_{\partial(U\cap\mathbb{C}_i)} f(s)ds_i F^R(s,q). \tag{9.2}$$

The integrals depend neither on U and nor on the imaginary unit $i \in \mathbb{S}$.

We now consider the formulations of the F-functional calculus in the quaternionic setting for right linear quaternionic operators. The same formulation holds also for left linear operators with a suitable interpretation of the symbols.

Definition 9.3. We will denote by $\mathcal{BC}(V)$ the subclass of $\mathcal{B}(V)$ that consists of those quaternionic operators T that can be written as

$$T = T_0 + e_1T_1 + e_2T_2 + e_3T_3$$

where the operators T_ℓ, $\ell = 0, 1, 2, 3$ commute among themselves.

Definition 9.4 (The F-spectrum and the F-resolvent sets). Let $T \in \mathcal{BC}(V)$. We define the F-spectrum $\sigma_F(T)$ of T as

$$\sigma_F(T) = \{s \in \mathbb{H} : \ s^2\mathcal{I} - s(T+\overline{T}) + T\overline{T} \quad \text{is not invertible}\}.$$

The S-resolvent set $\rho_S(T)$ is defined as

$$\rho_F(T) = \mathbb{H} \setminus \sigma_F(T).$$

Theorem 9.5 (Structure of the F-spectrum). *Let $T \in \mathcal{BC}(V)$ and let $p = p_0 + p_1 I \in p_0 + p_1\mathbb{S} \subset \mathbb{H}\setminus\mathbb{R}$, such that $p \in \sigma_F(T)$. Then all the elements of the sphere $p_0 + p_1\mathbb{S}$ belong to $\sigma_F(T)$.*

We point out that the F-spectrum is well defined just for operators T that belong to $\mathcal{BC}(V)$. The F-spectrum can be seen as the commutative version of the S-spectrum because

$$\sigma_S(T) = \sigma_F(T), \quad \text{for all } \; T \in \mathcal{BC}(V).$$

In the case the components of T do not commute then it is not true that $T\overline{T} = \overline{T}T = T_0^2 + T_1^2 + T_2^2 + T_3^2$ and so the F-spectrum is not well defined.

As a consequence if $T \in \mathcal{BC}(V)$, then the F-spectrum $\sigma_F(T)$ is a compact nonempty set.

Definition 9.6 (F-resolvent operators). Let $T \in \mathcal{BC}(V)$. For $s \in \rho_F(T)$ we define the left F-resolvent operator as

$$F^L(s,T) := -4(s\mathcal{I} - \overline{T})(s^2\mathcal{I} - s(T + \overline{T}) + T\overline{T})^{-2},$$

and the right F-resolvent operator as

$$F^R(s,T) := -4(s^2\mathcal{I} - s(T + \overline{T}) + T\overline{T})^{-2}(s\mathcal{I} - \overline{T}).$$

The definition of T-admissible set U and of locally left (resp. right) slice hyperholomorphic functions on the F-spectrum $\sigma_F(T)$ can be obtained by rephrasing the definition for the S-spectrum.

We will denote by $\mathcal{SH}^L_{\sigma_F(T)}$ (resp. right $\mathcal{SH}^R_{\sigma_F(T)}$) the set of locally left (resp. right) slice hyperholomorphic functions on $\sigma_F(T)$.

Definition 9.7 (The quaternionic F-functional calculus for bounded operators). Let $T \in \mathcal{BC}(V)$ and set $ds_i = ds/i$, for $i \in \mathbb{S}$. We define the formulations of the quaternionic F-functional calculus as

$$\breve{f}(T) := \frac{1}{2\pi} \int_{\partial(U \cap \mathbb{C}_i)} F^L(s,T)\, ds_i\, f(s), \qquad f \in \mathcal{SH}^L_{\sigma_F(T)}, \tag{9.3}$$

and

$$\breve{f}(T) := \frac{1}{2\pi} \int_{\partial(U \cap \mathbb{C}_i)} f(s)\, ds_i\, F^R(s,T), \qquad f \in \mathcal{SH}^R_{\sigma_F(T)}, \tag{9.4}$$

where U is T-admissible.

The definition of the quaternionic F-functional calculus for bounded operators is well posed since it was proved that the integrals (9.3) and (9.4) depend neither on the open set U (that contains the F-spectrum) nor on the imaginary unit $i \in \mathbb{S}$.

Theorem 9.8 (The quaternionic F-resolvent equation). *Let $T \in \mathcal{BC}(V)$. Then for $p, s \in \rho_F(T)$ the following equation holds*

$$\begin{aligned}
&F^R(s,T)S^{-1}_{C,L}(p,T) + S^{-1}_{C,R}(s,T)F^L(p,T) + \frac{1}{4}\Big(sF^R(s,T)F^L(p,T)p \\
&\quad - sF^R(s,T)TF^L(p,T) - F^R(s,T)TF^L(p,T)p + F^R(s,T)T^2F^L(p,T)\Big) \\
&\quad - \left[\left(F^R(s,T) - F^L(p,T)\right)p - \bar{s}\left(F^R(s,T) - F^L(p,T)\right)\right]\left(p^2 - 2s_0p + |s|^2\right)^{-1}.
\end{aligned}$$

where the quaternionic SC-resolvent operators are defined as

$$S_{C,L}^{-1}(s,T) := (s\mathcal{I} - \overline{T})(s^2\mathcal{I} - s(T+\overline{T}) + T\overline{T})^{-1}, \quad s \in \rho_F(T) \tag{9.5}$$

and

$$S_{C,R}^{-1}(s,T) := (s^2\mathcal{I} - s(T+\overline{T}) + T\overline{T})^{-1}(s\mathcal{I} - \overline{T}), \quad s \in \rho_F(T). \tag{9.6}$$

As a consequence of the quaternionic F-resolvent equations we can study the Riesz projectors associated with the quaternionic F-functional calculus.

Theorem 9.9. *Let $T \in \mathcal{BC}(V)$. Let $\sigma_F(T) = \sigma_{1F}(T) \cup \sigma_{2F}(T)$, with*

$$\text{dist}\,(\sigma_{1F}(T), \sigma_{2F}(T)) > 0.$$

Let U_1 and U_2 be two T-admissible sets such that $\sigma_{1F}(T) \subset U_1$ and $\sigma_{2F}(T) \subset U_2$, with $\overline{U}_1 \cap \overline{U}_2 = \emptyset$. Set

$$\breve{P}_\ell := \frac{C}{2\pi} \int_{\partial(U_\ell \cap \mathbb{C}_i)} F^L(s,T)\, ds_i s^2, \qquad \ell = 1,2,$$

where $C := \Delta q^2$. Then, for $\ell = 1,2$, the following properties hold:

(1) $\breve{P}_\ell^2 = \breve{P}_\ell$,
(2) $T\breve{P}_\ell = \breve{P}_\ell T, \quad \ell = 1,2.$

Even thought the F-resolvent equation seems to be very complicated it is allows one to prove that $\breve{P}_\ell^2 = \breve{P}_\ell$, for $\ell = 1,2$. We cannot expect a simpler F-resolvent equation because the F-functional calculus is based on an integral transform and not on a Cauchy formula.

References

[1] S. Adler, *Quaternionic Quantum Field Theory*, Oxford University Press, 1995.

[2] D. Albrecht, X. Duong, A. McIntosh, *Operator theory and harmonic analysis.* Instructional Workshop on Analysis and Geometry, Part III (Canberra, 1995), 77–136, Proc. Centre Math. Appl. Austral. Nat. Univ., 34, Austral. Nat. Univ., Canberra, (1996).

[3] D. Alpay, F. Colombo, T. Qian, I. Sabadini, *The H^∞ functional calculus based on the S-spectrum for quaternionic operators and for n-tuples of noncommuting operators*, Journal of Functional Analalysis, **271** (2016), 1544–1584.

[4] D. Alpay, F. Colombo, I. Sabadini, *Slice Hyperholomorphic Schur Analysis*, Operator Theory: Advances and Applications, Vol. 256 (2017) Springer Basel, 365pp.

[5] D. Alpay, F. Colombo, I. Sabadini, *Perturbation of the generator of a quaternionic evolution operator*, Anal. Appl. (Singap.), **13** (2015), 347–370.

[6] D. Alpay, F. Colombo, J. Gantner, D.P. Kimsey, *Functions of the infinitesimal generator of a strongly continuous quaternionic group*, arXiv:1502.02954, to appear in Anal. Appl. (Singap.) (2016), doi.org/10.1142/S021953051650007X.

[7] D. Alpay, F. Colombo, J. Gantner, I. Sabadini, *A new resolvent equation for the S-functional calculus*, J. Geom. Anal., **25** (2015), 1939–1968.

[8] D. Alpay, F. Colombo, D.P. Kimsey, *The spectral theorem for quaternionic unbounded normal operators based on the S-spectrum*, J. Mathematical Phys., **57** (2016), no. 2, 023503, 27 pp.

[9] D. Alpay, F. Colombo, D.P. Kimsey, I. Sabadini. *An extension of Herglotz's theorem to the quaternions*, J. Math. Anal. Appl., **421** (2015), no. 1, 754–778.

[10] D. Alpay, F. Colombo, D.P. Kimsey, I. Sabadini. *The spectral theorem for unitary operators based on the S-spectrum*, Milan J. Math., **84** (2016), 41–61.

[11] G. Birkhoff, J. von Neumann, *The logic of quantum mechanics*, Ann. of Math., **37** (1936), 823–843.

[12] J. L. Brenner, *Matrices of quaternions*, Pacific J. Math., **1** (1951), 329–335.

[13] F. Colombo, J. Gantner, *Formulations of the F-functional calculus and some consequences*, Proc. Roy. Soc. Edinburgh Sect. A, **146** (2016), 509–545.

[14] F. Colombo, R. Lávička, I. Sabadini, V. Souček, *Monogenic plane waves and the W-functional calculus*, Math. Methods Appl. Sci., **3** (2016), 412–424.

[15] F. Colombo, I. Sabadini, *On some properties of the quaternionic functional calculus*, J. Geom. Anal., **19** (2009), 601–627.

[16] F. Colombo, I. Sabadini, *On the formulations of the quaternionic functional calculus*, J. Geom. Phys., **60** (2010), 1490–1508.

[17] F. Colombo, I. Sabadini, *The quaternionic evolution operator*, Adv. Math., **227** (2011), 1772–1805.

[18] F. Colombo, I. Sabadini, *A structure formula for slice monogenic functions and some of its consequences*, Hypercomplex analysis, 101–114, Trends Math., Birkhäuser Verlag, Basel, 2009.

[19] F. Colombo, I. Sabadini, *The F-functional calculus for unbounded operators*, J. Geom. Phys., **86** (2014), 392–407.

[20] F. Colombo, I. Sabadini, F. Sommen, *The Fueter mapping theorem in integral form and the F-functional calculus*, Math. Methods Appl. Sci., **33** (2010), 2050–2066.

[21] F. Colombo, I. Sabadini, D.C. Struppa, *Noncommutative functional calculus. Theory and applications of slice regular functions*, volume 289 of *Progress in Mathematics*. Birkhäuser/Springer Basel AG, Basel, 2011.

[22] F. Colombo, J. Gantner, *On power series expansions of the S-resolvent operator and the Taylor formula*, to appear in Journal of Geometry and Physics. doi.org/10.1016/j.geomphys.2016.07.018.

[23] F. Colombo, J. Gantner, *Fractional powers of quaternionic operators and Kato's formula using slice hyperholomorphicity*, to appear in: Trans. Amer. Math. Soc., doi.org/10.1090/tran/7013.

[24] N. Dunford, J. Schwartz. *Linear Operators, part I: General Theory*, J. Wiley and Sons (1988).

[25] N. Dunford, J. Schwartz. *Linear Operators, part II: Spectral theory*, J. Wiley and Sons (1988).

[26] G. Emch, *Mécanique quantique quaternionienne et relativité restreinte*, I, Helv. Phys. Acta, **36** (1963), 739–769.

[27] D.R. Farenick, B.A.F. Pidkowich, *The spectral theorem in quaternions*, Linear Algebra Appl., **371** (2003), 75–102.

[28] D. Finkelstein, J.M. Jauch, S. Schiminovich, D. Speiser, *Foundations of quaternion quantum mechanics*, J. Mathematical Phys., **3** (1962), 207–220.

[29] R. Fueter, *Die Funktionentheorie der Differentialgleichungen $\Delta u = 0$ und $\Delta\Delta u = 0$ mit vier reellen Variablen*, Comm. Math. Helv., **7** (1934), 307–330.

[30] G. Gentili, C. Stoppato, D.C. Struppa, *Regular functions of a quaternionic variable.* Springer Monographs in Mathematics. Springer, Heidelberg, 2013.

[31] R. Ghiloni, V. Moretti, A. Perotti, *Continuous slice functional calculus in the quaternionic Hilbert spaces*, Rev. Math. Phys., **25** (2013), 1350006, 83 pp.

[32] R. Ghiloni, V. Moretti, A. Perotti, *Spectral properties of compact normal quaternionic operators*, in Hypercomplex Analysis: New Perspectives and Applications Trends in Mathematics, 133–143, (2014).

[33] R. Ghiloni, V. Recupero, *Semigroups over real alternative *-algebras: generation theorems and spherical sectorial operators*, Trans. Amer. Math. Soc. ISSN 0002-9947 (in press).

[34] L.P. Horwitz, L.C. Biedenharn, *Quaternion quantum mechanics: Second quantization and gauge fields*, Annals of Physics, **157** (1984), 432–488.

[35] B. Jefferies, *Spectral properties of noncommuting operators*, Lecture Notes in Mathematics, 1843, Springer-Verlag, Berlin, 2004.

[36] B. Jefferies, A. McIntosh, *The Weyl calculus and Clifford analysis*, Bull. Austral. Math. Soc., **57** (1998), 329–341.

[37] B. Jefferies, A. McIntosh, J. Picton-Warlow, *The monogenic functional calculus*, Studia Math., **136** (1999), 99–119.

[38] H.C. Lee, *Eigenvalues and canonical forms of matrices with quaternion coefficients*, Proc. Roy. Irish Acad. Sci. A, **52** (1949), 253–260.

[39] A. McIntosh, A. Pryde, *A functional calculus for several commuting operators*, Indiana Univ. Math. J., **36** (1987), 421–439.

[40] A. McIntosh, *Operators which have an H^∞ functional calculus.* Miniconference on operator theory and partial differential equations (North Ryde, 1986), 210–231, Proc. Centre Math. Anal. Austral. Nat. Univ., 14, Austral. Nat. Univ., Canberra, (1986).

[41] T. Qian, *Generalization of Fueter's result to $\mathbb{R}^{n+1}$*, Rend. Mat. Acc. Lincei, **8** (1997), 111–117.

[42] M. Sce, *Osservazioni sulle serie di potenze nei moduli quadratici*, Atti Acc. Lincei Rend. Fisica, **23** (1957), 220–225.

[43] C.S. Sharma, T.J. Coulson, *Spectral theory for unitary operators on a quaternionic Hilbert space*, J. Math. Phys., **28** (1987), 1941–1946.

[44] F. Sommen, *Clifford Analysis and integral geometry*, Clifford algebras and their applications in mathematical physics (Montpellier, 1989), 293–311, Fund. Theories Phys., 47, Kluwer Acad. Publ., Dordrecht, 1992.

[45] K. Viswanath, *Normal operators on quaternionic Hilbert spaces*, Trans. Amer. Math. Soc., **162** (1971), 337–350.

[46] F. Zhang, *Quaternions and matrices of quaternions*, Linear Algebra Appl., **251** (1997), 21–57.

Fabrizio Colombo
Politecnico di Milano
Dipartimento di Matematica
Via E. Bonardi, 9
I-20133 Milano, Italy
e-mail: `fabrizio.colombo@polimi.it`

David P. Kimsey
Newcastle University
School of Mathematics and Statistics
Newcastle upon Tyne
NE1 7RU, United Kingdom
e-mail: `dpkimsey@gmail.com`

Modern Trends in Hypercomplex Analysis
Trends in Mathematics, 143–159

Models for Some Irreducible Representations of $\mathfrak{so}(m,\mathbb{C})$ in Discrete Clifford Analysis

Hilde De Ridder and Tim Raeymaekers

Abstract. In this paper we work in the 'split' discrete Clifford analysis setting, i.e., the m-dimensional function theory concerning null-functions of the discrete Dirac operator ∂, defined on the grid $\mathbb{Z}^m$, involving both forward and backward differences. This Dirac operator factorizes the (discrete) Star-Laplacian ($\Delta^* = \partial^2$). We show how the space $\mathcal{H}_k$ of discrete k-homogeneous spherical harmonics, which is a reducible $\mathfrak{so}(m,\mathbb{C})$-representation, may explicitly be decomposed into 2^{2m} isomorphic copies of irreducible $\mathfrak{so}(m,\mathbb{C})$-representations with highest weight $(k,0,\ldots,0)$. The key element is the introduction of 2^{2m} idempotents, dividing the discrete Clifford algebra in 2^{2m} subalgebras of dimension $\binom{k+m-1}{k} - \binom{k+m-3}{k}$.

Mathematics Subject Classification (2010). 17B15, 47A67, 20G05, 15A66, 39A12.

Keywords. Discrete Clifford analysis, irreducible representation, orthogonal Lie algebra, harmonic functions.

In classical Clifford analysis, the infinitesimal 'rotations' are given by the angular momentum operators, in our function theoretical setting denoted by the differential operators $L_{a,b} = x_a\partial_{x_b} - x_b\partial_{x_a}$. These operators satisfy the commutation relations

$$[L_{a,b}, L_{c,d}] = \delta_{b,c}\, L_{a,d} - \delta_{b,d}\, L_{a,c} - \delta_{a,c}\, L_{b,d} + \delta_{a,d}\, L_{b,c},$$

which are the defining relations of the orthogonal Lie algebra $\mathfrak{so}(m)$. Since these are endomorphisms of the space $\mathcal{H}_k(m,\mathbb{C})$ of scalar-valued harmonic polynomials homogeneous of degree k, this polynomial space is a model for an (irreducible) $\mathfrak{so}(m,\mathbb{C})$-representation [see, e.g., [9, 1]]. Classically, to establish $\mathcal{M}_k$, the space of spinor-valued monogenics, homogeneous of degree k, as $\mathfrak{so}(m,\mathbb{C})$-representation, the following operators are considered

$$dR(e_{a,b}) : \mathcal{M}_k \to \mathcal{M}_k, \qquad M_k \mapsto \left(L_{a,b} + \frac{1}{2}\, e_a\, e_b\right) M_k.$$

The first author acknowledges the support of the Research Foundation – Flanders (FWO), grant no. FWO13\PDO\039.

These operators are endomorphisms of $\mathcal{M}_k$ which also satisfy the defining relations of $\mathfrak{so}(m,\mathbb{C})$:

$$[dR(e_{a,b}), dR(e_{c,d})] = \delta_{b,c}\, dR(e_{a,d}) - \delta_{b,d}\, dR(e_{a,c}) - \delta_{a,c}\, dR(e_{b,d}) + \delta_{a,d}\, dR(e_{b,c}).$$

In [5], we developed discrete counterparts of the operators $L_{a,b}$ and $dR(e_{a,b})$ in the discrete Clifford analysis setting.

Definition 1. The (discrete) angular momentum operators are discrete operators $L_{a,b} = \xi_a\, \partial_b + \xi_b\, \partial_a$, $1 \leqslant a \neq b \leqslant m$, acting on the discrete functions. Here ξ_a and ∂_a, $a = 1, \ldots, m$ are the discrete vector variables and co-ordinate differences in the discrete Clifford analysis setting (see Section 1). For $a = b$, we define $L_{a,a} = 0$. Furthermore, let the operator $\Omega_{a,b}$ act on discrete functions f as $\Omega_{a,b}\, f = L_{a,b}\, f\, e_b\, e_a$.

The discrete angular momentum operators also satisfy the defining relations of the orthogonal Lie algebra $\mathfrak{so}(m)$ (see, e.g., [11]):

$$[\Omega_{a,b}, \Omega_{c,d}] = \delta_{b,c}\, \Omega_{a,d} - \delta_{b,d}\, \Omega_{a,c} - \delta_{a,c}\, \Omega_{b,d} + \delta_{a,d}\, \Omega_{b,c}.$$

Furthermore, they are endomorphisms of the space $\mathcal{H}_k$ of Clifford algebra-valued k-homogeneous harmonics since $\Omega_{a,b}$ commutes with $\mathfrak{sl}_2 = \{\Delta, \xi^2, \mathbb{E} + \frac{m}{2}\}$, for all (a, b). We thus concluded that $\mathcal{H}_k$ is a representation of $\mathfrak{so}(m,\mathbb{C})$; however, this is not an irreducible representation, as will be shown in the following sections.

An important difference with the Euclidean Clifford setting is the addition of the basis elements $e_b\, e_a$ to the right of the considered function f. This will have consequences later on in this paper, when we describe irreducible representations by means of an idempotent; the action of $\mathfrak{so}(m,\mathbb{C})$ elements will affect this idempotent, in the sense that the representation can no longer be interpreted as a left ideal in the Clifford algebra. Another unexpected result was the possibility to rotate points of the grid $\mathbb{Z}^m$ over all real angles by rotation of the discrete delta functions (resp. distributions); we are thus not longer restricted to rotating over (integer multiples of) right angles.

In a similar manner, discrete operators $dR(e_{a,b})$ were constructed in [5], satisfying the defining relations of the orthogonal lie algebra $\mathfrak{so}(m)$ and commuting with $\mathfrak{osp}(1|2) = \{\partial, \xi, \mathbb{E} + \frac{m}{2}\}$ which makes them endomorphisms of the space $\mathcal{M}_k$ of k-homogeneous discrete monogenic polynomials. As such, $\mathcal{M}_k$ is a reducible $\mathfrak{so}(m,\mathbb{C})$-representation. The decomposition of $\mathcal{M}_k$ into irreducible representations will be the topic of an upcoming paper.

Describing the discrete harmonic spaces $\mathcal{H}_k$ as $\mathfrak{so}(m,\mathbb{C})$-representations will be most effective (from a representation-theoretic point of view) when the representations are irreducible. Only then will we be able to draw conclusions about for example Gelfand–Tsetlin bases (see, e.g., [12]). An accurate description of the decomposition is thus very important, and this will be done for $\mathcal{H}_k$ in the following sections.

To keep this paper self-contained, we start with a preliminary section on the discrete Clifford analysis framework. Throughout this paper, we will use concepts regarding the Cartan subalgebra and the positive roots of the orthogonal Lie algebra $\mathfrak{so}(m,\mathbb{C})$, hence in Section 2, we briefly summarize these concepts, based on [8]. In Section 3, we describe the decomposition of $\mathcal{H}_k$ in irreducible components, both for even and for odd dimension.

1. Preliminaries

Let $\mathbb{R}^m$ be the m-dimensional Euclidian space with orthonormal basis e_j, $j = 1,\ldots,m$ and consider the Clifford algebra $\mathbb{R}_{m,0}$ over $\mathbb{R}^m$, governed by the multiplication relations $e_ie_j + e_je_i = 2\delta_{ij}$. Passing to the so-called 'split' discrete setting [6, 2], we embed this Clifford algebra into the bigger complex one $\mathbb{C}_{2m}$, the underlying vector space of which has twice the dimension, and introduce forward and backward basis elements $\mathbf{e}_j^{\pm}$ satisfying the following anti-commutator rules:

$$\{\mathbf{e}_j^-,\mathbf{e}_\ell^-\} = \{\mathbf{e}_j^+,\mathbf{e}_\ell^+\} = 0, \qquad \{\mathbf{e}_j^+,\mathbf{e}_\ell^-\} = \delta_{j\ell}, \qquad j,\ell = 1,\ldots,m.$$

The connection to the original basis e_j is given by $\mathbf{e}_j^+ + \mathbf{e}_j^- = e_j$, $j = 1,\ldots,m$. This implies $e_j^2 = 1$, in contrast to the usual Clifford setting where traditionally $e_j^2 = -1$ is chosen.

Now consider the standard equidistant lattice $\mathbb{Z}^m$; the coordinates of a Clifford vector $\underline{x}$ will thus only take integer values. We construct a discrete Dirac operator factorizing the discrete Laplacian, using both forward and backward differences $\Delta_j^{\pm}$, $j = 1,\ldots,m$, acting on Clifford-valued functions f as follows:

$$\Delta_j^+[f](\underline{x}) = f(\underline{x}+\mathbf{e}_j) - f(\underline{x}), \qquad \Delta_j^-[f](\underline{x}) = f(\underline{x}) - f(\underline{x}-\mathbf{e}_j).$$

With respect to the $\mathbb{Z}^m$-grid, the usual definition of the discrete Laplacian in $\underline{x} \in \mathbb{Z}^m$ is

$$\Delta^*[f](\underline{x}) = \sum_{j=1}^m \Delta_j^+\Delta_j^-[f](\underline{x}) = \sum_{j=1}^m (f(\underline{x}+\mathbf{e}_j) + f(\underline{x}-\mathbf{e}_j)) - 2m\, f(\underline{x}).$$

This operator is also known as "Star-Laplacian", denoted from now on as Δ. An appropriate definition (see, e.g., [6, 7]) of a discrete Dirac operator ∂ factorizing Δ, i.e., satisfying $\partial^2 = \Delta$, is obtained by combining the forward and backward basis elements with the corresponding forward and backward differences, more precisely

$$\partial = \sum_{j=1}^m \left(\mathbf{e}_j^+\Delta_j^+ + \mathbf{e}_j^-\Delta_j^-\right).$$

In order to receive an analogue of the classical Weyl relations $\partial_{x_j}x_k - x_k\partial_{x_j} = \delta_{jk}$, the coordinate vector variable operators $\xi_j = \mathbf{e}_j^+X_j^- + \mathbf{e}_j^-X_j^+$ are defined by their interaction with the corresponding coordinate operators $\partial_j = \mathbf{e}_j^+\Delta_j^+ + \mathbf{e}_j^-\Delta_j^-$, according to the skew Weyl relations, cf. [2]:

$$\partial_j\,\xi_j - \xi_j\,\partial_j = 1,\ j = 1,\ldots,m,$$

which imply that $\partial_j \xi_j^k[1] = k\,\xi_j^{k-1}[1]$. The operators ξ_j and ∂_j furthermore satisfy the following anti-commutator relations:

$$\{\xi_j, \xi_k\} = \{\partial_j, \partial_k\} = \{\partial_j, \xi_k\} = 0, \qquad j \neq k,\ j,k = 1,\dots,m$$

implying that $\partial_\ell\, \xi_j^k[1] = 0$, $j \neq \ell$.

The natural powers $\xi_j^k[1]$ of the operator ξ_j acting on the ground state 1 are the basic discrete k-homogeneous polynomials of degree k in the variable x_j, i.e., $\mathbb{E}\,\xi_j^k[1] = k\,\xi_j^k[1]$, where $\mathbb{E} = \sum_{j=1}^m \xi_j\,\partial_j$ is the discrete Euler operator. They constitute a basis for all discrete polynomials. Explicit formulas for $\xi_j^k[1]$ are given for example in [2, 3]; furthermore $\xi_j^k[1](x_j) = 0$ if $k \geqslant 2\,|x_j| + 1$.

A discrete function is discrete harmonic in a domain $\Omega \subset \mathbb{Z}^m$ if $\Delta f(\underline{x}) = 0$ for all $\underline{x} \in \Omega$. The space of discrete harmonic homogeneous polynomials of degree k is denoted $\mathcal{H}_k$, while the space of all discrete harmonic homogeneous polynomials is denoted $\mathcal{H}$. It is clear that $\mathcal{H} = \bigoplus_{k=0}^{\infty} \mathcal{H}_k$. The dimensions over the discrete Clifford algebra are

$$\dim(\mathcal{H}_k) = \binom{k+m-1}{k} - \binom{k+m-3}{k}.$$

2. Orthogonal Lie algebras

We will start by briefly introducing the orthogonal Lie algebra $\mathfrak{so}(m,\mathbb{C})$; a detailed description can be found for example in, e.g., [8]. The orthogonal Lie algebra $\mathfrak{so}(m,\mathbb{C})$ is generated in even dimension $m = 2n$ by $\frac{m(m-1)}{2}$ basis elements H_a, $X_{a,b}$, $Y_{a,b}$ and $Z_{a,b}$ $(1 \leqslant a,b \leqslant n)$ and in odd dimension $m = 2n+1$ these basis elements are extended to a full basis of $\mathfrak{so}(m,\mathbb{C})$ by $2n$ extra elements U_j and V_j, $1 \leqslant j \leqslant n$:

$$\begin{aligned}\mathfrak{so}(2n,\mathbb{C}) &= \operatorname{span}_{\mathbb{C}}\{H_a, X_{a,b}, Y_{a,b}, Z_{a,b}, 1 \leqslant a,b \leqslant n,\ a \neq b\},\\ \mathfrak{so}(2n+1,\mathbb{C}) &= \operatorname{span}_{\mathbb{C}}\{H_a, X_{a,b}, Y_{a,b}, Z_{a,b}, U_a, V_a, 1 \leqslant a,b \leqslant n,\ a \neq b\}.\end{aligned}$$

These basis elements are the root vectors of the adjoint representation of $\mathfrak{so}(m,\mathbb{C})$ in accordance to [8, 10].

The Cartan subalgebra can be chosen as $\mathfrak{h} = \{H_a, 1 \leqslant a \leqslant n\}$, independent of the parity of the dimension, i.e., $\mathfrak{so}(2n,\mathbb{C})$ and $\mathfrak{so}(2n+1,\mathbb{C})$ are both Lie algebras of rank n. The roots of $\mathfrak{so}(m,\mathbb{C})$ (see also [11]) are determined by considering the adjoint representation:

$$\begin{aligned}[H_s, Y_{a,b}] &= (\delta_{sa} + \delta_{sb})\,Y_{a,b} = ((L_a + L_b)\,(H_s))\,Y_{a,b},\\ [H_s, X_{a,b}] &= (\delta_{sa} - \delta_{sb})\,X_{a,b} = ((L_a - L_b)\,(H_s))\,X_{a,b},\\ [H_s, Z_{a,b}] &= -\,(\delta_{sa} + \delta_{sb})\,Z_{a,b} = ((-L_a - L_b)\,(H_s))\,Z_{a,b},\\ [H_s, U_a] &= \delta_{sa}\,U_a = (L_a(H_s))\,U_a,\\ [H_s, V_a] &= -\delta_{sa}\,U_a = (-L_a(H_s))\,U_a,\end{aligned}$$

where $\{L_a, 1 \leqslant a \leqslant n\}$ is a basis of the dual vector space $\mathfrak{h}^*$ of the Cartan subalgebra $\mathfrak{h}$, i.e., $L_a(H_b) = \delta_{a,b}$. Note in particular that the Cartan subalgebra elements H_a can be calculated by taking the commutator of a positive root vector with its corresponding negative root vector:

$$[Y_{a,b}, Z_{a,b}] = -H_a - H_b, \qquad [X_{a,b}, X_{b,a}] = H_a - H_b.$$

We thus deduce the following roots and root vectors.

$m = 2n$

root	root vector
$L_a - L_b$	$X_{a,b}$
$L_a + L_b$	$Y_{a,b}$
$-L_a - L_b$	$Z_{a,b}$

$m = 2n + 1$

root	root vector
$L_a - L_b$	$X_{a,b}$
$L_a + L_b$	$Y_{a,b}$
$-L_a - L_b$	$Z_{a,b}$
L_a	U_a
$-L_a$	V_a

To make a distinction between positive and negative roots, we consider the linear functional $l : \mathfrak{h}^* \to \mathbb{R}$ defined by fixing n different real numbers c_i such that for all $a_i \in \mathbb{R}$:

$$l(a_1 L_1 + \cdots + a_n L_n) = a_1 c_1 + \cdots + a_n c_n.$$

We choose the constants c_i such that the ordering $c_1 > c_2 > \cdots > c_n > 0$ is satisfied. With this convention, the positive roots in even dimension, i.e., roots α for which $l(\alpha) > 0$, are given by

$$\{L_a + L_b : 1 \leqslant a \neq b \leqslant n\} \cup \{L_a - L_b : 1 \leqslant a < b \leqslant n\}.$$

The negative roots, i.e., roots α for which $l(\alpha) < 0$, are given by

$$\{-L_a - L_b : 1 \leqslant a \neq b \leqslant n\} \cup \{L_a - L_b : 1 \leqslant b < a \leqslant n\}.$$

In odd dimension, one finds positive roots

$$\{L_a + L_b : 1 \leqslant a \neq b \leqslant n\} \cup \{L_a - L_b : 1 \leqslant a < b \leqslant n) \cup \{L_a : 1 \leqslant a \leqslant n\}$$

and negative roots

$$\{-L_a - L_b : 1 \leqslant a \neq b \leqslant n\} \cup \{L_a - L_b : 1 \leqslant b < a \leqslant n) \cup \{-L_a : 1 \leqslant a \leqslant n\}.$$

As illustrative examples, we depict the root diagrams of $\mathfrak{so}(6, \mathbb{C})$:

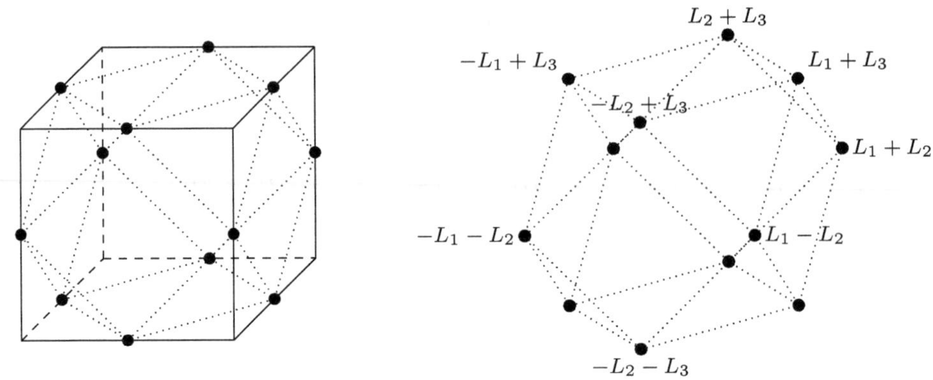

and of $\mathfrak{so}(7,\mathbb{C})$:

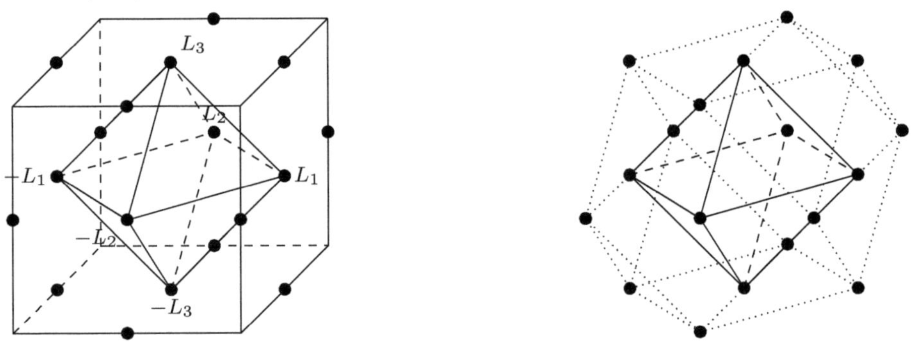

In [5], we introduced the algebra $\mathfrak{so}(m,\mathbb{C})$ (up to an isomorphism) in the discrete Clifford analysis context. The generators of $\mathfrak{so}(m,\mathbb{C})$ were not given in terms of the root vectors and Cartan subalgebra, but rather by the generators $\{\Omega_{a,b} : 1 \leqslant a \neq b \leqslant m\}$ introduced in Definition 1, satisfying the defining relations of $\mathfrak{so}(m,\mathbb{C})$:

$$[\Omega_{a,b}, \Omega_{c,d}] = \delta_{a,d}\,\Omega_{b,c} + \delta_{b,c}\,\Omega_{a,d} - \delta_{a,c}\,\Omega_{b,d} - \delta_{b,d}\,\Omega_{a,c}. \tag{2.1}$$

In the following sections, we will re-establish the orthogonal Lie algebra in the discrete Clifford analysis setting, but now by determining the explicit expressions of the root vectors and Cartan subalgebra.

3. Decomposition of $\mathcal{H}_k$ in irreducible representations

The space of discrete harmonic Clifford-valued homogeneous polynomials $\mathcal{H}_k$ is a representation for $\mathfrak{so}(m,\mathbb{C})$. To see this, we again consider the operators $\Omega_{a,b} : \mathcal{H}_k \to \mathcal{H}_k$:

$$\Omega_{a,b}(H_k) = L_{a,b}\,H_k\,e_b\,e_a = (\xi_b\,\partial_a + \xi_a\,\partial_b)\,H_k\,e_b\,e_a, \quad a \neq b \qquad \text{and} \qquad \Omega_{a,a} = 0.$$

By calculating the dimension, we immediately may conclude that this representation is not just a model for the irreducible representation with highest weight $(k, 0, \ldots, 0)$ like the classical case:

$$\dim_{\mathbb{C}}(\mathcal{H}_k) = \dim_{\mathbb{C}_{2m}}(\mathcal{H}_k) \dim_{\mathbb{C}}(\mathbb{C}_{2m}) = 2^{2m}\left(\binom{k+m-1}{k} - \binom{k+m-3}{k}\right),$$

while

$$\dim_{\mathbb{C}}(k, 0, \ldots, 0) = \left(\binom{k+m-1}{k} - \binom{k+m-3}{k}\right).$$

Here $(k, 0, \ldots, 0)$ represents the irreducible representation of $\mathfrak{so}(m, \mathbb{C})$ with highest weight $(k, 0, \ldots, 0)$. This means that $\mathcal{H}_k$ is probably reducible. The remainder of this article exactly deals with the decomposition of this space into irreducible representations.

Classically, one considers the scalar-valued harmonic polynomials as an irreducible representation of $\mathfrak{so}(m, \mathbb{C})$ within the space of Clifford-valued harmonic polynomials. However, note that in the discrete setting, due to the addition of the basis elements $e_b\, e_a$ in the definition of the operators $\Omega_{a,b}$, and the fact that $L_{a,b}$ itself is not scalar, the operators $\Omega_{a,b}$ are no longer scalar. Hence the subspace of $\mathcal{H}_k$ of scalar harmonics, i.e., harmonic functions that have scalar Taylor series coefficients, is *not* an invariant under their action. To arrive at irreducible representations of $\mathfrak{so}(m, \mathbb{C})$ within the space of *Clifford-valued* discrete harmonics $\mathcal{H}_k$, we must thus reconsider our approach. We will do this by introducing an appropriate idempotent with which we multiply our harmonics from the right. To determine which idempotent is appropriate, we first consider the analogues of the root system of $\mathfrak{so}(m, \mathbb{C})$ in our discrete Clifford setting.

3.1. Even dimension $m = 2n$

Definition 2. We define the operators H_a, $X_{a,b}$, $Y_{a,b}$ and $Z_{a,b} \in \mathfrak{so}(m, \mathbb{C})$:

$$\begin{aligned}
H_a &= i\,\Omega_{2a-1,2a}, \qquad 1 \leqslant a \leqslant n,\\
X_{a,b} &= \frac{1}{2}\left(\Omega_{2a-1,2b-1} + i\,\Omega_{2a-1,2b} - i\,\Omega_{2a,2b-1} + \Omega_{2a,2b}\right),\\
Y_{a,b} &= \frac{1}{2}\left(\Omega_{2a-1,2b-1} - i\,\Omega_{2a-1,2b} - i\,\Omega_{2a,2b-1} - \Omega_{2a,2b}\right),\\
Z_{a,b} &= \frac{1}{2}\left(\Omega_{2a-1,2b-1} + i\,\Omega_{2a-1,2b} + i\,\Omega_{2a,2b-1} - \Omega_{2a,2b}\right), \qquad 1 \leqslant a, b \leqslant n.
\end{aligned}$$

Note that, because $\Omega_{a,b} = -\Omega_{b,a}$, we find that $Y_{b,a} = -Y_{a,b}$. Furthermore, $\Omega_{a,a} = 0$ implies that $Y_{a,a} = 0$. The same holds for $Z_{a,b}$. For $X_{a,b}$, we find that $X_{b,a} \neq X_{a,b}$ and that $X_{a,a} = H_a$ hence we consider all couples (a, b) with $a \neq b$.

Remark 3. Note that we can reconstruct the original operators $\Omega_{a,b}$ as $\Omega_{2a-1,2a} = -i\,H_a$ and for $a \neq b$:

$$\begin{aligned} 2\,\Omega_{2a-1,2b-1} &= X_{a,b} - X_{b,a} + Y_{a,b} + Z_{a,b}, \\ -2i\,\Omega_{2a,2b-1} &= X_{a,b} + X_{b,a} + Y_{a,b} - Z_{a,b}, \\ 2i\,\Omega_{2a-1,2b} &= X_{a,b} + X_{b,a} - Y_{a,b} + Z_{a,b}, \\ 2\,\Omega_{2a,2b} &= X_{a,b} - X_{b,a} - Y_{a,b} - Z_{a,b}. \end{aligned}$$

We will now show that these operators indeed show the expected commutator relations associated with the root system of $\mathfrak{so}(m, \mathbb{C})$:

Lemma 4. *The following commutator relations hold:*

$$\begin{aligned} [H_j, Y_{a,b}] &= (\delta_{ja} + \delta_{jb})\,Y_{a,b} = (L_a + L_b)\,(H_j)\,Y_{a,b}, \\ [H_j, X_{a,b}] &= (\delta_{ja} - \delta_{jb})\,X_{a,b} = (L_a - L_b)\,(H_j)\,X_{a,b}, \\ [H_j, Z_{a,b}] &= -\,(\delta_{ja} + \delta_{jb})\,Z_{a,b} = -\,(L_a + L_b)\,(H_j)\,Z_{a,b}, \\ [Y_{a,b}, Z_{c,d}] &= \delta_{ad}\,X_{b,c} + \delta_{bc}\,X_{a,d} - \delta_{ac}\,X_{b,d} - \delta_{bd}\,X_{a,c}, \\ [X_{a,b}, X_{c,d}] &= \delta_{bc}\,X_{a,d} - \delta_{ad}\,X_{c,b}, \\ [X_{a,b}, Y_{c,d}] &= \delta_{bc}\,Y_{a,d} - \delta_{bd}\,Y_{a,c}, \\ [X_{a,b}, Z_{c,d}] &= \delta_{ad}\,Z_{b,c} - \delta_{ac}\,Z_{b,d}, \\ [Y_{a,b}, Y_{c,d}] &= 0, \\ [Z_{a,b}, Z_{c,d}] &= 0. \end{aligned}$$

In particular, $X_{a,b}$*,* $a < b$ *resp.* $Y_{a,b}$ *are root vectors corresponding to the positive roots* $L_a - L_b$*, resp.* $L_a + L_b$*. Furthermore,* $X_{a,b}$ *with* $a > b$ *and* $Z_{a,b}$ *are root vectors corresponding to the negative roots* $L_a - L_b$ *resp.* $-L_a - L_b$*.*

Proof. We will only write down the commutator relation $[H_j, Y_{a,b}]$ and $[X_{a,b}, Y_{c,d}]$ here, the other ones can be proven in a similar fashion.

$$[H_j, Y_{a,b}] = \frac{1}{2}\left[i\,\Omega_{2j-1,2j}, \Omega_{2a-1,2b-1} - i\,\Omega_{2a-1,2b} - i\,\Omega_{2a,2b-1} - \Omega_{2a,2b}\right].$$

Applying the commutator rule (2.1) results in:

$$\begin{aligned} [H_j, Y_{a,b}] &= \frac{i}{2}\,(\delta_{jb}\,\Omega_{2j,2a-1} - \delta_{ja}\,\Omega_{2j,2b-1}) + \frac{1}{2}\,(-\delta_{jb}\,\Omega_{2j-1,2a-1} - \delta_{ja}\,\Omega_{2j,2b}) \\ &\quad + \frac{1}{2}\,(\delta_{ja}\,\Omega_{2j-1,2b-1} + \delta_{jb}\,\Omega_{2j,2a}) - \frac{i}{2}\,(-\delta_{jb}\,\Omega_{2j-1,2a} + \delta_{ja}\,\Omega_{2j-1,2b}) \\ &= \delta_{ja}\,Y_{j,b} + \delta_{jb}\,Y_{a,j} = (\delta_{ja} + \delta_{jb})\,Y_{a,b}. \end{aligned}$$

For the second statement, we again apply (2.1):

$$\begin{aligned} 4\,[X_{a,b}, Y_{c,d}] &= 2\,\delta_{bc}\,(\Omega_{2a-1,2d-1} - i\,\Omega_{2a-1,2d} - i\,\Omega_{2a,2d-1} - \Omega_{2a,2d}) \\ &\quad - 2\,\delta_{bd}\,(\Omega_{2a-1,2c-1} - i\,\Omega_{2a-1,2c} - \Omega_{2a,2c} - i\,\Omega_{2a,2c-1}) \\ &= 2\,\delta_{bc}\,Y_{a,d} - 2\,\delta_{bd}\,Y_{a,c}. \end{aligned}$$

□

In this discrete setting, the Cartan subalgebra $\mathfrak{h} \subset \mathfrak{so}(m, \mathbb{C})$ is given by

$$\mathfrak{h} = \{H_a,\ 1 \leqslant a \leqslant n\}.$$

As the Cartan elements mutually commute, their action on any representation of $\mathfrak{so}(m, \mathbb{C})$ can be diagonalized simultaneously, since $\mathfrak{so}(m, \mathbb{C})$ is semi-simple, $m > 2$. Any finite-dimensional representation $\mathbb{V}_\mu$ of the Lie algebra $\mathfrak{so}(m, \mathbb{C})$ may thus be decomposed as eigenspaces for the subalgebra $\mathfrak{h}$. The set of n eigenvalues of such an eigenspace is also known as the weight of the considered eigenspace and the eigenspace itself is called weight space. We may decompose $\mathbb{V}_\mu$ according to a finite set of weights W:

$$\mathbb{V}_\mu = \bigoplus_{\lambda \in W} V_\lambda$$

where $V_\lambda = \{P \in \mathbb{V}_\mu : H_a\, P = \ell_i\, P,\ 1 \leqslant a \leqslant n\}$, for all $\lambda = (\ell_1, \ldots, \ell_n) \in W$.

In particular, we consider the decomposition of the representation $\mathcal{H}_k$. If we are thus to introduce an idempotent I and a harmonic polynomial P_k such that the space $\operatorname{span}_{\mathbb{C}} \{P_k\, I\}$ is a weight space of a representation of the orthogonal algebra with weight $(\ell_1, \ldots, \ell_n)$, it must certainly hold that $H_a\, P_k\, I = \ell_a\, P_k\, I$. Therefor, the necessary idempotents must satisfy

$$\forall\, a = 1, \ldots, n, \quad \exists\, c_a \in \mathbb{C}: \qquad I\, e_{2a-1}\, e_{2a} = c_a\, I.$$

Definition 5. For $1 \leqslant s \leqslant n$, we denote the following Clifford elements

$$\begin{aligned} I_{2s-1}^{\pm} &= \left(\mathbf{e}_{2s-1}^{+}\mathbf{e}_{2s-1}^{-} \pm \mathbf{e}_{2s-1}^{+}\right), & I_{2s}^{\pm} &= \left(\mathbf{e}_{2s}^{+}\mathbf{e}_{2s}^{-} \pm i\, \mathbf{e}_{2s}^{+}\right), \\ K_{2s-1}^{\pm} &= \left(\mathbf{e}_{2s-1}^{-}\mathbf{e}_{2s-1}^{+} \pm \mathbf{e}_{2s-1}^{-}\right), & K_{2s}^{\pm} &= \left(\mathbf{e}_{2s}^{-}\mathbf{e}_{2s}^{+} \pm i\, \mathbf{e}_{2s}^{-}\right). \end{aligned}$$

Lemma 6. *Let* $I_s = I_{2s-1}^{+} I_{2s}^{-} = \left(\mathbf{e}_{2s-1}^{+}\mathbf{e}_{2s-1}^{-} + \mathbf{e}_{2s-1}^{+}\right)\left(\mathbf{e}_{2s}^{+}\mathbf{e}_{2s}^{-} - i\mathbf{e}_{2s}^{+}\right)$ *for* $1 \leqslant s \leqslant n$. *Then the Clifford element*

$$I = \prod_{s=1}^{n} I_s = \prod_{s=1}^{n} \left(\mathbf{e}_{2s-1}^{+}\mathbf{e}_{2s-1}^{-} + \mathbf{e}_{2s-1}^{+}\right)\left(\mathbf{e}_{2s}^{+}\mathbf{e}_{2s}^{-} - i\, \mathbf{e}_{2s}^{+}\right)$$

is an idempotent ($I^2 = I$) *and it satisfies*

$$I\, e_{2s-1} e_{2s} = i\, I, \quad \forall\, 1 \leqslant s \leqslant n.$$

Proof. Note that

$$\begin{aligned} I_{2s-1}^{\pm}\, e_{2s-1} &= \left(\mathbf{e}_{2s-1}^{+} \pm \mathbf{e}_{2s-1}^{+}\mathbf{e}_{2s-1}^{-}\right) = \pm\left(\mathbf{e}_{2s-1}^{+}\mathbf{e}_{2s-1}^{-} \pm \mathbf{e}_{2s-1}^{+}\right) = \pm\, I_{2s-1}^{\pm}, \\ K_{2s-1}^{\pm}\, e_{2s-1} &= \left(\mathbf{e}_{2s-1}^{-} \pm \mathbf{e}_{2s-1}^{-}\mathbf{e}_{2s-1}^{+}\right) = \pm\left(\mathbf{e}_{2s-1}^{-}\mathbf{e}_{2s-1}^{+} \pm \mathbf{e}_{2s-1}^{-}\right) = \pm\, K_{2s-1}^{\pm}, \\ I_{2s}^{\pm}\, e_{2s} &= \left(\mathbf{e}_{2s}^{+} \pm i\, \mathbf{e}_{2s}^{+}\mathbf{e}_{2s}^{-}\right) = \pm i\left(\mathbf{e}_{2s}^{+}\mathbf{e}_{2s}^{-} \mp i\, \mathbf{e}_{2s}^{+}\right) = \pm i\, I_{2s}^{\mp}, \\ K_{2s}^{\pm}\, e_{2s} &= \left(\mathbf{e}_{2s}^{-} \pm i\, \mathbf{e}_{2s}^{-}\mathbf{e}_{2s}^{+}\right) = \pm i\left(\mathbf{e}_{2s}^{-}\mathbf{e}_{2s}^{+} \mp i\, \mathbf{e}_{2s}^{-}\right) = \pm i\, K_{2s}^{\mp}. \end{aligned}$$

We start with the second statement. We choose $s = 1$ (the general proof is similar). Since $I_s^{\pm}\, e_k = e_k\, I_s^{\mp}$, $\forall s \neq k$, the element $e_1 e_2$ commutes with all I_s,

$s = 2, \dots, n$:

$$\begin{aligned} I\, e_1 e_2 &= I_1^+\, I_2^-\, e_1\, e_2 \prod_{s=2}^{n} (I_{2s-1}^+\, I_{2s}^-) = I_1^+\, e_1\, I_2^+\, e_2 \prod_{s=2}^{n} (I_{2s-1}^+\, I_{2s}^-) \\ &= i\, I_1^+\, I_2^- \prod_{s=2}^{n} (I_{2s-1}^+\, I_{2s}^-) \\ &= i\, I. \end{aligned}$$

To check the idempotency, we consider:

$$I^2 = \left(\prod_{s=1}^{n} I_s \right)^2 = \left(\prod_{s=1}^{n} \left(\mathbf{e}_{2s-1}^+ \mathbf{e}_{2s-1}^- + \mathbf{e}_{2s-1}^+ \right) \left(\mathbf{e}_{2s}^+ \mathbf{e}_{2s}^- - i\, \mathbf{e}_{2s}^+ \right) \right)^2 .$$

For $s = 1$, the factor $I_1^+ = \left(\mathbf{e}_1^+ \mathbf{e}_1^- + \mathbf{e}_1^+ \right)$ has a term $\mathbf{e}_1^+ \mathbf{e}_1^-$ that commutes with all $I_j^\pm$, $j \neq 1$, and a term $\mathbf{e}_1^+$ that does not. Indeed, $\mathbf{e}_1^+\, I_s^\pm = I_s^\mp\, \mathbf{e}_1^+$, $s \neq 1$. However, as both terms will eventually be multiplied with I_1^+ from the right and as $\mathbf{e}_1^+\, I_1^+ = \mathbf{e}_1^+ \left(\mathbf{e}_1^+ \mathbf{e}_1^- + \mathbf{e}_1^+ \right) = 0$ (because of the isotropy of $\mathbf{e}_1^+$), the $\mathbf{e}_1^+$-term automatically vanishes. We only need to determine the result of the commuting part:

$$\begin{aligned} I^2 &= \prod_{p=1}^{n} \left(\mathbf{e}_{2p-1}^+ \mathbf{e}_{2p-1}^-\, \mathbf{e}_{2p}^+ \mathbf{e}_{2p}^- \right) \prod_{s=1}^{n} \left(\mathbf{e}_{2s-1}^+ \mathbf{e}_{2s-1}^- + \mathbf{e}_{2s-1}^+ \right) \left(\mathbf{e}_{2s}^+ \mathbf{e}_{2s}^- - i\, \mathbf{e}_{2s}^+ \right) \\ &= \prod_{s=1}^{n} \mathbf{e}_{2s-1}^+ \mathbf{e}_{2s-1}^- \left(\mathbf{e}_{2s-1}^+ \mathbf{e}_{2s-1}^- + \mathbf{e}_{2s-1}^+ \right) \mathbf{e}_{2s}^+ \mathbf{e}_{2s}^- \left(\mathbf{e}_{2s}^+ \mathbf{e}_{2s}^- - i\, \mathbf{e}_{2s}^+ \right) \\ &= \prod_{s=1}^{n} \left(\mathbf{e}_{2s-1}^+ \mathbf{e}_{2s-1}^- + \mathbf{e}_{2s-1}^+ \right) \left(\mathbf{e}_{2s}^+ \mathbf{e}_{2s}^- - i\, \mathbf{e}_{2s}^+ \right) \\ &= I. \end{aligned}$$

□

Lemma 7. *If we replace any (or multiple) elements I_{2s-1}^+ with I_{2s-1}^-, K_{2s-1}^+ or K_{2s-1}^-, the resulting Clifford element is still an idempotent. The same holds if we replace any (or multiple) elements I_{2s}^- by I_{2s}^+ or $K_{2s}^\pm$. Furthermore*

$$\begin{aligned} I_{2s-1}^\pm I_{2s}^\pm\, e_{2s-1}\, e_{2s} &= \pm(\mp i)\, I_{2s-1}^\pm I_{2s}^\pm, \\ I_{2s-1}^\pm\, K_{2s}^\pm\, e_{2s-1}\, e_{2s} &= \pm(\mp i)\, I_{2s-1}^\pm\, K_{2s}^\pm, \\ K_{2s-1}^\pm\, I_{2s}^\pm\, e_{2s-1}\, e_{2s} &= \pm(\mp i)\, K_{2s-1}^\pm\, I_{2s}^\pm, \\ K_{2s-1}^\pm\, K_{2s}^\pm\, e_{2s-1}\, e_{2s} &= \pm(\mp i)\, K_{2s-1}^\pm\, K_{2s}^\pm. \end{aligned}$$

Proof. We here only consider the combination of $I_{2s-1}^\pm$ with $I_{2s}^\pm$. The proofs of the other combinations are similar. So let

$$I = \prod_{s=1}^{n} I_{2s-1}^\pm I_{2s}^\pm = \prod_{s=1}^{n} \left(\mathbf{e}_{2s-1}^+ \mathbf{e}_{2s-1}^- \pm \mathbf{e}_{2s-1}^+ \right) \left(\mathbf{e}_{2s}^+ \mathbf{e}_{2s}^- \pm i\, \mathbf{e}_{2s}^+ \right).$$

Again because of the isotropy of $\mathbf{e}_i^+$ (and of $\mathbf{e}_i^-$), in each case, we only need to consider the commuting part:

$$\begin{aligned} I^2 &= \prod_{p=1}^{n} \left(\mathbf{e}_{2p-1}^+\mathbf{e}_{2p-1}^- \, \mathbf{e}_{2p}^+\mathbf{e}_{2p}^-\right) \prod_{s=1}^{n} \left(\mathbf{e}_{2s-1}^+\mathbf{e}_{2s-1}^- \pm \mathbf{e}_{2s-1}^+\right) \left(\mathbf{e}_{2s}^+\mathbf{e}_{2s}^- \pm i\, \mathbf{e}_{2s}^+\right) \\ &= \prod_{s=1}^{n} \mathbf{e}_{2s-1}^+\mathbf{e}_{2s-1}^- \left(\mathbf{e}_{2s-1}^+\mathbf{e}_{2s-1}^- \pm \mathbf{e}_{2s-1}^+\right) \mathbf{e}_{2s}^+\mathbf{e}_{2s}^- \left(\mathbf{e}_{2s}^+\mathbf{e}_{2s}^- \pm i\, \mathbf{e}_{2s}^+\right) \\ &= \prod_{s=1}^{n} \left(\mathbf{e}_{2s-1}^+\mathbf{e}_{2s-1}^- \pm \mathbf{e}_{2s-1}^+\right) \left(\mathbf{e}_{2s}^+\mathbf{e}_{2s}^- \pm i\, \mathbf{e}_{2s}^+\right) = I. \end{aligned}$$

Furthermore $I_1^\pm I_2^\pm e_1 e_2 = I_1^\pm e_1 I_2^\mp e_2 = \pm(\mp i)\, I_1^\pm I_2^\pm$ and similar for other $1 \leqslant s \leqslant n$. □

Remark 8. By combining the different idempotents, we can form all basis elements of the Clifford algebra. Indeed

$$\begin{aligned} I_{2s-1}^+ + I_{2s-1}^- &= \left(\mathbf{e}_{2s-1}^+\mathbf{e}_{2s-1}^- + \mathbf{e}_{2s-1}^+\right) + \left(\mathbf{e}_{2s-1}^+\mathbf{e}_{2s-1}^- - \mathbf{e}_{2s-1}^+\right) = 2\, \mathbf{e}_{2s-1}^+\mathbf{e}_{2s-1}^-, \\ K_{2s-1}^+ + K_{2s-1}^- &= \left(\mathbf{e}_{2s-1}^-\mathbf{e}_{2s-1}^+ + \mathbf{e}_{2s-1}^-\right) + \left(\mathbf{e}_{2s-1}^-\mathbf{e}_{2s-1}^+ - \mathbf{e}_{2s-1}^-\right) = 2\, \mathbf{e}_{2s-1}^-\mathbf{e}_{2s-1}^+, \\ I_{2s-1}^+ - I_{2s-1}^- &= \left(\mathbf{e}_{2s-1}^+\mathbf{e}_{2s-1}^- + \mathbf{e}_{2s-1}^+\right) - \left(\mathbf{e}_{2s-1}^+\mathbf{e}_{2s-1}^- - \mathbf{e}_{2s-1}^+\right) = 2\, \mathbf{e}_{2s-1}^+, \\ K_{2s-1}^+ - K_{2s-1}^- &= \left(\mathbf{e}_{2s-1}^-\mathbf{e}_{2s-1}^+ + \mathbf{e}_{2s-1}^-\right) - \left(\mathbf{e}_{2s-1}^-\mathbf{e}_{2s-1}^+ - \mathbf{e}_{2s-1}^-\right) = 2\, \mathbf{e}_{2s-1}^- \end{aligned}$$

and, since $1 = \mathbf{e}_{2s-1}^+\mathbf{e}_{2s-1}^- + \mathbf{e}_{2s-1}^-\mathbf{e}_{2s-1}^+$, we can also produce scalars. Similar combinations are used to arrive at $\mathbf{e}_{2s}^+\mathbf{e}_{2s}^-$, $\mathbf{e}_{2s}^-\mathbf{e}_{2s}^+$ and $\mathbf{e}_{2s}^\pm$.

We find that $\mathcal{H}_k$ decomposes into the direct sum of $4^m = 2^{2m}$ subspaces $\mathcal{H}_k\, I$, where the idempotent I runs over all idempotents mentioned above. In the following sections, we will always denote the idempotent

$$\prod_{s=1}^{n} \left(\mathbf{e}_{2s-1}^+\mathbf{e}_{2s-1}^- + \mathbf{e}_{2s-1}^+\right) \left(\mathbf{e}_{2s}^+\mathbf{e}_{2s}^- - i\, \mathbf{e}_{2s}^+\right)$$

by I.

We already established that $\mathcal{H}_{2k}$ is a representation of $\mathfrak{so}(2n, \mathbb{C})$. This representation is reducible. A mutual eigenspace of all generators of $\mathfrak{h}$ is given by $\operatorname{span}_{\mathbb{C}} \{f_{2k}\, I\}$ where

$$f_{2k} = \left((\xi_2 + \xi_1)(\xi_2 - \xi_1)\right)^k .$$

If we consider the representation $\mathcal{H}_{2k+1}$, then the corresponding weight space of all elements of the Cartan subalgebra is given by $\operatorname{span}_{\mathbb{C}} \{f_{2k+1}\, I\}$ where

$$f_{2k+1} = (\xi_2 + \xi_1)\left((\xi_2 - \xi_1)(\xi_2 + \xi_1)\right)^k .$$

Lemma 9. *The subspace* $\operatorname{span}_{\mathbb{C}} \{f_{2k}\, I\}$ *resp.* $\operatorname{span}_{\mathbb{C}} \{f_{2k+1}\, I\}$ *is an eigenspace of all generators of* $\mathfrak{h}$*, and can hence be seen as (part of) a weight space of an* $\mathfrak{so}(m, \mathbb{C})$*-representation with weight* $(2k, 0, \ldots, 0)$*, resp.* $(2k+1, 0, \ldots, 0)$*.*

Proof. Since the generating elements f_{2k} and f_{2k+1} only contain ξ_1 and ξ_2, we only need to consider $H_1 = i\,\Omega_{1,2}$. The generating elements will automatically vanish under the action of the other $\mathfrak{h}$-elements. It was previously established (see [4], proof of Prop. 2) that

$$\begin{aligned}
\partial_1 f_{2k} = \partial_2\, f_{2k} &= 2k\,(\xi_2 - \xi_1)\,((\xi_2 + \xi_1)\,(\xi_2 - \xi_1))^{k-1}\,,\\
\partial_1 f_{2k+1} = \partial_2 f_{2k+1} &= (2k+1)\,((\xi_2 - \xi_1)\,(\xi_2 + \xi_1))^{k}\,.
\end{aligned}$$

From this, the action of $H_1 = i\,\Omega_{12}$ follows easily:

$$\begin{aligned}
H_1\, f_{2k}\, I &= i\,(\xi_1\,\partial_2 + \xi_2\,\partial_1)\, f_{2k}\, I\, e_2 e_1\\
&= i\,(2k)\,(\xi_1 + \xi_2)\,(\xi_2 - \xi_1)\,((\xi_2 + \xi_1)\,(\xi_2 - \xi_1))^{k-1}\,(-i\,I)\\
&= (2k)\,((\xi_2 + \xi_1)\,(\xi_2 + \xi_1))^{k}\, I\\
&= 2k\, f_{2k}\, I,
\end{aligned}$$

while

$$\begin{aligned}
H_1\, f_{2k+1}\, I &= i\,(\xi_1\,\partial_2 + \xi_2\,\partial_1)\, f_{2k+1}\, I\, e_2 e_1\\
&= i\,(2k+1)\,(\xi_1 + \xi_2)\,((\xi_2 - \xi_1)\,(\xi_2 + \xi_1))^{k}\,(-i\,I)\\
&= (2k+1)\, f_{2k+1}\, I.
\end{aligned}$$

□

Lemma 10. *The* 1*-dimensional spaces* $\mathrm{span}_{\mathbb{C}}\,\{f_{2k}\, I\}$ *and* $\mathrm{span}_{\mathbb{C}}\,\{f_{2k+1}\, I\}$ *vanish under the action of the positive roots, i.e.,*

$$\begin{aligned}
Y_{a,b}\,(f_{2k}\, I) &= 0,\ \forall (a,b),\ a \neq b,\\
X_{a,b}\,(f_{2k}\, I) &= 0,\ \forall (a,b),\ a < b,\\
Y_{a,b}\,(f_{2k+1}\, I) &= 0,\ \forall (a,b),\ a \neq b,\\
X_{a,b}\,(f_{2k+1}\, I) &= 0,\ \forall (a,b),\ a < b.
\end{aligned}$$

Proof. Since f_{2k} and f_{2k+1} only contain ξ_2 and ξ_1, we only need to consider (a,b) with $b = 1$ and $a > 1$:

$$\begin{aligned}
2\,Y_{a,1}\,(f_{2k}\, I) = {} & (\Omega_{2a-1,1} - i\,\Omega_{2a-1,2} - i\,\Omega_{2a,1} - \Omega_{2a,2})\, f_{2k}\, I\\
= {} & \xi_{2a-1}\,(\partial_1 f_{2k})\, I\, e_1 e_{2a-1} - i\,\xi_{2a-1}\,(\partial_2 f_{2k})\, I\, e_2 e_{2a-1}\\
& - i\,\xi_{2a}\,(\partial_1 f_{2k})\, I\, e_1 e_{2a} - \xi_{2a}\,(\partial_2 f_{2k})\, I\, e_2 e_{2a}.
\end{aligned}$$

We thus consider

$$\begin{aligned}
I\, e_1 e_{2a} &= I_1^+\, I_2^- \ldots I_{2a-1}^+\, I_{2a}^-\; e_1\, e_{2a}\; I_{2a+1}^+\, I_{2a+2}^- \cdots I_{2n-1}^+ I_{2n}^-\\
&= I_1^+\, e_1\; I_2^+\, I_3^- \ldots I_{2a-2}^+\, I_{2a-1}^-\, I_{2a}^+\, e_{2a}\; I_{2a+1}^+\, I_{2a+2}^- \cdots I_{2n-1}^+ I_{2n}^-\\
&= i\, I_1^+\, I_2^+\, I_3^- \ldots I_{2a-2}^+\, I_{2a-1}^-\, I_{2a}^-\, I_{2a+1}^+\, I_{2a+2}^- \cdots I_{2n-1}^+ I_{2n}^-.
\end{aligned}$$

Denote for now

$$I^{1,a} = I_1^+ \quad \overbrace{I_2^+\, I_3^- \ldots I_{2a-2}^+\, I_{2a-1}^-}^{\text{changed sign of factor } 2\,\ldots\,2a-1} \quad I_{2a}^-\, I_{2a+1}^+\, I_{2a+2}^- \cdots I_{2n-1}^+ I_{2n}^-.$$

Then $I\, e_1\, e_{2a} = i\, I^{1,a}$ and

$$\begin{aligned}
I\, e_1\, e_{2a-1} &= I_1^+\, I_2^- \dots I_{2a-1}^+\, e_1\, e_{2a-1}\, I_{2a}^- \dots I_{2n-1}^+\, I_{2n}^- \\
&= I_1^+\, e_1\, I_2^+ \dots I_{2a-1}^-\, e_{2a-1}\, I_{2a}^- \dots I_{2n-1}^+\, I_{2n}^- \\
&= -\, I_1^+\, I_2^+ \dots I_{2a-2}^+\, I_{2a-1}^-\, I_{2a}^- \dots I_{2n-1}^+\, I_{2n}^- \\
&= -I^{1,a}, \\
I\, e_2\, e_{2a} &= I_1^+\, I_2^- \dots I_{2a-1}^+\, I_{2a}^-\, e_2\, e_{2a}\, I_{2a+1}^+ \dots I_{2n-1}^+\, I_{2n}^- \\
&= I_1^+\, I_2^-\, e_2\, I_3^- \dots I_{2a-1}^-\, I_{2a}^+\, e_{2a}\, I_{2a+1}^+ \dots I_{2n-1}^+\, I_{2n}^- \\
&= (-i)\, i\, I_1^+\, I_2^+\, I_3^- \dots I_{2a-1}^-\, I_{2a}^-\, I_{2a+1}^+ \dots\, I_{2n-1}^+\, I_{2n}^- \\
&= I^{1,a}, \\
I\, e_2\, e_{2a-1} &= I_1^+\, I_2^- \dots I_{2a-1}^+\, e_2\, e_{2a-1}\, I_{2a}^- \dots I_{2n-1}^+\, I_{2n}^- \\
&= I_1^+\, I_2^-\, e_2\, I_3^- \dots I_{2a-1}^-\, e_{2a-1}\, I_{2a}^- \dots I_{2n-1}^+\, I_{2n}^- \\
&= (-1)\,(-i)\, I_1^+\, I_2^+\, I_3^- \dots I_{2a-2}^+\, I_{2a-1}^-\, I_{2a}^- \dots I_{2n-1}^+\, I_{2n}^- \\
&= i\, I^{1,a}.
\end{aligned}$$

We recall that $\partial_1 f_{2k} = \partial_2 f_{2k}$ and combining all this, we find that

$$\begin{aligned}
2\, Y_{a,1}\,(f_{2k}\, I) = &-\xi_{2a-1}\,(\partial_1 f_{2k})\, I^{1,a} - i\, \xi_{2a-1}\,(\partial_1 f_{2k})\,\big(i\, I^{1,a}\big) \\
&- i\, \xi_{2a}\,(\partial_1 f_{2k})\,\big(i\, I^{1,a}\big) - \xi_{2a}\,(\partial_1 f_{2k})\, I^{1,a} = 0.
\end{aligned}$$

The case of f_{2k+1} is completely similar.

For $X_{a,b}$ with $a < b$ we only need to consider the case where $a = 1$ and $1 < b$:

$$\begin{aligned}
2\, X_{1,b}\,(f_{2k}\, I) = &\;(\Omega_{1,2b-1} + i\, \Omega_{1,2b} - i\, \Omega_{2,2b-1} + \Omega_{2,2b})\, f_{2k}\, I \\
= &\;\xi_{2b-1}\, \partial_1 f_{2k}\, I\, e_{2b-1} e_1 + i\, \xi_{2b}\, \partial_1 f_{2k}\, I\, e_{2b} e_1 \\
&- i\, \xi_{2b-1}\, \partial_2 f_{2k}\, I\, e_{2b-1} e_2 + \xi_{2b}\, \partial_2 f_{2k}\, I\, e_{2b} e_2 \\
= &\;\xi_{2b-1}\, \partial_1 f_{2k}\, I^{1,b} - i^2\, \xi_{2b}\, \partial_1 f_{2k}\, I^{1,b} \\
&+ (-i)^2\, \xi_{2b-1}\, \partial_1 f_{2k}\, I^{1,b} - \xi_{2b}\, \partial_1 f_{2k}\, I^{1,b} = 0.
\end{aligned}$$

The case of f_{2k+1} is again completely similar. □

Corollary 11. *The space* $\mathrm{span}_{\mathbb{C}}\,\{f_k\, I\}$ *is a 1-dimensional highest weight space with weight* $(k, 0, \dots, 0)$. *As such, it generates an irreducible representation of* $\mathfrak{so}(m, \mathbb{C})$, *see for example* [8].

We will from now on denote $(k) = (k, 0, \dots, 0)$.

Remark 12. The space $\mathcal{H}_k\, I$ is not a left $\mathfrak{so}(m, \mathbb{C})$-module, i.e., the image of the space $\mathcal{H}_k\, I$ under the action of a rotations $\Omega_{a,b}$ does not belong to $\mathcal{H}_k\, I$, but to some $\mathcal{H}_k\, J$ with J a different idempotent. As such, direct calculations with the given irreducible representations may become somewhat trickier (although not impossible) than in the classical Clifford setting.

Each space $\text{span}_{\mathbb{C}} \{f_k J\}$, where J runs over all possible idempotents such that the highest weight is (k), generates an independent isomorphic irreducible representation since the highest weight space of such an irreducible representation is one-dimensional. The element $f_k J$ of weight (k) can thus not be found in the irreducible representation spanned by another element $f_k J'$ (with J' a different idempotent).

Note that not every combination of f_k with an idempotent J delivers the weight vector (k). Half of all idempotents J delivers a weight space $\text{span}_{\mathbb{C}} \{f_k J\}$ with weight (k); the other half delivers a weight space $\text{span}_{\mathbb{C}} \{f_k J\}$ with weight $(-k)$ (which is in fact a lowest weight space, see remark below). However, in those cases, the vector $g_k J$ where we denoted

$$g_{2k} = ((\xi_2 - \xi_1)(\xi_2 + \xi_1))^k,$$
$$g_{2k+1} = (\xi_2 - \xi_1)((\xi_2 + \xi_1)(\xi_2 - \xi_1))^k,$$

is a highest weight vector with weight (k) for our choice of positive root system. Again, these weight vectors will all generate independent isomorphic $\mathfrak{so}(m, \mathbb{C})$-representations. We thus get 2^{2m} different highest weight vectors which generate 2^{2m} different isomorphic irreducible representations.

Remark 13. When the space $\text{span}_{\mathbb{C}} \{f_k J\}$ spans a weight space with weight $(-k)$, the negative roots act trivially on this space, as one would expect.

Each of these representations has dimension $\binom{k+m-1}{k} - \binom{k+m-3}{k}$ (see, e.g., [8]). By considering all 2^{2m} idempotents and as

$$\begin{aligned}\dim_{\mathbb{C}} (\mathcal{H}_k(m, \mathbb{C}_{2m})) &= \left(\binom{k+m-1}{k} - \binom{k+m-3}{k}\right) \dim_{\mathbb{C}}(\mathbb{C}_{2m}) \\ &= 2^{2m} \left(\binom{k+m-1}{k} - \binom{k+m-3}{k}\right),\end{aligned}$$

we find that we can decompose the space $\mathcal{H}_k(m, \mathbb{C}_{2m})$ into 2^{2m} irreducible isomorphic representations of $\mathfrak{so}(m, \mathbb{C})$.

3.2. Odd dimension $m = 2n + 1$

In odd dimension, we extend the set of generators H_a, $X_{a,b}$, $Y_{a,b}$ and $Z_{a,b}$ of the root system with the $2n$ mappings:

$$U_s = \frac{1}{\sqrt{2}} (\Omega_{2s-1,m} - i\,\Omega_{2s,m}), \qquad V_s = \frac{1}{\sqrt{2}} (\Omega_{2s-1,m} + i\,\Omega_{2s,m}),$$

where $1 \leqslant s \leqslant n$. With the addition of these $2n$ mappings, we are again able to reconstruct all original Ω_{ab}'s since $\sqrt{2}\,\Omega_{2s-1,m} = U_s + V_s$ and $-\sqrt{2}i\,\Omega_{2s,m} = U_s - V_s$.

Lemma 14. *For* $1 \leqslant s, j \leqslant n$, *it holds that*

$$[H_s, U_j] = \delta_{sj}\, U_j = L_j(H_s)\, U_j,$$
$$[H_s, V_j] = -\delta_{sj}\, V_j = -L_j(H_s)\, V_j.$$

In particular, U_j is a root vector corresponding to the positive root L_j and V_j is a root vector corresponding with the negative root $-L_j$, $\forall\, 1 \leqslant j \leqslant n$.

Proof. Take $1 \leqslant s, j \leqslant n$:

$$\begin{aligned}\sqrt{2}\,[H_s, U_j] &= i\,[\Omega_{2s-1,2s}, \Omega_{2j-1,m}] + [\Omega_{2s-1,2s}, \Omega_{2j,m}]\\ &= i\,(-\delta_{sj}\,\Omega_{2s,m}) + (\delta_{sj}\,\Omega_{2s-1,m})\\ &= \sqrt{2}\,\delta_{sj}\,U_s,\\ \sqrt{2}\,[H_s, V_j] &= i\,[\Omega_{2s-1,2s}, \Omega_{2j-1,m}] - [\Omega_{2s-1,2s}, \Omega_{2j,m}]\\ &= i\,(-\delta_{sj}\,\Omega_{2s,m}) - (\delta_{sj}\,\Omega_{2s-1,m})\\ &= -\sqrt{2}\,\delta_{sj}\,V_s.\end{aligned}$$ □

Lemma 15. *The operators U_j and V_j satisfy the following additional commutator relations with $X_{a,b}$, $Y_{a,b}$ and $Z_{a,b}$:*

$$\begin{aligned}[U_j, X_{a,b}] &= -\delta_{jb}\,U_a, & [V_j, X_{a,b}] &= \delta_{ja}\,V_b,\\ [U_j, Y_{a,b}] &= 0, & [V_j, Y_{a,b}] &= \delta_{ja}\,U_b - \delta_{jb}\,U_a,\\ [V_j, Z_{a,b}] &= 0, & [U_j, Z_{a,b}] &= -\delta_{jb}\,V_a + \delta_{ja}\,V_b,\\ [U_j, U_\ell] &= -Y_{j,\ell},\ j \neq \ell, & [U_j, V_j] &= -H_j,\\ [V_j, V_\ell] &= -Z_{j,\ell},\ j \neq \ell & [U_j, V_\ell] &= -X_{j,\ell},\ j \neq \ell.\end{aligned}$$

Proof. We only show the first proof, as the other relations use similar arguments.

$$\begin{aligned}\sqrt{2}\,[U_j, X_{a,b}] &= \delta_{jb}\,(\Omega_{m,2a-1} - i\,\Omega_{m,2a}) = \delta_{jb}\,(-\Omega_{2a-1,m} + i\,\Omega_{2a,m})\\ &= -\sqrt{2}\,\delta_{jb}\,U_a.\end{aligned}$$ □

To establish highest weight vectors in the odd-dimensional case, we further introduce

$$I_m^{\pm} = \left(\mathbf{e}_m^{+}\mathbf{e}_m^{-} \pm \mathbf{e}_m^{+}\right), \qquad K_m^{\pm} = \left(\mathbf{e}_m^{-}\mathbf{e}_m^{+} \pm \mathbf{e}_m^{-}\right)$$

and denote $I = \prod_{s=1}^{n}\left(I_{2s-1}^{+}\,I_{2s}^{-}\right) I_m^{+}$. Then the elements $f_{2k}\,I$ and $f_{2k+1}\,I$ are still weight vectors with weights $(2k, 0, \ldots, 0)$ resp. $(2k+1, 0, \ldots, 0)$ which vanish under the positive roots $Y_{a,b}$ and $X_{a,b}$ $(a < b)$. We only need to consider the positive roots involving the extra factor m.

Lemma 16. *The generators of the representation $f_{2k}\,I$ and $f_{2k+1}\,I$ vanish under the operator U_j, $1 \leqslant j \leqslant n$, i.e., $f_{2k}\,I$ and $f_{2k+1}\,I$ are highest weight vectors:*

$$U_j\,(f_{2k}\,I) = 0, \qquad U_j\,(f_{2k+1}\,I) = 0, \quad \forall\, 1 \leqslant j \leqslant n.$$

Proof. Since

$$\sqrt{2}\,U_j\,(f_{2k}\,I) - (\Omega_{2j-1,m} - i\,\Omega_{2j,m})\,f_{2k}\,I$$

and f_{2k} contains only ξ_1 and ξ_2, we find that $U_j(f_{2k}\,I)$ will immediately be zero unless $j = 1$. Then

$$\begin{aligned}\sqrt{2}\,U_1(f_{2k}\,I) &= \xi_m\,\partial_1\,f_{2k}\,I\,e_m\,e_1 - i\,\xi_m\,\partial_2 f_{2k}\,I\,e_m\,e_2\\ &= 2k\,\xi_m\,g_{2k-1}\,(I\,e_m\,e_1 - i\,I\,e_m\,e_2)\,.\end{aligned}$$

We complete the proof by noting that

$$\begin{aligned}I\,e_m\,e_1 &= -I_1^+\,e_1\,I_2^+\,I_3^-\ldots I_{2n-1}^-\,I_{2n}^+\,I_m^-\,e_m = I_1^+\,I_2^+\,I_3^-\ldots I_{2n-1}^-\,I_{2n}^+\,I_m^-,\\ I\,e_m\,e_2 &= -I_1^+\,I_2^-\,e_2\,I_3^-\ldots I_{2n-1}^-\,I_{2n}^+\,I_m^-\,e_m = (-i)\,I_1^+\,I_2^+\,I_3^-\ldots I_{2n-1}^-\,I_{2n}^+\,I_m^-.\end{aligned}$$

The proof for $f_{2k+1}\,I$ is completely similar. □

Remark 17. Let $m = 2n+1$. We compare the dimensions:

$$\dim_{\mathbb{C}}\mathcal{H}_k = 2^{2m}\left(\binom{k+m-1}{k} - \binom{k+m-3}{k}\right)$$

On the other hand, we have found $4^{2n+1} = 2^{2m}$ different highest weight vectors and thus 2^{2m} isomorphic irreducible representations with combined dimension

$$2^{2m}\left(\binom{k+m-1}{k} - \binom{k+m-3}{k}\right).$$

We may thus conclude that $\mathcal{H}_k$ decomposes into 2^{2m} isomorphic irreducible representations of $\mathfrak{so}(m,\mathbb{C})$.

4. Conclusion and future research

The space $\mathcal{H}_k$ of discrete k-homogeneous harmonic polynomials is a reducible representation of $\mathfrak{so}(m,\mathbb{C})$, which can be decomposed into 2^{2m} isomorphic copies of irreducible $\mathfrak{so}(m,\mathbb{C})$-representations with highest weight $(k,0,\ldots,0)$. This is done by means of 2^{2m} idempotents. Let

$$\begin{aligned}f_{2k} &= ((\xi_2+\xi_1)(\xi_2-\xi_1))^k\,,\\ g_{2k} &= ((\xi_2-\xi_1)(\xi_2+\xi_1))^k\,,\\ f_{2k+1} &= (\xi_2+\xi_1)((\xi_2-\xi_1)(\xi_2+\xi_1))^k\,,\\ g_{2k+1} &= (\xi_2-\xi_1)((\xi_2+\xi_1)(\xi_2-\xi_1))^k\,,\end{aligned}$$

be discrete homogeneous harmonic functions of degree $2k$ (resp. $2k+1$). For each choice of idempotent I from the set of 2^{2m} idempotents

$$\prod_{s=1}^{n} F_s, \qquad F_s \in \{I_{2s-1}^{\pm}I_{2s}^{\pm},\ I_{2s-1}^{\pm}K_{2s}^{\pm},\ K_{2s-1}^{\pm}I_{2s}^{\pm},\ K_{2s-1}^{\pm}K_{2s}^{\pm}\}\,,$$

where

$$\begin{aligned}I_{2s-1}^{\pm} &= \left(\mathbf{e}_{2s-1}^{+}\mathbf{e}_{2s-1}^{-} \pm \mathbf{e}_{2s-1}^{+}\right), & I_{2s}^{\pm} &= \left(\mathbf{e}_{2s}^{+}\mathbf{e}_{2s}^{-} \pm i\,\mathbf{e}_{2s}^{+}\right),\\ K_{2s-1}^{\pm} &= \left(\mathbf{e}_{2s-1}^{-}\mathbf{e}_{2s-1}^{+} \pm \mathbf{e}_{2s-1}^{-}\right), & K_{2s}^{\pm} &= \left(\mathbf{e}_{2s}^{-}\mathbf{e}_{2s}^{+} \pm i\,\mathbf{e}_{2s}^{-}\right),\end{aligned}$$

either the subspace $\text{span}_{\mathbb{C}}\{f_j I\}$ or the subspace $\text{span}_{\mathbb{C}}\{g_j I\}$ generates an irreducible $\mathfrak{so}(m, \mathbb{C})$-representation with highest weight $(j, 0, \ldots, 0)$, $j \in \{2k, 2k+1\}$, under the action of the negative roots.

In an upcoming paper, we decompose the space of discrete k-homogeneous monogenic polynomials in irreducible $\mathfrak{so}(m, \mathbb{C})$-representation, creating in this way a notion of discrete spinors.

References

[1] D. Constales, F. Sommen and P. Van Lancker, *Models for irreducible representations of Spin(m)*, Adv. Appl. Clifford Algebras **11 (S1)**, 2001, pp. 271–289.

[2] H. De Ridder, H. De Schepper, U. Kähler and F. Sommen, *Discrete function theory based on skew Weyl relations*, Proc. Amer. Math. Soc. **138**, 2010, pp. 3241–3256.

[3] H. De Ridder, H. De Schepper and F. Sommen, *The Cauchy–Kovalevskaya Extension Theorem in Discrete Clifford Analysis*, Comm. Pure Appl. Math. **10 (4)**, 2011, pp. 1097–1109.

[4] H. De Ridder, H. De Schepper and F. Sommen, *Fueter Polynomials in Discrete Clifford Analysis*, Mathematische Zeitschrift **272 (1-2)**, 2012, pp. 253–268.

[5] H. De Ridder, T. Raeymaekers and F. Sommen, *Rotations in discrete Clifford analysis*, Appl. Math. Comput. **285 (C)**, July 2016, pp. 114–140.

[6] H. De Schepper, F. Sommen and L. Van de Voorde, *A basic framework for discrete Clifford analysis*, Experimental Mathematics **18 (4)**, 2009, pp. 385–395.

[7] N. Faustino, U. Kähler and F. Sommen, *Discrete Dirac operators in Clifford analysis*, Adv. in Appl. Cliff. Alg., **17 (3)**, 2007, pp. 451–467.

[8] W. Fulton and J. Harris, *Representation theory*, Graduate Texts in Mathematics **129**, 1991, Springer-Verlag, New York. A first course, Readings in Mathematics.

[9] J. Gilbert and M.A.M. Murray, *Clifford algebras and Dirac operators in harmonic analysis*, 1991, Cambridge University Press, Cambridge.

[10] J. Humphreys, *Introduction to Lie algebra and representation theory*, 1972, Springer-Verlag, New York.

[11] A.W. Knapp, *Lie Groups Beyond an Introduction*, Progress in Mathematics **140**, 1996, Birkhäuser Boston, Inc., Boston, MA.

[12] A.I. Molev, *Yangians and classical Lie algebras*, Mathematical surveys and monographs **143**, 2007, AMS Bookstore.

Hilde De Ridder and Tim Raeymaekers
Department of Mathematical Analysis
Ghent University
Building S22, Galglaan 2
B-9000 Gent, Belgium
e-mail: `hdr@cage.ugent.be`
`tr@cage.ugent.be`

Modern Trends in Hypercomplex Analysis
Trends in Mathematics, 161–174

Gegenbauer Type Polynomial Solutions for the Higher Spin Laplace Operator

David Eelbode and Tim Janssens

Abstract. In this paper we study special polynomial solutions to the higher spin Laplace operator, which is a conformally invariant second-order operator acting on fields taking values in the space of symmetric tensors. We will consider a particular subalgebra of the conformal symmetry algebra and use a ladder formalism to generate special solutions for this operator. For the normal Laplace operator this leads to harmonic polynomials expressed in terms of Gegenbauer polynomials, in the higher spin case the resulting solutions are more complicated.

Mathematics Subject Classification (2010). 42B37, 33C45, 30G35.

Keywords. Higher spin operators, special functions, Gegenbauer polynomials.

1. Introduction

Classical harmonic and Clifford analysis are function theories in m dimensions in which conformally invariant operators are studied using a unified framework. We refer to some of the standard textbooks, see [2, 6, 15, 16]. Traditionally, most of the attention was aimed at the Laplace operator Δ_x and the Dirac operator ∂_x, but it seems that various higher spin generalizations of both operators have recently gained their place in the aforementioned function theories. Far from claiming completeness, we refer to, e.g., [3, 7, 9–11, 13] for papers in the context of Clifford analysis.

In the present paper we continue the investigation of a particular higher spin generalization: we focus our attention on the operators $\mathcal{D}_k$, indexed by $k \in \mathbb{N}$ and introduced in [5], which are connected to both the Laplace operator ($\mathcal{D}_0 = \Delta_x$) and the Rarita–Schwinger operators $\mathcal{R}_k$, see [3]. These second-order operators are conformally invariant and defined on functions taking values in the space $\mathcal{H}_k(\mathbb{R}^m, \mathbb{C})$ of k-homogeneous harmonics in a dummy variable (our model for higher spin fields, hereby drawing inspiration from [4], which reduce to scalar values in $\mathbb{C}$ for $k = 0$). In particular, we construct special solutions using a method which for $k = 0$ yields harmonics expressed in terms of Gegenbauer polynomials. This method uses a special conformal symmetry which gives rise to a subalgebra $\mathfrak{sl}(2)$ of the full conformal symmetry algebra $\mathfrak{so}(1, m+1)$; this was heavily exploited in [12] to provide a representation theoretical proof for the celebrated Fueter theorem in Clifford analysis.

In the classical case $k = 0$, the role of these solutions cannot be underestimated: they are related to the reproducing kernel for the spaces $\mathcal{H}_l(\mathbb{R}^m, \mathbb{C})$ and they are crucial in the construction of Gelfand–Tsetlin bases due to their connection with the so-called branching problem from $\mathfrak{so}(m)$ to $\mathfrak{so}(m-1)$. However, the situation is more complicated for $\mathcal{D}_k$ with $k > 0$. The reason for this is the following: whereas the spaces $\mathcal{H}_l(\mathbb{R}^m, \mathbb{C})$ define an irreducible representation for $\mathfrak{so}(m)$, the space of l-homogeneous solutions for $\mathcal{D}_k$ is no longer irreducible. In a sense, this leads to a certain ambiguity in the search for a higher spin version of these Gegenbauer type solutions: one can either generalize them using the conformal inversion (the approach adopted in the present paper), or one can focus on the fact that they should for instance reproduce certain solution spaces (which is the path we have pursued in our paper [8]).

So in this paper, we will again consider a particular subalgebra $\mathfrak{sl}(2)$ of the full conformal symmetry algebra and use the associated ladder formalism to generate solutions for the operator $\mathcal{D}_k$ (see Section 3). In contrast to the classical case, recognizing these solutions in terms of well-known special functions turns out to be a difficult problem; we believe that this stems from the fact that the solution spaces are not irreducible. This will be explained in Section 4, invoking the branching problem. Finally, in the last sections we will give an explicit example (Section 5) and an overview of future interests (Section 6).

2. The higher spin Laplace operator $\mathcal{D}_k$

First of all we give a quick introduction into higher spin Clifford analysis. The operators $\mathcal{D}_k$ will be defined on the space $\mathcal{C}^\infty(\mathbb{R}^m, \mathcal{H}_k(\mathbb{R}^m, \mathbb{C}))$, which consists of functions $f(x, u)$ such that

$$f(x,u) := f_x(u) \in \mathcal{H}_k(\mathbb{R}^m, \mathbb{C}), \quad \forall x \in \mathbb{R}^m.$$

There is a natural action of the group $\mathrm{Spin}(m)$ on these functions (the so-called regular action), given by $H(s)[f](x,u) = f(\overline{s}xs, \overline{s}us)$. The derived action of the corresponding Lie algebra $\mathfrak{so}(m)$ is given by

$$dH(e_{ij})[f](x,u) = (L^x_{ij} + L^u_{ij})f(x,u),$$

with for instance $L^x_{ij} := x_i\partial_{x_j} - x_j\partial_{x_i}$ the angular operators in the x-variable. Let $\mathcal{P}_{l,k}(\mathbb{R}^{2m}, \mathbb{C})$ be the space of polynomials in two variables x and u with degree of homogeneity in x (resp. u) equal to l (resp. k).

Definition 2.1. For all integers $l \geq k$, the space of simplicial harmonics is defined as

$$\mathcal{H}_{l,k}(\mathbb{R}^{2m}, \mathbb{C}) := \mathcal{P}_{l,k}(\mathbb{R}^{2m}, \mathbb{C}) \cap \ker(\Delta_x, \Delta_u, \langle x, \partial_u\rangle, \langle \partial_x, \partial_u\rangle)\ ,$$

where we adopted the notation $\ker(D_1, \ldots, D_n) := \ker(D_1) \cap \cdots \cap \ker(D_n)$.

We hereby list some properties of the space of simplicial harmonics $\mathcal{H}_{l,k}(\mathbb{R}^{2m}, \mathbb{C})$:

- If $m > 4$, the spaces $\mathcal{H}_{l,k}$ form an irreducible $\mathfrak{so}(m)$-representation.
- In this case their highest weight is equal to $(l, k, 0, \ldots, 0)$, with highest weight vector

$$(x_1 - ix_2)^{l-k}((x_1 - ix_2)(u_3 - iu_4) - (x_3 - ix_4)(u_1 - iu_2))^k.$$

- If $l = k$ then by symmetry $\mathcal{H}_{k,k}(\mathbb{R}^{2m}, \mathbb{C}) \subset \ker(\langle u, \partial_x \rangle)$.

Next we will give the definition of the higher spin Laplace operator as well as some basic properties, for more details we refer to [5]. The generalization of the Laplace operator to the higher spin case is defined as the operator

$$\mathcal{D}_k : \mathcal{C}^\infty(\mathbb{R}^m, \mathcal{H}_k) \to \mathcal{C}^\infty(\mathbb{R}^m, \mathcal{H}_k)$$

with

$$\mathcal{D}_k = \Delta_x - \frac{4}{2k + m - 2} \left(\langle u, \partial_x \rangle - \frac{|u|^2}{2k + m - 4} \langle \partial_u, \partial_x \rangle \right) \langle \partial_u, \partial_x \rangle .$$

This operator is conformally invariant with respect to the following inversion:

$$f(x, u) \mapsto \mathcal{J}_R[f](x, u) := |x|^{2-m} f\left(\frac{x}{|x|^2}, \frac{xux}{|x|^2}\right) .$$

The following operator will be crucial for our purposes, as it will play the role of the ladder operator mentioned in the introduction (the proof follows from straightforward calculations):

Lemma 2.2. *One has that*

$$\mathcal{J}_R \partial_{x_i} \mathcal{J}_R = |x|^2 \partial_{x_i} + 2 \langle x, u \rangle \partial_{u_i} - 2u_i \langle x, \partial_u \rangle - x_i(2\mathbb{E}_x + m - 2).$$

Using this operator we can realize a copy of the Lie algebra $\mathfrak{sl}(2)$ inside the full conformal Lie algebra $\mathfrak{so}(1, m+1)$ of (generalized) symmetries for the operators $\mathcal{D}_k$ (see [5] for more details).

Proposition 2.3. *We have the following realization of* $\mathfrak{sl}(2)$*:*

$$\mathfrak{sl}(2) \cong \mathrm{Alg}(\mathcal{J}_R \partial_{x_j} \mathcal{J}_R, \partial_{x_j}, 2\mathbb{E}_x + m - 2).$$

As mentioned earlier, one of the main differences between the Laplace operator Δ_x and the operators $\mathcal{D}_k$ for $k \neq 0$ is the fact that the polynomial kernel for the latter operator is *not* irreducible under the action of $\mathfrak{so}(m)$. As a matter of fact, in [5] is was shown that one can decompose the l-homogeneous kernel of $\mathcal{D}_k$ as follows:

$$\ker_l \mathcal{D}_k = \bigoplus_{i=0}^{k} \bigoplus_{j=0}^{k-i} (\mathcal{J}_R \Delta_x \mathcal{J}_R)^i \langle u, \partial_x \rangle^{i+j} \mathcal{H}_{l-i+j,k-i-j}.$$

Note that these embedding operators are explicitly given as follows:

Lemma 2.4. *The operator* $\mathcal{J}_R\Delta_x\mathcal{J}_R$ *is given by:*

$$\mathcal{J}_R\Delta_x\mathcal{J}_R = |x|^4\Delta_x + 4\left((2\mathbb{E}_u + m - 4)\langle u, x\rangle + |u|^2\langle x, \partial_u\rangle\right)\langle x, \partial_u\rangle + 4|x|^2\left(\langle u, x\rangle\langle \partial_u, \partial_x\rangle - \langle u, \partial_x\rangle\langle x, \partial_u\rangle\right).$$

3. Invariant polynomial solutions

Similar to the construction of the harmonic Gegenbauer polynomials, we will construct special solutions for $\mathcal{D}_k$ by repeated action of the operator from Lemma 2.2. In the harmonic case ($k = 0$), this was done by letting it act on the constant 1 but here we need the raising operator to act on a k-homogeneous polynomial in the dummy variable $u \in \mathbb{R}^m$. This polynomial must belong to the kernel of the operator Δ_u and should be $\mathfrak{so}(m-1)$-invariant, so as not to violate the invariance built into the raising operator. Therefore, the only possibility is the harmonic Gegenbauer polynomial in the variable u. We will implement the following notation: for $j \leq k$ we put

$$P_k^j(u) = |u|^{k-j}C_{k-j}^{\frac{m}{2}-1+j}\left(\frac{u_1}{|u|}\right) .$$

For $j > k$ we adopt the convention that $P_k^j(u) = 0$. Note that for $j = 0$, this is precisely the harmonic polynomial in $u \in \mathbb{R}^m$ we use as a starting point. Although this is no longer true for $j > 0$, the resulting Gegenbauer polynomials still have a special meaning: they occur as embedding factors for the branching problem for harmonic polynomials in $u \in \mathbb{R}^m$. Indeed, one has that

$$P_k^j(u) : \mathcal{H}_j(\mathbb{R}^{m-1}) \to \mathcal{H}_k(\mathbb{R}^m) : H_j(u_2, \ldots, u_m) \mapsto P_k^j(u)H_j(u_2, \ldots, u_m) .$$

In other words: $P_k^j(u)$ can be interpreted as a multiplication operator which gives harmonics on $\mathbb{R}^m$ when acting on a harmonic of a certain degree in a space of one dimension less. The reason for this is the following: these embeddings can also be written in terms of the classical Kelvin inversion $\mathcal{J}_\Delta$ for harmonic functions given by

$$\mathcal{J}_\Delta[f(x)] := |u|^{2-m}f\left(\frac{u}{|u|^2}\right) .$$

Indeed, one has that

$$(\mathcal{J}_\Delta\partial_{u_1}\mathcal{J}_\Delta)^{k-j} = \mathcal{J}_\Delta\partial_{u_1}^{k-j}\mathcal{J}_\Delta : \mathcal{H}_j(\mathbb{R}^{m-1}) \to \mathcal{H}_k(\mathbb{R}^m) .$$

To lighten the notation we denote the raising operator by means of

$$X := \mathcal{J}_R\partial_{x_1}\mathcal{J}_R .$$

We are interested in finding an expression for $X[P_k^j(u)]$, as this often occurs in what follows. We first prove a Leibniz type rule:

Lemma 3.1. *For* $f(x, u)$ *and* $g(x, u)$ *in* $\mathcal{C}^\infty(\mathbb{R}^{2m}, \mathbb{C})$, *one has:*

$$X[fg] = X[f]g + fX[g] + (m-2)x_1fg .$$

Proof. Each term in the formula for X satisfies the Leibniz rule, except for the multiplication with $(2-m)x_1$. To compensate this we have to add the additional last term in the formula above. □

Lemma 3.2. *For every $j \in \mathbb{N}$ we have that:*

$$X[P_k^j(u)] = 2(m-2+2j)\left(\langle x,u\rangle - u_1x_1\right)P_k^{j+1}(u) - (m-2)x_1P_k^j(u).$$

Proof. As we are acting on polynomials that are independent of x, our raising operator reduces to $X = 2\langle x,u\rangle\,\partial_{u_1} - 2u_1\langle x,\partial_u\rangle - x_1(m-2)$. Using the fact that

$$\frac{d}{dt}C_n^\mu(t) = 2\mu C_{n-1}^{\mu+1}(t)$$

the desired result follows from straightforward calculations. □

Next, we show that the action of the raising operator yields polynomial solutions for $\mathcal{D}_k$ of a special form (linear combinations of Gegenbauer polynomials in u):

Theorem 3.3. *For each $k,l \in \mathbb{N}$ we have that:*

$$X^l[P_k^0(u)] = \sum_{i=0}^{\min(k,l)} f_i^{(l)}(x_1,|x|,u_1,\langle u,x\rangle)P_k^i(u).$$

Proof. We take an arbitrary k fixed and proceed via induction on l. If $l=0$ the result is trivial since $f_0^{(0)} = 1$. Assume it holds for all values up to and including $(l-1)$. This means that:

$$\begin{aligned} X^l[P_k^0(u)] &= XX^{l-1}[P_k^0(u)] \\ &= \sum_{i=0}^{\min(k,l-1)} X[f_i^{(l-1)}(x_1,|x|,u_1,\langle u,x\rangle)P_k^i(u)] \end{aligned}$$

For each $0 \le i \le \min(k,l-1)$ we have that

$$\begin{aligned} X[f_i^{(l-1)}(x_1,|x|,u_1,\langle u,x\rangle)P_k^i(u)] &= X[f_i^{(l-1)}(x_1,|x|,u_1,\langle u,x\rangle)]P_k^i(u) \\ &\quad + f_i^{(l-1)}(x_1,|x|,u_1,\langle u,x\rangle)X[P_k^i(u)] \\ &\quad + (m-2)x_1f_i^{(l-1)}(x_1,|x|,u_1,\langle u,x\rangle)P_k^i(u). \end{aligned}$$

We know from the previous lemma that $X[P_k^i(u)]$ looks as follows:

$$X[P_k^j(u)] = 2(m-2+2i)\left(\langle x,u\rangle - u_1x_1\right)P_k^{i+1}(u) - (m-2)x_1P_k^i(u)$$

and thus this is of the correct form. All that we have to check is that $X[f_i^{(l-1)}]$ only depends on the given parameters to complete the proof which follows from the chain rule and straightforward calculations. □

From this proof we can extract a recursive relation for the functions $f_i^{(l)}$:

Proposition 3.4. *For $0 \leq i \leq \min(k, l)$ we have that:*

$$f_i^l(x_1, |x|, u_1, \langle u, x\rangle) = X[f_i^{(l-1)}(x_1, |x|, u_1, \langle u, x\rangle)] \\ + f_{i-1}^{(l-1)}(x_1, |x|, u_1, \langle u, x\rangle) 2(m-4+2i)\left(\langle x, u\rangle - u_1 x_1\right).$$

An explicit formula for $f_0^{(l)}$ is obtained in the following lemma:

Lemma 3.5. *For each $l \in \mathbb{N}$ we have that:*

$$f_0^{(l)}(x_1, |x|, u_1, \langle u, x\rangle) = (-1)^l l! |x|^l C_l^{\frac{m}{2}-1}\left(\frac{x_1}{|x|}\right).$$

Proof. As the recursive relation reduces to $f_0^{(l)} = X[f_0^{(l-1)}]$ for all $l \in \mathbb{N}$, and the fact that

$$X^l[1] = \left(|x|^2 \partial_{x_1} - x_1(2\mathbb{E}_x + m - 2)\right)^l [1],$$

the result follows from the harmonic case ($k = 0$). □

Note that this shows that $X^l[P_k^0(u)]$ can in fact be seen as 'a deformation' of a harmonic polynomial in both x and $u \in \mathbb{R}^m$ (i.e., up to a remaining polynomial to make it a solution for $\mathcal{D}_k$):

$$X^l[P_k^0(u)] = (-1)^l l! |x|^l |u|^k C_l^{\frac{m}{2}-1}\left(\frac{x_1}{|x|}\right) C_k^{\frac{m}{2}-1}\left(\frac{u_1}{|u|}\right) + \text{Rest}_{l,k}(x, u).$$

We can illustrate the previous results with a scheme, where the arrows show which coefficients contribute to a specific term:

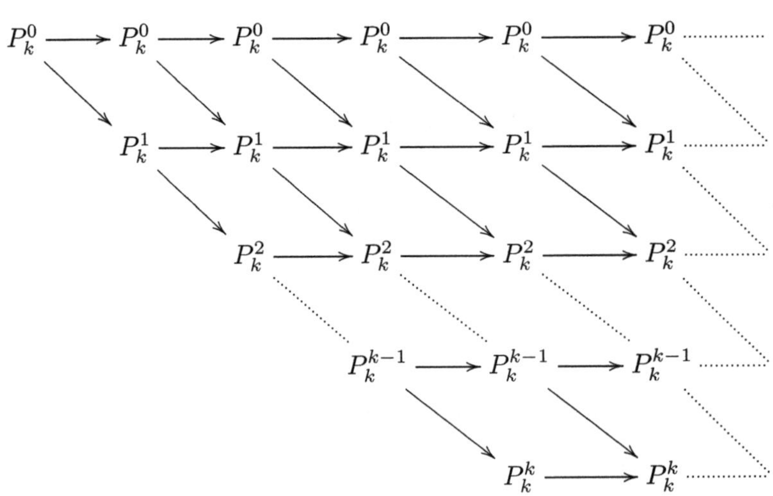

In the harmonic case, the repeated action of the raising operator leads to a polynomial depending on the squared norm of the vector variable (an invariant)

and the inner product with a fixed unit vector (reducing the invariance by one dimension). In the present setting, the previous results suggest a result depending on the variables

$$(r,s,t) := \left(\frac{x_1}{|x|}, \frac{u_1}{|u|}, \left\langle \frac{x}{|x|}, \frac{u}{|u|} \right\rangle\right)$$

which is shown in the next theorem:

Theorem 3.6. *For all $l,k \in \mathbb{N}$ there exists a function $f_{l,k}(r,s,t) : \mathbb{R}^3 \to \mathbb{C}$ such that:*

$$X^l[P_k^0(u)] = |x|^l |u|^k f_{l,k}(r,s,t)\,.$$

Proof. Take arbitrary k and l. Using Theorem 3.3 we know that:

$$\begin{aligned} X^l[P_k^0(u)] &= \sum_{i=0}^{\min\{k,l\}} f_i^{(l)}(x_1,|x|,u_1,\langle u,x\rangle) P_k^i(u) \\ &= \sum_{i=0}^{\min\{k,l\}} |u|^{k-i} f_i^{(l)}(x_1,|x|,u_1,\langle u,x\rangle) C_{k-i}^{\frac{m}{2}-1+i}(s)\,. \end{aligned}$$

Using Proposition 3.4 we can conclude that

$$\begin{aligned} \mathbb{E}_x f_i^{(l)} &= l f_i^{(l-1)} \\ \mathbb{E}_u f_i^{(l)} &= i f_i^{(l-1)} \end{aligned}$$

which means that $f_i^{(l)}(x_1,|x|,u_1,\langle u,x\rangle) = |x|^l |u|^i g_i^{(l)}(r,s,t)$ for some function $g_i^{(l)}$. This finishes the proof. □

One can rewrite the action of the raising operator X, using the variables (r,s,t). We want to find an operator Q_l such that:

$$X[|x|^l |u|^k f_{l,k}(r,s,t)] = |x|^{l+1} |u|^k Q_l f_{l,k}(r,s,t).$$

Using Lemma 3.1 and the chain rule, straightforward calculations give us that:

$$Q_l := (1-r^2)\partial_r + 2(t-sr)\partial_s - (s-rt)\partial_t - (l+m-2).$$

We can also use our $\mathfrak{sl}(2)$-realization to find an inverse. Recall that we have that

$$0 \xleftarrow[\partial_{x_1}]{} \underset{\underset{2\mathbb{E}_x+m-2}{\circlearrowright}}{P_k^0(u)} \underset{\partial_{x_1}}{\overset{\mathcal{J}_R\partial_{x_1}\mathcal{J}_R}{\rightleftarrows}} \underset{\underset{2\mathbb{E}_x+m-2}{\circlearrowright}}{(\mathcal{J}_R\partial_{x_1}\mathcal{J}_R)[P_k^0(u)]} \underset{\partial_{x_1}}{\overset{\mathcal{J}_R\partial_{x_1}\mathcal{J}_R}{\rightleftarrows}} \underset{\underset{2\mathbb{E}_x+m-2}{\circlearrowright}}{(\mathcal{J}_R\partial_{x_1}\mathcal{J}_R)^2[P_k^0(u)]} \underset{\partial_{x_1}}{\overset{\mathcal{J}_R\partial_{x_1}\mathcal{J}_R}{\rightleftarrows}} \bullet \cdots\cdots$$

It is also well known that for $\mathfrak{sl}(2) \cong \mathrm{Alg}(X,Y,H)$ the following commutation relation holds (with $a \in \mathbb{N}$):

$$[Y, X^{a+1}] = -(a+1)X^a(H+a).$$

If we apply this to our situation we find that:

$$\begin{aligned}\partial_{x_1}(\mathcal{J}_R\partial_{x_1}\mathcal{J}_R)^{l+1}[P_k^0(u)] &= [\partial_{x_1},(\mathcal{J}_R\partial_{x_1}\mathcal{J}_R)^{l+1}][P_k^0(u)]\\ &= -(l+1)(m-2+l)(\mathcal{J}_R\partial_{x_1}\mathcal{J}_R)^{l}[P_k^0(u)].\end{aligned}$$

Moreover it should be easy to see that $\partial_{x_1}|x|^{l+1}|u|^k f_{l+1,k}(r,s,t)$ can be written as:

$$|x|^l|u|^k\left((l+1)r+(1-r^2)\partial_r+(s-rt)\partial_t\right)f_{l+1,k}(r,s,t).$$

Defining the operator L_l by means of

$$L_l=-\frac{1}{(l+1)(m-2+l)}\left((l+1)r+(1-r^2)\partial_r+(s-rt)\partial_t\right),$$

we can now say that

$$\partial_{x_1}|x|^{l+1}|u|^k f_{l+1,k}(r,s,t)=-(l+1)(m-2+l)|x|^l|u|^k L_l(\partial_r,0,\partial_t)f_{l+1}(r,s,t).$$

This operator serves as the inverse, as can be observed from direct calculations. Indeed, we have that both

$$\begin{aligned}&|x|^l|u|^k L_l(\partial_r,0,\partial_t)Q_l(\partial_r,\partial_s,\partial_t)f_{l,k}(r,s,t)\\ &=-\frac{1}{(l+1)(m-2+l)}\partial_{x_1}|x|^{l+1}|u|^kQ_l(\partial_r,\partial_s,\partial_t)f_{l,k}(r,s,t)\\ &=-\frac{1}{(l+1)(m-2+l)}\partial_{x_1}(\mathcal{J}_R\partial_{x_1}\mathcal{J}_R)^{l+1}[P_k^0(u)]\\ &=(\mathcal{J}_R\partial_{x_1}\mathcal{J}_R)^{l}[P_k^0(u)]\\ &=|x|^l|u|^kf_{l,k}(r,s,t)\end{aligned}$$

and

$$\begin{aligned}&|x|^{l+1}|u|^kQ_l(\partial_r,\partial_s,\partial_t)L_l(\partial_r,0,\partial_t)f_{l+1,k}(r,s,t)\\ &=(\mathcal{J}_R\partial_{x_1}\mathcal{J}_R)|x|^l|u|^kL_l(\partial_r,0,\partial_t)f_{l+1,k}(r,s,t)\\ &=-\frac{1}{(l+1)(m-2+l)}(\mathcal{J}_R\partial_{x_1}\mathcal{J}_R)\partial_{x_1}|x|^{l+1}|u|^kf_{l+1,k}(r,s,t)\\ &=-\frac{1}{(l+1)(m-2+l)}(\mathcal{J}_R\partial_{x_1}\mathcal{J}_R)\partial_{x_1}(\mathcal{J}_R\partial_{x_1}\mathcal{J}_R)^{l+1}[P_k^0(u)]\\ &=(\mathcal{J}_R\partial_{x_1}\mathcal{J}_R)^{l+1}[P_k^0(u)]\\ &=|x|^{l+1}|u|^kf_{l+1,k}(r,s,t)\ .\end{aligned}$$

The motivation for introducing this inverse is encoded in the scheme at the top of the next page, where we have also defined the polynomial:

$$S_k=-\frac{1}{k+1}\left((1-s^2)\partial_s-(k+m-2)s\right).$$

Given any $f_{l,k}$, we can thus complete the scheme.

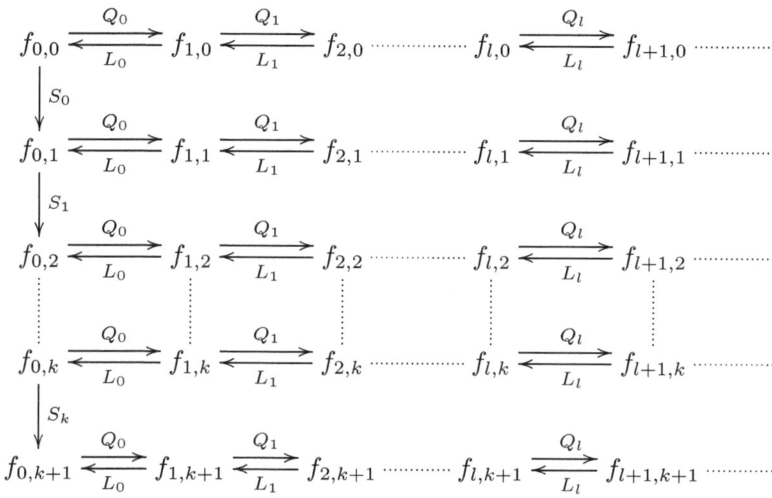

Proposition 3.7. *The functions* $f_{l,k}(r,s,t)$*, defined by*

$$X^l\left[|u|^k C_k^{\frac{m}{2}-1}\left(\frac{u_1}{|u|}\right)\right] = |x|^l|u|^k f_{l,k}(r,s,t)$$

can be written as:

$$f_{l,k}(r,s,t) = \sum_{a=0}^{l}\sum_{b=0}^{k}\sum_{c=0}^{\min\{l-a,k-b\}} \alpha_{l,k}(a,b,c) r^a s^b t^c$$

where the $\alpha_{l,k}(a,b,c)$ *satisfy the following recursive relation:*

$$\begin{aligned}\alpha_{l,k}(a,b,c) &= (a+1)\alpha_{l-1,k}(a+1,b,c) + (c-a-2b-l-m+4)\alpha_{l-1,k}(a-1,b,c)\\ &\quad - 2(b+1)\alpha_{l-1,k}(a,b+1,c-1) - (c+1)\alpha_{l-1,k}(a,b-1,c+1).\end{aligned}$$

We have adopted the convention that: $\alpha_{l,k}(a,b,c) = 0$ *if any of the indices are out of bounds. Moreover, because* $|x|^l|u|^k f_{l,k}(r,s,t) \in \mathcal{P}_{l,k}$*, if:*

$$l-a-c \not\equiv 0 \mod 2 \text{ or } k-b-c \not\equiv 0 \mod 2$$

then $\alpha_{l,k}(a,b,c) = 0$.

Proof. This follows from the fact that $f_{l,k}(r,s,t) = Q_{l-1}f_{l-1,k}(r,s,t)$ and direct calculations. □

Despite the existence of this recursive relation, it proves difficult to find a general expression for the coefficients. We expected the coefficients to be a rational function involving polynomials in the dimension m of degree one. However, at some point in the calculation an irreducible (over $\mathbb{Q}$) second degree polynomial in m appears (even when restricting to low k values) which makes it impossible to recognize a product of Gamma functions (something which could lead to hypergeometric coefficients). For instance, let $k = 2$ and look at the coefficient of the term $r^2 s^2$ for the first values for l:

l-values	$\alpha_{l,2}(2,2,0)$
$l = 2$	$\frac{1}{2}(m-2)m(m+2)(m+4)$
$l = 4$	$-3(m-2)m(m+2)(m+4)(m+10)$
$l = 6$	$\frac{45}{2}(m-2)m(m+2)(m+4)(m^2+22m+104)$
$l = 8$	$-210(m-2)m(m+2)(m+4)(m+6)(m+10)(m+20)$
$l = 10$	$\frac{4725}{2}(m-2)m(m+2)(m+4)(m+6)(m+8)(m^2+38m+328)$

The reason for this unexpected term could be the following: in the classical case we found a unique invariant when looking at the repeated action of the raising operator. In our current setting this is no longer the case: the space of the $\mathfrak{so}(m-1)$-invariant polynomial solutions to $\mathcal{D}_k$ is $(k+1)$-dimensional (provided that $l \geq k$), which means that we are dealing with a certain linear combination. Fortunately we can find a suitable basis for this space.

4. Branching rules

In the previous section, we have found special solutions for $\mathcal{D}_k$ which can be written as $|x|^l |u|^k f_{lk}(r,s,t)$. As they are polynomials on $\mathbb{R}^{2m}$, this implies that

$$f_{l,k}(r,s,t) = \sum_{a=0}^{l} \sum_{b=0}^{k} \sum_{c=0}^{\min\{l-a,k-b\}} \alpha_{l,k}(a,b,c) r^a s^b t^c.$$

Since multiplying with $|x|^l |u|^k$ has to give a polynomial, we can conclude that

l even (resp. odd) $\Longrightarrow$ $\alpha_{lk}(a,b,c) = 0$ if $a+c$ is odd (resp. even)
k even (resp. odd) $\Longrightarrow$ $\alpha_{lk}(a,b,c) = 0$ if $b+c$ is odd (resp. even)

These polynomials belong to the $(k+1)$-dimensional space of $\mathfrak{so}(m-1)$-invariant polynomials in $\ker_l \mathcal{D}_k$ and the following theorem provides us with a suitable basis:

Theorem 4.1. *Let $m > 4$, and let $P_{l,k}(x,u) \in \ker_l \mathcal{D}_k$ be an $\mathfrak{so}(m-1)$-invariant solution for $\mathcal{D}_k$ with $l \geq k$. In that case there exist constants $c_i \in \mathbb{C}$ $(i = 0, \ldots, k)$ such that:*

$$P_{l,k}(x,u) = \sum_{i=0}^{k} c_i (\mathcal{J}_R \Delta_x \mathcal{J}_R)^i \langle u, \partial_x \rangle^k |x|^{l+k-2i} C_{l+k-2i}^{\frac{m}{2}-1}(r)$$

Proof. We recall the fact that $\mathcal{H}_{l,k}$ is an irreducible $\mathfrak{so}(m)$-representation with highest weight $(l,k,0,\ldots,0)$. This means that:

$$\mathcal{H}_{l,k}\Big|_{\mathfrak{so}(m-1)}^{\mathfrak{so}(m)} \cong \bigoplus_{i=k}^{l} \bigoplus_{j=0}^{k} (i,j,0,\ldots,0)$$

and thus there is no scalar component to be found unless $k = 0$. From the classical harmonic result we know that the $\mathfrak{so}(m-1)$-invariant subspace in $\mathcal{H}_{l+k-2i}$ is generated by the harmonic Gegenbauer polynomial of degree $l+k-2i$. Using the embedding factors for the simplicial harmonics in $\ker_l \mathcal{D}_k$ (see [5]), we arrive at the proof. □

From this theorem we can also conclude that each $\mathfrak{so}(m-1)$-invariant polynomial in $\ker_l \mathcal{D}_k$ has to be of the form $|x|^l|u|^k g(r,s,t)$ since it can be shown that both the operators $\langle u, \partial_x\rangle$ and $\mathcal{J}_R\Delta_x\mathcal{J}_R$ preserve this form.

5. Example

We will find an explicit formula for one of the $\mathfrak{so}(m-1)$-invariants in $\ker_l \mathcal{D}_k$ namely:

$$\langle u, \partial_x\rangle^k |x|^{l+k} C_{l+k}^{\frac{m}{2}-1}\left(\frac{x_1}{|x|}\right).$$

This is a rather special solution, as it is *not* induced by the solutions for $\ker \mathcal{D}_{k-1}$. By this we mean the following: a special class of solutions for $\mathcal{D}_k$ contains polynomials in (x,u) which belong to the kernel of both Δ_x and $\langle \partial_x, \partial_u\rangle$ (see the definition of $\mathcal{D}_k$ in Section 2). In physics, these solutions are important as they satisfy certain gauge conditions (they are harmonic and satisfy the condition $\langle \partial_x, \partial_u\rangle f(x,u) = 0$). As was shown in [5], this operator $\langle \partial_x, \partial_u\rangle$ also maps solutions for $\mathcal{D}_k$ surjectively to solutions for $\mathcal{D}_{k-1}$, although the inversion is a non-trivial operator. The component we are about to describe does not come from such an inversion procedure, as it is killed by the operator $\langle \partial_x, \partial_u\rangle$. To obtain an explicit expression we calculate the repeated action of $\langle u, \partial_x\rangle$ on

$$|x|^n C_n^{\mu}(r) = \sum_{j=0}^{\lfloor \frac{n}{2}\rfloor}(-1)^j 2^{n-2j}\frac{\Gamma(n-j+\mu)}{\Gamma(\mu)j!(n-2j)!}x_1^{n-2j}|x|^{2j} = \sum_{j=0}^{\lfloor \frac{n}{2}\rfloor}\alpha_n(j,\mu)x_1^{n-2j}|x|^{2j}.$$

We need to following lemmas, all of which are easily proven by straightforward calculations and induction.

Lemma 5.1. *Let $k \in \mathbb{N}$ and f, g k-times differentiable functions on $\mathbb{R}^m$ then*

$$\langle u, \partial_x\rangle^k (fg) = \sum_{i=0}^{k}\binom{k}{i}\left(\langle u, \partial_x\rangle^{k-i} f\right)\left(\langle u, \partial_x\rangle^{i} g\right).$$

Lemma 5.2. *Let $a, b \in \mathbb{N}$ then*

$$\langle u, \partial_x\rangle^a x_1^b = (b)_a u_1^a x_1^{b-a}$$

where $(b)_a = b(b-1)\cdots(b-a+1)$ is the lowering factorial.

Lemma 5.3. *Let $a, b \in \mathbb{N}$ then*

$$\langle u, \partial_x \rangle^a |x|^{2b} = \sum_{i=0}^{\lfloor \frac{a}{2} \rfloor} \frac{2^{a-2i}(a)_{2i}(b)_{a-i}}{i!} \langle u, x \rangle^{a-2i} |u|^{2i}|x|^{2b-2a+2i}.$$

This means that $\langle u, \partial_x \rangle^k |x|^n C_n^\mu(r)$ is given by:

$$\sum_{j=0}^{\lfloor \frac{n}{2} \rfloor} \alpha_n(j,\mu) \sum_{i=0}^{k} \binom{k}{i} \left(\langle u, \partial_x \rangle^{k-i} x_1^{n-2j} \right) \left(\langle u, \partial_x \rangle^{i} |x|^{2j} \right)$$

$$= \sum_{j=0}^{\lfloor \frac{n}{2} \rfloor} \alpha_n(j,\mu) \sum_{i=0}^{k} \binom{k}{i} \left((n-2j)_{k-i} u_1^{k-i} x_1^{n-2j-k+i} \right)$$

$$\times \sum_{h=0}^{\lfloor \frac{i}{2} \rfloor} \frac{2^{i-2h}(i)_{2h}(j)_{i-h}}{h!} \langle u, x \rangle^{i-2h} |u|^{2h}|x|^{2j-2i+2h}$$

$$= \sum_{j=0}^{\lfloor \frac{n}{2} \rfloor} \sum_{i=0}^{k} \sum_{h=0}^{\lfloor \frac{i}{2} \rfloor} \gamma_{n,k,\mu}(i,j,h) u_1^{k-i} x_1^{n-2j-k+i} \langle u, x \rangle^{i-2h} |u|^{2h}|x|^{2j-2i+2h}$$

$$= |x|^{n-k}|u|^k \sum_{j=0}^{\lfloor \frac{n}{2} \rfloor} \sum_{i=0}^{k} \sum_{h=0}^{\lfloor \frac{i}{2} \rfloor} \gamma_{n,k,\mu}(i,j,h) \left(\frac{u_1}{|u|}\right)^{k-i} \left(\frac{x_1}{|x|}\right)^{n-2j-k+i} \left(\left\langle \frac{u}{|u|}, \frac{x}{|x|} \right\rangle\right)^{i-2h}$$

with

$$\gamma_{n,k,\mu}(i,j,h) := \alpha_n(j,\mu) \binom{k}{i} (n-2j)_{k-i} \frac{2^{i-2h}(i)_{2h}(j)_{i-h}}{h!}.$$

If we want to write this into our chosen standard form then we would have to do the following substitutions:

$$a := n - 2j - k + i$$
$$b := k - i$$
$$c := i - 2h.$$

It is here that the parity conditions on our coefficients will appear. Since $c = k - b - 2h$ we know that $c + b \equiv k \mod 2$. Completely analogue one can use the fact that $a = n - 2j - k + c + 2h$ to conclude that $a + c \equiv n - k \mod 2$. Also we can see that $c \leq k - b$ and $c \leq n - k - a$ to end up with:

$$\sum_{a=0}^{n-k} \sum_{b=0}^{k} \sum_{c=0}^{\min(n-k-a,k-b)} \epsilon_{n,k}(a,b,c) \gamma_{n,k,\mu} \left(k - b, \frac{n-a-b}{2}, \frac{k-b-c}{2} \right) r^a s^b t^c$$

where $\epsilon_{n,k}(a,b,c) = 0$ if the parity conditions are not met, and equal to 1 otherwise. These conditions also guarantee that the arguments of the $\gamma_{n,k,\mu}$ are positive integers. There is however a way to get rid of the factor $\epsilon_{n,k}(a,b,c)$ in the summation if we slightly change our summation indices. If we use the fact that, for each

a, b, c, there have to exist i, j such that $a = n - k - c - 2i$ and $b = k - c - 2j$ we can write our expression as:

$$\sum_{c=0}^{\min(n-k,k)} \sum_{i=0}^{\lfloor \frac{n-k-c}{2} \rfloor} \sum_{j=0}^{\lfloor \frac{k-c}{2} \rfloor} \gamma_{n,k,\mu} (c + 2j, c + i + j, j)\, r^{n-k-c-2i} s^{k-c-2j} t^c.$$

Combining all of the above gives us the following theorem:

Theorem 5.4. *For each $l, k \in \mathbb{N}$ the following $\mathfrak{so}(m-1)$-invariant polynomial belongs to $\ker_l \mathcal{D}_k$:*

$$\sum_{a=0}^{l} \sum_{b=0}^{k} \sum_{c=0}^{\min(l-a,k-b)} \epsilon_{l,k}(a, b, c) \gamma_{l,k} \left(k - b, \frac{l + k - a - b}{2}, \frac{k - b - c}{2} \right) r^a s^b t^c$$

where

$$\gamma_{l,k}(i, j, h) := (-1)^j 2^{l+k+i-2j-2h} \binom{k}{i} \frac{(i)_{2h} (j)_{i-h}}{h! j! (l + i - 2j)!} \frac{\Gamma(l + k - j + \frac{m}{2} - 1)}{\Gamma(\frac{m}{2} - 1)}$$

and $\epsilon_{l,k}(a, b, c)$ is equal to 1 when both $a + c \equiv l \mod 2$ and $b + c \equiv k \mod 2$, and equal to zero otherwise.

6. Further research

It is clear that knowing invariants in the spaces $\mathcal{H}_{i,j}(\mathbb{R}^{2m}, \mathbb{C})$ under a certain subalgebra is crucial in order to understand the invariants in the kernel of $\mathcal{D}_k$. There is however one particular important subspace of $\ker_l \mathcal{D}_k$ we are interested in: namely $\mathcal{H}_{l,k}(\mathbb{R}^{2m}, \mathbb{C})$. To find an invariant in the latter space a branching to $\mathfrak{so}(m-1)$ will be insufficient, but when branching to $\mathfrak{so}(m-2)$ one has to deal with multiplicities (see Section 4). In [8] we further explore this problem to find an arguably more suitable generalisation for the harmonic Gegenbauer polynomials.

References

[1] Askey, R., *Orthogonal Polynomials and Special Functions*, Society for Industrial and Applied Mathematics, 1975

[2] Brackx, F. and Delanghe, R. and Sommen, F., *Clifford analysis*, Pitman Advanced Publishing Program, 1982

[3] Bureš, J., Sommen, F., Souček, V., Van Lancker, P., *Rarita–Schwinger type operators in Clifford analysis*, Journal of Funct. Anal. **185** (2001), pp. 425–456.

[4] Constales, D., Sommen. F., Van Lancker, P., *Models for irreducible representations of* Spin(m), Adv. Appl. Clifford Algebras **11** No. S1 (2001), pp. 271–289.

[5] De Bie, H., Eelbode, D., Roels, M., *The higher spin Laplace operator*, submitted.

[6] Delanghe, R., Sommen, F., Souček, V., *Clifford analysis and spinor valued functions*, Kluwer Academic Publishers, Dordrecht (1992).

[7] De Schepper, H., Eelbode, D., Raeymaekers, T., *On a special type of solutions of arbitrary higher spin Dirac operators*, J. Phys. A **43** No. 32 (2010), pp. 1–13.

[8] Eelbode, D., Janssens, T., *Higher spin generalisation of the Gegenbauer polynomials*, in preparation.

[9] Eelbode, D., Raeymaekers, T., Van der Jeugt, J., *Decomposition of the polynomial kernel of arbitrary higher spin Dirac operators*, J. Math. Phys. **56** No. 10 (2015).

[10] Eelbode, D., Raeymaekers, T., Van Lancker, P., *On the fundamental solution for higher spin Dirac operators*, J. Math. Anal. Appl. **405** No. 2 (2013), pp. 555–564.

[11] Eelbode, D., Šmíd, D., *Factorization of Laplace operators on higher spin representations*, Compl. Anal. Oper. Theo **6** No. 5 (2012), pp. 1011–1023.

[12] Eelbode, D., Souček, V., Van Lancker, P., *Gegenbauer polynomials and the Fueter theorem*, Compl. Var. Ell. Equa. **59** No. 6 (2014), pp. 826–840.

[13] Eelbode, D., Van Lancker, P., *Total Rarita–Schwinger operators in Clifford analysis*, Ann. Glob. Anal. Geom. **42** No. 4 (2012), pp. 473–493.

[14] Fulton, W. and Harris, J., *Representation theory: A first course*, Springer-Verlag, 1991

[15] Gilbert, J., Murray, M., *Clifford algebras and Dirac operators in harmonic analysis*, Cambridge University Press, Cambridge (1991).

[16] Gürlebeck, K. Sprössig, W., *Quaternionic Analysis and Elliptic Boundary Value Problems*, ISNM 89, Birkhäuser Verlag, Basel (1990).

[17] Lavička, R. and Souček, V. and Van Lancker, P., *Orthogonal basis for spherical monogenics by step two branching*, Annals of Global Analysis and Geometry, volume 41 No. 2, pp. 161–186, 2012

David Eelbode and Tim Janssens
University of Antwerp
Department of Mathematics
Middelheimlaan, 1
B-2020 Antwerp, Belgium
e-mail: `david.eelbode@uantwerpen.be`
`tim.janssens4@uantwerpen.be`

Modern Trends in Hypercomplex Analysis
Trends in Mathematics, 175–189

A New Cauchy Type Integral Formula for Quaternionic k-hypermonogenic Functions

Sirkka-Liisa Eriksson and Heikki Orelma

Abstract. In complex function theory holomorphic functions are conjugate gradient of real harmonic functions. We may build function theories in higher dimensions based on this idea if we generalize harmonic functions and define the conjugate gradient operator. We study this type of function theory in $\mathbb{R}^3$ connected to harmonic functions with respect to the Laplace–Beltrami operator of the Riemannian metric $ds^2 = x_2^{-2k}\left(\sum_{i=0}^{2} dx_i^2\right)$. The domain of the definition of our functions is in $\mathbb{R}^3$ and the image space is the associative algebra of quaternions $\mathbb{H}$ generated by $1, e_1, e_2$ and $e_{12} = e_1 e_2$ satisfying the relation $e_i e_j + e_j e_i = -2\delta_{ij}$, $i, j = 1, 2$. The complex field $\mathbb{C}$ is identified by the set $\{x_0 + x_1 e_1 \mid x_0, x_1 \in \mathbb{R}\}$. The conjugate gradient is defined in terms of modified Dirac operator, introduced by $M_k f = Df + k x_2^{-1}\overline{Qf}$, where Qf is given by the decomposition $f(x) = Pf(x) + Qf(x)\, e_2$ with $Pf(x)$ and $Qf(x)$ in $\mathbb{C}$ and $\overline{Qf}$ is the usual complex conjugation.

Leutwiler noticed around 1990 that if the usual Euclidean metric is changed to the hyperbolic metric of the Poincaré upper half-space model ($k = 1$), then the power function $(x_0 + x_1 e_1 + x_2 e_2)^n$, calculated using quaternions, is the conjugate gradient of the a hyperbolic harmonic function. We study functions, called k-hypermonogenic, satisfying $M_k f = 0$. Monogenic functions are 0-hypermonogenic. Moreover, 1-hypermonogenic functions are hypermonogenic defined by H. Leutwiler and the first author.

We prove a new Cauchy type integral formulas for k-hypermonogenic functions where the kernels are calculated using the hyperbolic distance and are k-hypermonogenic functions. This formula gives the known formulas in case of monogenic and hypermonogenic functions. It also produces new Cauchy and Teodorescu type integral operators investigated in the future research.

Mathematics Subject Classification (2010). Primary 30G35; Secondary 30A96.

Keywords. k-hypermonogenic, k-hyperbolic harmonic, Laplace–Beltrami, monogenic, Clifford algebra, hyperbolic metric, hyperbolic Laplace.

1. Introduction

We study generalized function theory connected to the Riemannian metric

$$ds^2 = \frac{dx_0^2 + dx_1^2 + dx_2^2}{x_2^{2k}}.$$

When $k = 0$ the metrics is Euclidean and when $k = 1$ the metric is the hyperbolic metric of Poincaré upper half-space. We are studying functions whose domain is in the Euclidean space $\mathbb{R}^3$ and the image space the associative real division algebra of quaternions. In this case we define generalized holomorphic functions, called k-hypermonogenic. Our theory combines together the theory of monogenic functions $(k = 0)$ and hypermonogenic functions $(k = 1)$. Moreover, it is also connected to the eigenfunctions of the hyperbolic Laplace operator of the Poincaré upper half-space model.

Hypermonogenic functions were introduced by H. Leutwiler and the first author in [6]. Overview to the theory is written in [8] or [10]. Two types of Cauchy integral formulas for hypermonogenic functions were proved in [7] and the total formula with two hypermonogenic kernels in [2]. The formulas were improved to contain just one single kernel in [9] and [4]. Later in [5] it was invented the surprising result that the kernel is the Cauchy kernel of monogenic function shifted to the Euclidean center of the hyperbolic ball.

The general complicated integral formulas for k-hypermonogenic functions were proved in [3]. In our main result of this paper we present in $\mathbb{R}^3$ a new Cauchy formula for k-hypermongenic functions.This formula corrects and improves the formula presented in [14, Theorems 3.22 and 3.23] where the function g_k has a calculation error. Moreover the kernels are vanishing at the infinity. When $k = 0$, this formula is just the Cauchy formula of monogenic functions. In case $k = 1$, we obtain the Cauchy formula of hypermonogenic functions. Moreover our formula also gives an integral operators of boundary functions producing a k-hypermongenic function. The generalization of these results to higher dimensions is under investigations.

2. Preliminaries

To make the reading easier, we recall the notations and main concepts used in this paper. The real associated division algebra of quaternions is denoted by $\mathbb{H}$. Its generating elements are denoted by e_1, e_2, e_3 and they satisfy the properties $e_1e_2 = e_3$ and $e_ie_j + e_je_i = -2\delta_{ij}$ where δ_{ij} is the usual Kronecker delta. Elements $x = x_0 + x_1e_1 + x_2e_2$ for $x_0, x_1, x_2 \in \mathbb{R}$ are called *paravectors*. The vector space $\mathbb{R}^3$ is identified with the real vector space of paravectors and therefore elements $x_0 + x_1e_1 + x_2e_2$ and (x_0, x_1, x_2) are identified. The field of complex numbers is identified with the field $\{x_0 + x_1e_1 \mid x_0, x_1 \in \mathbb{R}\}$. Our general assumption is that the domain of our functions is an open subset Ω of $\mathbb{R}^3$ and their image space is the set of quaternions. We also assume that our functions are continuously differentiable.

We use three common involutions. If $q = x_0 + x_1e_1 + x_2e_2 + x_3e_3$ and $x_0, x_1, x_2, x_3 \in \mathbb{R}$ they are defined by

$$\begin{aligned} q' &= x_0 - x_1e_1 - x_2e_2 + x_3e_3, && \text{(the main involution)} \\ q^* &= x_0 + x_1e_1 + x_2e_2 - x_3e_3, && \text{(the reversion)} \\ \bar{q} &= x_0 - x_1e_1 - x_2e_2 - x_3e_3. && \text{(the conjugation)} \end{aligned}$$

The conjugation satisfies $\overline{q} = (q')^* = (q^*)'$.

Using the unique decomposition $q = u + ve_2$ for $u, v \in \mathbb{C}$ we define the mappings $P : \mathbb{H} \to \mathbb{C}$ and $Q : \mathbb{H} \to \mathbb{C}$ by $Pq = u$ and $Qq = v$ (see [6]). In order to compute the P- and Q-parts easily, we define also a new involution $q \to \widehat{q}$ by

$$\widehat{q} = q_0 + q_1e_1 - q_2e_2 - q_3e_3.$$

The simple observations

$$a'e_2 = e_2\widehat{a} \qquad \text{and} \qquad \widehat{a} = -e_2a'e_2$$

hold for all $a \in \mathbb{H}$. We also obtain the formulas

$$Pq = \frac{1}{2}(q + \widehat{q}) = \frac{1}{2}(q - e_2q'e_2) = -\frac{1}{2}(qe_2 + e_2q')\,e_2, \tag{2.1}$$

$$Qq = -\frac{1}{2}(q - \widehat{q})\,e_2 = -\frac{1}{2}(qe_2 - e_2q'). \tag{2.2}$$

The involutions satisfy the following product rules

$$\begin{aligned} (ab)' &= a'b', & (ab)^* &= b^*a^*, \\ \overline{ab} &= \overline{b}\,\overline{a}, & \widehat{ab} &= \widehat{a}\widehat{b} \end{aligned}$$

for all quaternions a and b. The mappings P and Q have the product rules (see [6])

$$P(ab) = (Pa)\,Pb - (Qa)\,Q'(b)\,, \tag{2.3}$$

$$Q(ab) = (Pa)\,Qb + (Qa)\,P'(b) = aQb + (Qa)\,b'. \tag{2.4}$$

Moreover if $a \in \mathbb{C}$ then

$$a' = \overline{a} \tag{2.5}$$

$$\widehat{a} = a \tag{2.6}$$

and

$$a'e_2 = e_2a.$$

The topology in $\mathbb{H}$ is introduced by the norm

$$|q| = \sqrt{x_0^2 + x_1^2 + x_2^2 + x_3^2} = \sqrt{q\overline{q}}.$$

The left and right Dirac (or Cauchy–Riemann) operators in $\mathbb{H}$ are defined by

$$D_lf = \frac{\partial f}{\partial x_0} + e_1\frac{\partial f}{\partial x_1} + e_2\frac{\partial f}{\partial x_2}, \quad D_rf = \frac{\partial f}{\partial x_0} + \frac{\partial f}{\partial x_1}e_1 + \frac{\partial f}{\partial x_2}e_2$$

and their conjugate operators $\overline{D}_l$ and $\overline{D}_r$ by

$$\overline{D}_l f = \frac{\partial f}{\partial x_0} - e_1 \frac{\partial f}{\partial x_1} - e_2 \frac{\partial f}{\partial x_2}, \quad \overline{D}_r f = \frac{\partial f}{\partial x_0} - \frac{\partial f}{\partial x_1} e_1 - \frac{\partial f}{\partial x_2} e_2.$$

The modified Dirac operators M_k^l, $\overline{M}_k^l$, M_k^r and $\overline{M}_k^r$ in $\mathbb{H}$ and for $k \in \mathbb{R}$ are introduced (see [1]) by

$$M_k^l f(x) = D_l f(x) + k \frac{\overline{Qf(x)}}{x_2}, \qquad M_k^r f(x) = D_r f(x) + k \frac{Qf(x)}{x_2},$$

$$\overline{M}_k^l f(x) = \overline{D}_l f(x) - k \frac{\overline{Qf(x)}}{x_2}, \qquad \overline{M}_k^r f(x) = \overline{D}_r f(x) + k \frac{Qf(x)}{x_2}$$

for $x \in \{x \in \Omega \mid x_2 \neq 0\}$. The operator M_1^l is abbreviated by M (see [6]).

Definition 2.1. Let $\Omega \subset \mathbb{R}^3$ be an open set. Let $k \in \mathbb{R}$. A mapping $f : \Omega \to \mathbb{H}$ is called left ***k*-hypermonogenic**, if $f \in C^1(\Omega)$ and $M_k^l f(x) = 0$ for any $x \in \{x \in \Omega \mid x_2 \neq 0\}$. The 0-hypermonogenic functions are called **monogenic**. The 1-hypermonogenic functions are called briefly **hypermonogenic**. The right k-hypermonogenic functions are defined similarly. A twice continuously differentiable function $f : \Omega \to \mathbb{H}$ is called ***k*-hyperbolic harmonic** if $\overline{M}_k^l M_k^l f = 0$ for any $x \in \{x \in \Omega \mid x_2 \neq 0\}$.

We have the following characterization of k-hyperbolic harmonic functions.

Proposition 2.2 (cf. [1]). *Let $f : \Omega \to \mathbb{H}$ be twice continuously differentiable. Then*

$$x_2^2 M_k \overline{M}_k f = x_2^2 \Delta f - k x_2 \frac{\partial f}{\partial x_2} + kQf e_2.$$

Moreover f is k-hyperbolic if and only if

$$x_2^2 \Delta P f - k x_2 \frac{\partial P f}{\partial x_2} = 0,$$

$$x_2^2 \Delta Q f - k x_2 \frac{\partial Q f}{\partial x_2} + kQf = 0.$$

We directly obtain the following corollaries.

Corollary 2.3. *A twice twice continuously differentiable function $f : \Omega \to \mathbb{H}$ is k-hyperbolic harmonic for any $k \in \mathbb{R}$ if nd only if it has the presentation*

$$f(x) = g(x) + c x_2 e_2,$$

where $g : \Omega \to \mathbb{C}$ is a harmonic function independent of x_2 and $c \in \mathbb{C}$.

Corollary 2.4. *The identity function $f(x) = x$ is k-hyperbolic harmonic for any value of k.*

We recall two main relations between k-hyperbolic harmonic functions and k-hypermonogenic functions.

Theorem 2.5 (cf. [6]). *Let $\Omega \subset \mathbb{R}^3$ be an open set and $f : \Omega \to \mathbb{H}$ be twice continuously differentiable. Then f is k-hypermonogenic if and only if f and $xf(x)$ are k-hyperbolic harmonic functions.*

Proposition 2.6 (cf. [1]). *Let $\Omega \subset \mathbb{R}^3$ be an open set. If a function $h : \Omega \to \mathbb{H}$ is k-hyperbolic harmonic then $\overline{M}_k^l h$ is k-hypermonogenic. Conversely, if a mapping $f : \Omega \to \mathbb{H}$ is k-hypermonogenic there exists locally a complex k-hyperbolic harmonic function h satisfying $f = \overline{D}h$.*

A key observation is the following relation between k- and $-k$-hypermonogenic functions.

Theorem 2.7 (cf. [2]). *Let Ω be an open subset of $\mathbb{R}^3 \backslash \{x_2 = 0\}$ and $f : \Omega \to \mathbb{H}$ be a $C^1(\Omega, \mathbb{H})$ function. A function $f : \Omega \to \mathbb{H}$ is k-hypermonogenic if and only if the function $x_2^{-k} f e_n$ is $-k$-hypermonogenic.*

3. Cauchy formula for k-hypermonogenic functions

The hyperbolic metric of the Poincaré upper half-space model is the Riemannian metric

$$ds^2 = \frac{dx_0^2 + dx_1^2 + dx_2^2}{x_2^2}$$

and its Laplace–Beltrami operator is

$$\Delta_h f = x_2^2 \Delta f - x_2 \frac{\partial f}{\partial x_2},$$

which is also called the hyperbolic Laplace operator. The hyperbolic distance may be computed as follows (see [16]).

Lemma 3.1. *The hyperbolic distance $d_h(x, a)$ between the points x and a in $\mathbb{R}^3_+$ is*

$$d_h(x, a) = \operatorname{arcosh} \lambda(x, a) = \ln\left(\lambda(x, a) + \sqrt{\lambda(x, a)^2 - 1}\right),$$

where

$$\lambda(x, a) = \frac{|x-a|^2 + |x-\hat{a}|^2}{4x_2 a_2} = \frac{|x-a|^2}{2x_2 a_2} + 1$$

and $|x - a|$ is the usual Euclidean distance in $\mathbb{R}^3$ between the points a and x.

We recall the following important relation between the Euclidean and hyperbolic balls.

Proposition 3.2 (cf. [16]). *The hyperbolic ball $B_h(a, r_h)$ in $\mathbb{R}^3_+$ with the hyperbolic center $a = a_0 + a_1 e_1 + a_2 e_2$ and the radius r_h is the same as the Euclidean ball with the Euclidean center*

$$c_a(r_h) = a_0 + a_1 e_1 + a_2 \cosh r_h e_2$$

and the Euclidean radius $r_e = a_2 \sinh r_h$.

We use the following calculation rules, proved in [11].

Lemma 3.3. *If* $c(x,a) = Pa + a_2\lambda(x,a)e_2$ *then*

$$\overline{D}^x \lambda(x,a) = \frac{\overline{x - c(x,a)}}{a_2 x_2},$$

$$\overline{D}^x d_h(x,a) = \frac{\overline{x - c(x,a)}}{a_2 x_2 \sinh d_h(x,a)} = \frac{\overline{x - c(x,a)}}{x_2 |x - c(x,a)|}.$$

We remark that applying the previous proposition we notice that $c(x,a)$ is the Euclidean center of the hyperbolic ball $B_h(a, d_h(x,a))$ and the value $|x - c(x,a)|$ is its Euclidean radius.

A key tool is the relation between k-hyperbolic harmonic functions and the eigenfunctions of the hyperbolic Laplace–Beltrami operator, stated slightly more general form as follows.

Proposition 3.4 (cf. [12]). *If* u *is a real-valued solution of the equation*

$$x_2^2 \triangle h(x) - k x_2 \frac{\partial h}{\partial x_2}(x) + l h(x) = 0 \tag{3.1}$$

in an open subset $\Omega \subset \mathbb{R}^3_+$, *then* $f(x) = x_2^{\frac{1-k}{2}} u(x)$ *is the eigenfunction of the hyperbolic Laplace operator corresponding to the eigenvalue* $\frac{1}{4}\left(k^2 + 2k - 3 - 4l\right)$. *Conversely, if* f *is an eigenfunction of the hyperbolic Laplace operator corresponding to the eigenvalue* l *in an open subset* $\Omega \subset \mathbb{R}^3_+$ *then* $u(x) = x_2^{\frac{k-1}{2}} f(x)$ *is the solution of the equation* (3.1) *in* Ω *with* $\gamma = \frac{1}{4}\left(k^2 + 2k - 3 - 4l\right)$.

The hyperbolic Laplace for functions depending only on $d_h(x, e_2) = r_h$ is

$$\triangle_h f(r_h) = \frac{\partial^2 f}{\partial r_h^2} + 2 \coth r_h \frac{\partial f}{\partial r_h}$$

(see [11] and [13]). The general solution depending on the hyperbolic distance is computed in [14].

Theorem 3.5 (cf. [14]). *The general solution of the equation*

$$\frac{\partial^2 f}{\partial r_h^2} + 2 \coth r_h \frac{\partial f}{\partial r_h} + \gamma f = 0 \tag{3.2}$$

is

$$f(r_h) = \begin{cases} \frac{C \cosh\left(\sqrt{1-\gamma} r_h\right)}{\sinh r_h} + \frac{C_0 \sinh\left(\sqrt{1-\gamma} r_h\right)}{\sinh r_h}, & \text{if } \gamma < 1, \\ \frac{C}{\sinh r_h} + \frac{C_0 r_h}{\sinh r_h}, & \text{if } \gamma = 1, \\ \frac{C \cos\left(\sqrt{\gamma-1} r_h\right)}{\sinh r_h} + \frac{C_0 \sin\left(\sqrt{\gamma-1} r_h\right)}{\sinh r_h}, & \text{if } \gamma > 1, \end{cases}$$

for real constants C *and* C_0.

In our case $\gamma = \frac{1}{4}(4 - (k+1)^2)$, hence there are bounded solutions if $-3 \leq k \leq 1$.

Corollary 3.6. *If $-3 \leq k \leq 1$ the function*

$$f(r_h) = \frac{C_0 \sinh\left(\frac{|k+1|}{2} r_h\right)}{\sinh r_h}$$

is a bounded solution of the equation (3.2) *vanishing when $r_h(x,a) \to \infty$.*

A kernel function in not unique but we later see that the following functions produce k-hypermonogenic functions possessing the nicest symmetry properties.

Proposition 3.7. *If $\gamma = \frac{1}{4}(4-(k+1)^2)$ then one solution of* (3.2) *with a singularity at a vanishing when $d_h(x,a) \to \infty$ is*

$$F_k(x,a) = \begin{cases} \frac{\cosh\left(\frac{d_h(x,a)(k+1)}{2}\right)}{\sinh d_h(x,a)}, & \text{if } -1 < k < 1 \\ \frac{e^{-\frac{|k+1| d_h(x,a)}{2}}}{\sinh d_h(x,a)}, & \text{if } k \leq -1 \text{ or } k \geq 1. \end{cases}$$

Moreover the function $g_k(x,a) = x_2^{\frac{k-1}{2}} a_2^{\frac{k-1}{2}} F_k(x,a)$ is k-hyperbolic harmonic with respect to the both variables x and a outside the point $x = a$ vanishing when $x_2 \to \infty$.

Proof. The first statement is clear. We look more carefully the second statement. Without loosing the generality, we may pick $a = e_2$ and abbreviate $\lambda = \lambda(x, e_2)$. If $-1 < k < 1$ the result holds. If $k \leq -1$ or $k \geq 1$ applying Lemma 3.1 we obtain

$$\lambda^2 - 1 = (\lambda - 1)(\lambda + 1) = \frac{|x - e_2|^2 |x + e_2|^2}{2x_2}$$

and therefore

$$\begin{aligned} \frac{x_2^{\frac{k-1}{2}} e^{-\frac{|k+1| d_h(x,a)}{2}}}{\sinh d_h(x,a)} &= \frac{x_2^{\frac{k-1}{2}}}{\left(\lambda + \sqrt{\lambda^2 - 1}\right)^{\frac{|k+1|}{2}} \sqrt{\lambda^2 - 1}} \\ &= \frac{2^{\frac{|k+1|+2}{2}} x_2^{\frac{k+1}{2} + \frac{|k+1|}{2}}}{\left(1 + |Px|^2 + x_2^2 + |x - e_2||x + e_2|\right)^{\frac{|k+1|}{2}} |x - e_2||x + e_2|} \\ &\leq \frac{2^{\frac{|k+1|+2}{2}}}{x_2^{\frac{|k+1|}{2} - \frac{k+1}{2}} |x_2^2 - 1|}, \end{aligned}$$

completing the proof of the first case.

Similarly in cases $k \leq -1$ and $k \geq 1$ we compute

$$h_k(x,a) = a_2^{s+1} x_2^{s-1} w_k(x,a) p(x,a),$$

where

$$w_k(x,a) = (1-s)\,e^{-|s|r_h} e_2 \frac{x - c_a(r_h)}{a_2} - |s| \sinh r_h e^{-|s|r_h} - e^{-|s|r_h} \cosh r_h$$
$$= \begin{cases} (1-s)\,e^{-sr_h} e_2 \frac{x-Pa}{a_2} - se^{-(s-1)r_h}, \text{ if } k \geq 1, \\ (1-s)\,e^{sr_h} e_2 \frac{x-Pa}{a_2} + se^{(s-1)r_h}, \text{ if } k \leq -1. \end{cases}$$

Applying [11], we infer that the function $p(x,a)$ is hypermonogenic, completing the proof. □

The kernel functions have been computed earlier in the classical harmonic case $k = 0$ and in the hyperbolic case $k = -1$ and $k = 1$ (see [15] and in the general case [2]). Our formulas are the same up to multiplying constants.

Corollary 3.8.

$$F_k(x,a) = \begin{cases} \frac{1}{|x-a|}, \ \textit{if } k = 0, \\ \frac{1}{a_2 x_2 \sinh d_h(x,a)} = \frac{2}{|x-a_2||x+a_2|}, \ \textit{if } k = -1, \\ \coth d_h(x,a) - 1, \ \textit{if } k = 1. \end{cases}$$

Proof. If $k = 0$ then using hyperbolic identities we compute

$$F_0(x,a) = \frac{\cosh\left(\frac{d_h(x,a)}{2}\right)}{\sqrt{x_2 a_2} \sinh d_h(x,a)}$$
$$= \frac{\sqrt{\cosh d_h(x,a) + 1}}{\sqrt{2x_2 a_2}\sqrt{\lambda^2 - 1}}$$
$$= \frac{1}{\sqrt{2x_2 a_2}\sqrt{\lambda - 1}} = \frac{1}{|x-a|}.$$

Similarly we calculate the case $k = -1$ as follows

$$F_{-1}(x,a) = \frac{1}{a_2 x_2 \sinh d_h(x,a)}$$
$$= \frac{1}{a_2 x_2 \sqrt{\lambda^2 - 1}} = \frac{2}{|x - a_2|\,|x + a_2|}.$$

The case $k = 1$ is obtained from

$$F_1(x,a) = \coth d_h(x,a) - 1.$$ □

Using Proposition 2.6, we may directly compute the corresponding k-hypermonogenic function.

Theorem 3.9. *Set* $r_h = d_h(x,a)$ *and* $s = \frac{k+1}{2}$. *If we denote*

$$w_k(x,a) = \begin{cases} (1-s)\,v(sr_h)\,e_2 \frac{x-Pa}{a_2} - sv((s-1)\,r_h), \ \textit{if } \ -1 < k, \\ v(sr_h)\left((1-s)\,e_2 \frac{x-Pa}{a_2} + se^{-r_h}\right), \ \textit{if } \ k \leq -1, \end{cases}$$

$$v(sr_h) = \begin{cases} \cosh\left(\frac{d_h(x,a)(k+1)}{2}\right), & \text{if } -1 < k < 1, \\ e^{-\frac{|k+1| d_h(x,a)}{2}}, & \text{if } k \le -1 \text{ or } k \ge 1. \end{cases}$$

and

$$p(x,a) = \frac{(x - c_a(r_h))^{-1}}{x_2 |x - c_a(r_h)|}$$

then the function

$$h_k(x,a) = a_2^{s+1} x_2^{s-1} w_k(x,a) p(x,a)$$

is paravector-valued k-hypermonogenic outside $x = a$ with respect to x and $p(x,a)$ is hypermonogenic with respect to x.

Proof. Denote $s = \frac{k+1}{2}$. Applying the previous corollary and Proposition 2.6, we note that the function

$$g_k(x_2, r_h) = \frac{a_2^{s-1} x_2^{s-1} v(sr_h)}{\sinh r_h}$$

is k-hyperbolic harmonic and $h_k = \overline{D}^x g(x_2, r_h)$ is k-hypermonogenic. Assume first that $-1 < k < 1$. We just make simple calculations

$$\frac{h_k(x,a)}{a_2^{s-1} x_2^{s-1}} = -(s-1) \frac{v(sr_h) e_2}{x_2 \sinh r_h} + \left(\frac{\sinh r_h \sinh(sr_h) - v(sr_h) \cosh r_h}{\sinh^2 r_h} \right) \overline{D}^x r_h.$$

Applying Lemma 3.3, we obtain

$$\frac{\overline{D}^x r_h}{a_2^2 \sinh^2 r_h} = \frac{\overline{x - c_a(r_h)}}{x_2 |x - c_a(r_h)|^3} = \frac{(x - c_a(r_h))^{-1}}{x_2 |x - c_a(r_h)|}$$

and

$$\begin{aligned} \frac{x - c(x,a)}{a_2} \frac{(x - c_a(r_h))^{-1}}{x_2 |x - c_a(r_h)|} &= \frac{1}{a_2 x_2 |x - c_a(r_h)|} \\ &= \frac{1}{a_2^2 x_2 \sinh r_h}. \end{aligned}$$

Hence we obtain

$$\frac{h_k(x,a)}{a_2^{s+1} x_2^{s-1}} = w_k(x,a) \frac{(x - c_a(r_h))^{-1}}{x_2 |x - c_a(r_h)|}$$

and

$$\begin{aligned} w_k(x,a) &= (1-s) v(sr_h) e_2 \frac{x - c_a(r_h)}{a_2} + s \sinh r_h \sinh(sr_h) - v(sr_h) \cosh r_h \\ &= (1-s) v(sr_h) e_2 \frac{x - Pa}{a_2} + s(\sinh r_h \sinh(sr_h) - v(sr_h) \cosh r_h) \\ &= (1-s) v(sr_h) e_2 \frac{x - Pa}{a_2} - sv((s-1) r_h). \end{aligned}$$

□

Denote the real surface measure by dS and a real weighted volume measure by

$$dm_k = \frac{1}{x_2^k} dm.$$

Using [14] we obtain the integral formula for the P-part.

Theorem 3.10. *Let Ω be an open subset of $\mathbb{R}^3_+$ (or $\mathbb{R}^3_-$) and $\overline{K} \subset \Omega$ be a smoothly bounded compact set with the outer unit normal field ν. Let $s = \frac{k+1}{2}$ and*

$$h_k(x,a) = a_2^{s+1} x_2^{s-1} w_k(x,a)\, p(x,a)$$

be the same function as in Theorem 3.9. If f is k-hypermonogenic in Ω and $a \in K$, then

$$Pf(a) = \frac{1}{4\pi} \int_{\partial K} P\left(h_k(x,a)\, \nu f(x)\right) \frac{dS}{x_2^k}.$$

Similarly using [14] we obtain the integral formula also for the Q-part.

Theorem 3.11. *Let Ω be an open subset of $\mathbb{R}^3_+$ (or $\mathbb{R}^3_-$) and $\overline{K} \subset \Omega$ be a smoothly bounded compact set with the outer unit normal field ν. The function*

$$v_{-k}(x,a) = a_2^k h_{-k}(x,a)$$

is paravector-valued $-k$-hypermonogenic outside $x = a$ with respect to x. If f is k-hypermonogenic in Ω and $a \in K$, then

$$Qf(a) = \frac{1}{4\pi} \int_{\partial K} Q\left(v_{-k}(x,a)\, \nu f(x)\right) dS.$$

We recall the Cauchy-type kernel of hypermonogenic functions.

Lemma 3.12 (cf. [5]). *If x and a belong to $\mathbb{R}^3_+$ then the function*

$$\begin{aligned}\frac{\left(x - c(x,a)\right)^{-1}}{x_2 \left|x - c(x,a)\right|} &= 4x_2 \frac{(x-a)^{-1}}{|x-a|} e_2 \frac{(x-\widehat{a})^{-1}}{|x-\widehat{a}|} \\ &= 4x_2 \frac{(x-\widehat{a})^{-1}}{|x-\widehat{a}|} e_2 \frac{(x-a)^{-1}}{|x-a|}\end{aligned}$$

is hypermonogenic with respect to x outside the point $x = a$.

We verify important symmetry properties of the kernels.

Lemma 3.13.

$$\begin{aligned}Ph_k(x,a) &= \overline{D}_1^x g_k(x,a) \\ &= -\overline{D}_1^a g_k(a,x) = -Ph_k(a,x),\end{aligned}$$

where

$$h_k(a,x) = \overline{D}^a g_k(a,x)$$

and

$$\overline{D}_1^a f = \frac{\partial f}{\partial a_0} + e_1 \frac{\partial f}{\partial a_1}.$$

Proof. We recall that

$$\overline{D}^x d_h(x,a) = \frac{\overline{x - c(x,a)}}{a_2 x_2 \sinh d_h(x,a)} = \frac{\overline{x - c(x,a)}}{x_2 |x - c(x,a)|}$$

and

$$\overline{D}^a d_h(x,a) = \frac{\overline{a - c(a,x)}}{a_2 |a - c(a,x)|} = \frac{\overline{a - c(a,x)}}{a_2 x_2 \sinh d_h(x,a)}.$$

Denoting g $s = \frac{k+1}{2}$ and $d_h(x,a) = r_h$, we just compute

$$\begin{aligned}
\overline{D_1}^x g(x_2, r_h) &= a_2^{\frac{k-1}{2}} x_2^{\frac{k-1}{2}} \left(\frac{sv'(sr)\sin r_h - v(sr)\cosh r_h}{\sinh^2 r_h} \right) \overline{D_1}^x r_h \\
&= a_2^{\frac{k-3}{2}} x_2^{\frac{k-3}{2}} \left(\frac{sv'(sr)\sin r_h - v(sr)\cosh r_h}{\sinh^2 r_h} \right) \frac{\overline{Px - Pa}}{\sinh r_h} \\
&= a_2^{\frac{k-3}{2}} x_2^{\frac{k-3}{2}} \left(\frac{sv'(sr)\sin r_h - v(sr)\cosh r_h}{\sinh^3 r_h} \right) \left(\overline{Px - Pa}\right).
\end{aligned}$$

Hence

$$\begin{aligned}
\overline{D_1}^a g(x_2, r_h) &= a_2^{\frac{k-3}{2}} x_2^{\frac{k-3}{2}} \left(\frac{sv'(sr)\sin r_h - v(sr)\cosh r_h}{\sinh^3 r_h} \right) \left(\overline{Pa - Px}\right) \\
&= -\overline{D_1}^x g(x_2, r_h) = -Ph_k.
\end{aligned}$$

□

There is a surprising symmetry relation between

$$x_2^{-k} Qh_k(a,x) \quad \text{and} \quad a_2^k Qh_{-k}(x,a)$$

stated next. We need this result in order to simplify the formula of the kernels in the Cauchy-type integral formula.

Lemma 3.14.

$$x_2^{-k} Qh_k(a,x) + a_2^k Qh_{-k}(x,a) = 0,$$
$$x_2^{-k} Qh_k(x,a) + a_2^k Qh_{-k}(a,x) = 0.$$

Proof. Assume first that $-1 < k < 1$. We just start to compute

$$g_k(x_2, r_h) = \frac{a_2^{\frac{k-1}{2}} x_2^{\frac{k-1}{2}} \cosh\left(\left(\frac{k+1}{2}\right) r_h\right)}{\sinh r_h}$$

and

$$\begin{aligned}
g_{-k}(x_2, r_h) &= \frac{a_2^{\frac{-k-1}{2}} x_2^{\frac{-k-1}{2}} \cosh\left(\left(\frac{-k+1}{2}\right) r_h\right)}{\sinh r_h} \\
&= \frac{a_2^{\frac{-k-1}{2}} x_2^{\frac{-k-1}{2}} \cosh\left(\left(\frac{k-1}{2}\right) r_h\right)}{\sinh r_h}.
\end{aligned}$$

Since

$$h_k(a,x) - \overline{D_1}^a g_k(a_2, r_h) - \frac{\partial g_k(a_2, r_h) e_2}{\partial a_2}$$

applying $\frac{\partial r_h}{\partial a_2} = \frac{a_2 - x_2 \cosh r_h}{a_2 x_2 \sinh r_h}$ and the hyperbolic identities we obtain the formula

$$\frac{Qh_k(a,x)}{x_2^{\frac{k-1}{2}} a_2^{\frac{k-1}{2}}} = -\frac{(k+1)\cosh(sr_h)}{2a_2 \sinh r_h} + \frac{(k+1)\sinh(sr_h)\cosh r_h}{2a_2 \sinh^2 r_h} - \frac{\cosh(sr_h)}{a_2 \sinh^3 r_h} - \frac{(k+1)\sinh(sr_h)}{2x_2 \sinh^2 r_h} + \frac{\cosh(sr_h)\cosh r_h}{x_2 \sinh^3 r_h}.$$

Hence we have

$$\frac{Qh_{-k}(x,a)}{a_2^{\frac{-1-k}{2}} x_2^{\frac{-1-k}{2}}} = \frac{(k-1)\cosh((s-1)r_h)}{2x_2 \sinh r_h} + \frac{(k-1)\sinh((s-1)r_h)\cosh r_h}{2x_2 \sinh^2 r_h} - \frac{\cosh((s-1)r_h)}{x_2 \sinh^3 r_h} - \frac{(k-1)\sinh((s-1)r_h)}{2a_2 \sinh^2 r_h} + \frac{\cosh((s-1)r_h)\cosh r_h}{a_2 \sinh^3 r_h}.$$

Moreover, applying hyperbolic identities of the sum, we infer

$$\frac{Qh_k(a,x)}{x_2^{\frac{k-1}{2}} a_2^{\frac{k-1}{2}}} = -\frac{(k-1)\cosh((s-1)r_h)}{2x_2 \sinh r_h} - \frac{(k-1)\sinh((s-1)r_h)\cosh r_h}{2x_2 \sinh^2 r_h} + \frac{\cosh((s-1)r_h)}{x_2 \sinh^3 r_h} + \frac{(k-1)\sinh((s-1)r_h)}{2a_2 \sinh^2 r_h} - \frac{\cosh((s-1)r_h)\cosh r_h}{a_2 \sinh^3 r_h}.$$

Hence we conclude the first assertion.

In the final case assume that $k \geq 1$. Since $\frac{\partial r_h}{\partial x_2} = \frac{x_2 - a_2 \cosh r_h}{a_2 x_2 \sinh r_h}$ after elementary calculations we may simplify as follows

$$\begin{aligned}\frac{e^{\frac{k-1}{2} r_h} Qh_{-k}(x,a)}{a_2^{-\frac{k-1}{2}} x_2^{\frac{-k-1}{2}}} &= \frac{k+1}{2x_2 \sinh r_h} + \frac{x_2 - \cosh r_h a_2}{x_2 a_2 \sinh^3 r_h} \cosh r_h + \frac{k-1}{2}\left(\frac{x_2 - \cosh r_h a_2}{x_2 a_2 \sinh^2 r_h}\right) \\ &= \frac{k-1}{2x_2 \sinh r_h} + \frac{x_2 \cosh r_h - a_2}{x_2 a_2 \sinh^3 r_h} + \frac{k-1}{2} \frac{x_2 - \cosh r_h a_2}{x_2 a_2 \sinh^2 r_h}.\end{aligned}$$

Moreover it holds

$$\begin{aligned}\frac{e^{\frac{k+1}{2} r_h} Qh_{-k}(x,a)}{a_2^{-\frac{k-1}{2}} x_2^{\frac{-k-1}{2}}} &= \frac{1-k}{2a_2 \sinh r_h} + \frac{a_2 - \cosh r_h x_2}{x_2 a_2 \sinh^3 r_h} \cosh r_h + \frac{k+1}{2}\left(\frac{a_2 - \cosh r_h x_2}{x_2 a_2 \sinh r_h}\right) \\ &= -\frac{k+1}{2} \frac{1}{a_2 \sinh r_h} + \frac{a_2 \cosh r_h - x_2}{x_2 a_2 \sinh^3 r_h} + \frac{k+1}{2} \frac{a_2 - \cosh r_h x_2}{x_2 a_2 \sinh^2 r_h}.\end{aligned}$$

Applying the formula $e^{-r_h} = \cosh r_h - \sinh r_h$ and the previous result we compute

$$\frac{e^{((k-1)/2)r_h} Qh_k(a,x)}{a_2^{(k-1)/2} x_2^{(k-1)/2}} = \frac{k-1}{2}\frac{a_2 \cos r_h - x_2}{x_2 a_2 \sinh^2 r_h} + \frac{a_2 - \cosh r_h x_2}{x_2 a_2 \sinh^3 r_h} - \frac{k-1}{2x_2 \sinh r_h}. \quad \square$$

Theorem 3.15. *Let Ω be an open subset of $\mathbb{R}^3_+$ and K be an open set whose closer $\overline{K} \subset \Omega$ is a smoothly bounded compact set with the outer unit normal field ν. If f is k-hypermonogenic in Ω and $a \in \int K$ then*

$$f(a) = \frac{1}{4\pi}\int_{\partial K}\left(r_1(a,x)\frac{P(\nu(x) f(x))}{x_2^k} + r_2(a,x) Q'(\nu(x) f(x))\right) dS(x)$$

where the functions

$$r_1(a,x) = -h_k(a,x) = -\overline{D}^a g_k(a,x),$$
$$r_2 a, x) = -a_2^k h_{-k}(a,x) e_2 = -a_2^k \overline{D}^a g_{-k}(a,x) e_2$$

are k-hypermonogenic with respect to a.

Proof. Applying Theorems 3.10 and 3.11 plus the formulas (2.1) and (2.2), we deduce

$$\begin{aligned} Pf(a) + Qf(a) e_2 &= \frac{1}{4\pi}\int_{\partial K} x_2^{-k} P(h_k(x,a)\nu f(x)) + Q(v_{-k}(x,a)\nu f(x)) e_2 \frac{dS}{x_2^k} \\ &= \frac{1}{8\pi}\int_{\partial K} x_2^{-k}\left(h_k(x,a)\nu f(x) + \widehat{h_k(x,a)}\widehat{\nu f(x)}\right) dS \\ &\quad + \frac{1}{8\pi}\int_{\partial K} a_2^k\left(h_{-k}(x,a)\nu f(x) - \widehat{h_{-k}(x,a)}\widehat{\nu f(x)}\right) dS. \end{aligned}$$

Collecting the similar parts and using (2.1), (2.2) and Lemma 3.14, we obtain

$$f(a) = \frac{1}{4\pi}\int_{\partial K}\left(x_2^{-k} A(x,y) P(\nu(x) f(x)) + B(x,y) Q'(\nu(x) f(x))\right) dS(x)$$

where the abbreviated functions are

$$\begin{aligned} A(x,y) &= 2^{-1} x_2^{-k}\left(h_k(x,a) + \widehat{h}_k(x,a)\right) + 2^{-1} a_2^k\left(h_{-k}(x,a) - \widehat{h}_{-k}(x,a)\right) \\ &= x_2^{-k} Ph_k(x,a) + a_2^k Qh_{-k}(x,a) e_2 \\ &= -x_2^{-k} Ph_k(a,x) - x_2^{-k} Qh_k(a,x) e_2 + \left(x_2^{-k} Qh_k(a,x) + a_2^k Qh_{-k}(x,a)\right) e_2 \\ &= -x_2^{-k} h_k(a,x) \end{aligned}$$

and

$$\begin{aligned} B(x,y) &= 2^{-1} x_2^{-k}\left(h_k(x,a) - \widehat{h}_k(x,a)\right) e_2 + 2^{-1} a_2^k\left(h_{-k}(x,a) + \widehat{h}_{-k}(x,a)\right) e_2 \\ &= -x_2^{-k} Qh_k(x,a) + a_2^k Ph_{-k}(x,a) e_2 \\ &= -a_2^k Ph_{-k}(a,x) e_2 + a_2^k Qh_{-k}(a,x) - x_2^{-k} Qh_k(x,a) - a_2^k Qh_{-k}(a,x) \\ &= -a_2^k h_{-k}(a,x) e_2, \end{aligned}$$

finishing the proof. $\square$

In the special cases $k = 0$ and $k = 1$, we have the known Cauchy formulas.

Theorem 3.16. *Let Ω be an open subset of $\mathbb{R}^3_+$ and K be an open set whose closer $\overline{K} \subset \Omega$ is a smoothly bounded compact set with the outer unit normal field ν. If f is monogenic in Ω and $a \in \int K$ then*

$$f(a) = \frac{1}{4\pi} \int_{\partial K} \frac{(x-a)^{-1}}{|x-a|} \nu(x) f(x)\, dS(x).$$

Theorem 3.17 (cf. [9]). *Let Ω be an open subset of $\mathbb{R}^3_+$ and $K \subset \Omega$ a smoothly bounded compact set with outer unit normal field ν. If f is hypermonogenic in Ω and $a \in \int K$, then*

$$f(a) = \frac{1}{4\pi} \int_{\partial K} (-k(a,x) P(\nu(y) f(y)) + k_1(a,y) Q'(\nu(y) f(y)))d\sigma,$$

where the kernels

$$\begin{aligned} k(a,y) &= 4a_2 \frac{(a-y)^{-1}}{|a-y|^1} e_n \frac{(a-\widehat{y})^{-1}}{|a-y|} = -\frac{1}{y_2} \overline{D}^a \left(\int_{\frac{|a-y|}{|a-\widehat{y}|}}^1 \frac{(1-s^2)}{s^2} ds \right) \\ &= \frac{(a-c(a,y))^{-1}}{a_2 |a-c(a,y)|} \end{aligned} \tag{3.3}$$

and

$$k_1(a,y) = a_2 \overline{D}^x \left(\frac{1}{|a-y|\,|a-\widehat{y}|} \right) e_2$$

are both hypermonogenic with respect to a in $\mathbb{R}^3 \setminus \{y, \widehat{y}\}$.

Using the standard arguments we may also verify that we obtain two integral operators producing k-hypermonogenic functions.

Theorem 3.18. *Let Ω be an open subset of $\mathbb{R}^3_+$ and K be an open set whose closer $\overline{K} \subset \Omega$ is a smoothly bounded compact set with the outer unit normal field ν. If f is continuous in Ω and $a \in \int K$ then the functions*

$$g_1(a) = \frac{1}{4\pi} \int_{\partial K} h_k(a,x) P(\nu(x) f(x))\, dS(x)$$

$$g_2(a) = \frac{1}{4\pi} \int_{\partial K} a_2^k h_{-k}(a,x) e_2 Q'(\nu(x) f(x))\, dS(x)$$

are k-hypermonogenic.

References

[1] Eriksson-Bique, S.-L., k-hypermonogenic functions. In Progress in Analysis, Vol. I, World Scientific (2003), 337–348.

[2] Eriksson, S.-L., Integral formulas for hypermonogenic functions, Bull. Bel. Math. Soc. 11 (2004), 705–717.

[3] Eriksson, S.-L., Cauchy-type integral formulas for k-hypermonogenic functions, in Proceedings of the 5th ISAAC conference, Catania, World Scientific, 2009.

[4] Eriksson, S.-L., Hyperbolic Extensions of Integral Formulas, Adv. appl. Clifford alg. 20, Numbers 3-4, (2010), 575–586.

[5] Eriksson, S.-L., *A Hyperbolic Dirac Operator and its Kernels*, Complex Variables and Elliptic Equations. Complex Variables and Elliptic Equations, Volume 58, Issue 6 (2013), 767–781.

[6] Eriksson-Bique, S.-L. and Leutwiler, H., Hypermonogenic functions. In Clifford Algebras and their Applications in Mathematical Physics, Vol. 2, Birkhäuser, Boston, 2000, 287–302.

[7] Eriksson, S.-L. and Leutwiler, H., Hypermonogenic functions and their Cauchy-type theorems. In Trends in Mathematics: Advances in Analysis and Geometry, Birkhäuser, Basel/Switzerland, 2004, 97–112.

[8] Eriksson, S.-L. and Leutwiler, H., Hyperbolic Function Theory, Adv. Appl. Clifford alg. 17 (2007), 437–450.

[9] Eriksson, S.-L. and Leutwiler, H., An Improved Cauchy Formula for Hypermonogenic Functions, Adv. Appl. Clifford Alg. 19 (2009), 269–282.

[10] Eriksson, S.-L. and H. Leutwiler, *Introduction to Hyperbolic Function Theory*, in Proceedings of Clifford Algebras and Inverse Problems, Research Report 90 (2009), 1–28.

[11] Eriksson, S.-L., H. Orelma, *On Hypermonogenic Functions*, Complex Variables and Elliptic Equations: An International Journal, Volume 58, Issue 7, (2013), 975–990.

[12] S.-L. Eriksson and H. Orelma, Mean Value Properties for the Weinstein Equation Using the Hyperbolic Metric, 1–10, Complex Analysis and Operator Theory© Springer Basel 2012 10.1007/s11785-012-0280-4.

[13] Eriksson, S.-L. and H. Orelma, Hyperbolic Laplace operator and the Weinstein equation in $\mathbb{R}^3$.

[14] Eriksson, S.-L., H. Orelma and N. Vieira, *Integral Formulas for k-hypermonogenic functions in* $\mathbb{R}^3$, Hypercomplex Analysis: New Perspectives and Applications, Trends in Mathematics 2014, pp. 119–132.

[15] Leutwiler, H., Quaternionic analysis in R^3versus its hyperbolic modification. In F. Brackx et al. (eds.) Clifford Analysis and its Applications, Kluwer, Dordrecht 2001, 193–211.

[16] Leutwiler, H., Appendix: Lecture notes of the course "Hyperbolic harmonic functions and their function theory". Clifford algebras and potential theory, 85–109, Univ. Joensuu Dept. Math. Rep. Ser., 7, Univ. Joensuu, Joensuu, 2004.

Sirkka-Liisa Eriksson
Department of Mathematics
and Statistics
University of Helsinki
PL 68 00014 Helsingin yliopisto
Finland
e-mail: sirkka-liisa.eriksson@professori.fi

Heikki Orelma
Department of Mathematics
Tampere University of Technology
P.O.Box 553
FI-33101 Tampere, Finland
e-mail: Heikki.Orelma@tut.fi

Modern Trends in Hypercomplex Analysis
Trends in Mathematics, 191–202

Eigenfunctions and Fundamental Solutions of the Caputo Fractional Laplace and Dirac Operators

Milton Ferreira and Nelson Vieira

Abstract. In this paper, by using the method of separation of variables, we obtain eigenfunctions and fundamental solutions for the three parameter fractional Laplace operator defined via fractional Caputo derivatives. The solutions are expressed using the Mittag-Leffler function and we show some graphical representations for some parameters. A family of fundamental solutions of the corresponding fractional Dirac operator is also obtained. Particular cases are considered in both cases.

Mathematics Subject Classification (2010). Primary: 35R11. Secondary: 35A08; 26A33; 30G35; 35C05.

Keywords. Fractional partial differential equations; fractional Laplace and Dirac operators; Caputo derivative; eigenfunctions; fundamental solution.

1. Introduction

In the last decades the interest in fractional calculus increased substantially. Among all the subjects there is a considerable interest in the study of ordinary and partial fractional differential equations regarding the mathematical aspects and methods of their solutions, and their applications in diverse areas such as physics, chemistry, engineering, optics or quantum mechanics (see, for example, [7–12, 14, 16]).

Here we consider a fractional Laplace operator in 3-dimensional space using Caputo derivatives with different order of differentiation for each direction. Previous approaches for this type of operators where considered in [15], [3], and [4]. In [15] the author studied eigenfunctions and fundamental solutions for the two-parameter fractional Laplace operator defined with Riemann–Liouville fractional derivatives. In [3] the authors extended the results for three dimensions and derived also fundamental solutions for the fractional Dirac operator which factorizes the fractional Laplace operator. Since there is a duality relation between

Caputo and Riemann–Liouville fractional derivatives presented in the formula of fractional integration by parts, there is need to study also fractional Laplace and Dirac operators with fractional derivatives defined in the Caputo sense. The aim of this paper is to use the method of separation of variables to present a formula for the family of eigenfunctions and fundamental solutions of the three-parameter fractional Laplace operator defined by Caputo fractional derivatives, as well as a family of fundamental solutions of the associated fractional Dirac operator. For the sake of simplicity we restrict ourselves to the three-dimensional case, however the results can be generalized for an arbitrary dimension. We observe that these operators were considered in [4] where the authors applied an operational approach via Laplace transform to construct general families of eigenfunctions and fundamental solutions.

The structure of the paper reads as follows: in the Preliminaries we recall some basic facts about fractional calculus, special functions and Clifford analysis, which are necessary for the development of this work. In Subsection 3.1 we use the method of separation of variables to describe a complete family of eigenfunctions and fundamental solutions of the fractional Laplace operator. In Subsection 3.2 we compute a family of fundamental solutions for the fractional Dirac operator. Finally, we point out that for the particular case of $\alpha = \beta = \gamma = 1$ the obtained formulas coincide with the correspondents classical formulas.

2. Preliminaries

2.1. Fractional calculus and special functions

Let $\left({}^C D^{\alpha}_{a^+} f\right)(x)$ denote the Caputo fractional derivative of order $\alpha > 0$ (see [10])

$$\left({}^C D^{\alpha}_{a^+} f\right)(x) = \frac{1}{\Gamma(n-\alpha)} \int_a^x \frac{f^{(n)}(t)}{(x-t)^{\alpha+1-n}}\, dt, \ n = [\alpha]+1, \ x > a, \tag{2.1}$$

where $[\alpha]$ means the integer part of α. When $0 < \alpha < 1$ then (2.1) takes the form

$$\left({}^C D^{\alpha}_{a^+} f\right)(x) = \frac{1}{\Gamma(1-\alpha)} \int_a^x \frac{f'(t)}{(x-t)^{\alpha}}\, dt. \tag{2.2}$$

The Riemann–Liouville fractional integral of order $\alpha > 0$ is given by (see [10])

$$\left(I^{\alpha}_{a^+} f\right)(x) = \frac{1}{\Gamma(\alpha)} \int_a^x \frac{f(t)}{(x-t)^{1-\alpha}}\, dt, \quad x > a. \tag{2.3}$$

We denote by $I^{\alpha}_{a^+}(L_1)$ the class of functions f represented by the fractional integral (2.3) of a summable function, that is $f = I^{\alpha}_{a^+}\varphi$, $\varphi \in L_1(a,b)$. A description of this class of functions was given in [13].

Theorem 2.1. *A function $f \in I^{\alpha}_{a^+}(L_1), \alpha > 0$ if and only if $I^{n-\alpha}_{a^+} f \in AC^n([a,b])$, $n = [\alpha]+1$ and $(I^{n-\alpha}_{a^+} f)^{(k)}(a) = 0$, $k = 0, \ldots, n-1$.*

In Theorem 2.1 $AC^n([a,b])$ denotes the class of functions f, which are continuously differentiable on the segment $[a,b]$ up to order $n-1$ and $f^{(n-1)}$ is absolutely continuous on $[a,b]$. Removing the last condition in Theorem 2.1 we obtain the class of functions that admits a summable fractional derivative.

Definition 2.2 ([13]). A function $f \in L_1(a,b)$ has a summable fractional derivative $\left(D^{\alpha}_{a+} f\right)(x)$ if $\left(I^{n-\alpha}_{a+} f\right)(x) \in AC^n([a,b])$, where $n = [\alpha] + 1$.

If a function f admits a summable fractional derivative, then the composition of (2.1) and (2.3) can be written in the form (see, e.g., [12])

$$\left(I^{\alpha}_{a+}\ {}^{C}D^{\alpha}_{a+} f\right)(x) = f(x) - \sum_{k=0}^{n-1} \frac{f^{(k)}(a)}{k!}(x-a)^k, \qquad n = [\alpha] + 1. \tag{2.4}$$

If $f \in I^{\alpha}_{a+}(L_1)$ then (2.4) reduces to $\left(I^{\alpha}_{a+}\ {}^{C}D^{\alpha}_{a+} f\right)(x) = f(x)$. Nevertheless we note that $\left({}^{C}D^{\alpha}_{a+}\ I^{\alpha}_{a+} f\right)(x) = f(x)$ in both cases. We observe that, in general, the semigroup property for the composition of Caputo fractional does not hold. We present three sufficient conditions under which the law of exponents hold. They can be applied in different situations accordingly with the conditions assumed to the function f.

Theorem 2.3 ([2, p. 56]). *Let $f \in C^k[a,b], a > 0$ and $k \in \mathbb{N}$. Moreover, let $\alpha, \beta > 0$ be such that there exists $l \in \mathbb{N}$ with $l \leq k$ and $\alpha, \alpha+\beta \in [l-1, l]$. Then*

$${}^{C}D^{\alpha}_{a+}\,{}^{C}D^{\beta}_{a+} f(x) = {}^{C}D^{\alpha+\beta}_{a+} f(x). \tag{2.5}$$

This theorem highlights a constraint on the applicability of the semigroup both with respect to the request of smoothness of the function and with respect to the ranges of the real orders of differentiation α and β. This means, for example, that, if $\alpha \in (0,1]$, then the law of exponents is applicable if $\beta \in [0, 1-\alpha)$ and $f \in C^k$, with $k = 1$. Here we notice that in most cases the law of exponents is not applicable for fractional Caputo derivatives, but anyhow there are different techniques to handle sequential fractional derivatives (see for example [12]). Since for $f \in C^{[\alpha]+1}([a,b])$ the Caputo derivative is a special case of the Grünwald–Letnikov fractional derivative (see [12, § 2.2.3]) then we have the following theorem:

Theorem 2.4 ([12, § 2.2.6]). *Let $\alpha, \beta > 0$ and $f \in C^n([a,b]), a > 0, n = [\alpha] + 1$. Then*

$${}^{C}D^{\alpha}_{a+}\,{}^{C}D^{\beta}_{a+} f(x) = {}^{C}D^{\alpha+\beta}_{a+} f(x) \tag{2.6}$$

holds for arbitrary β if the function f satisfies the conditions

$$f^{(k)}(a) = 0, \quad \textit{for} \quad k = 0, 1, \ldots, n-2. \tag{2.7}$$

For functions $f(x)$ that have a locally integrable singularity at $x = a$ we have the following result.

Theorem 2.5 ([5]). *Suppose that* $f(x) = (x-a)^\lambda g(x)$, *where* $a, \lambda > 0$ *and* $g(x)$ *has the generalized power series expansion* $g(x) = \sum_{n=0}^{\infty} a_n (x-a)^{n\gamma}$ *with radius of convergence* $R > 0$, $0 < \gamma \leq 1$. *Then*

$$ {}^C D_{a+}^{\alpha}\, {}^C D_{a+}^{\beta} f(x) = {}^C D_{a+}^{\alpha+\beta} f(x) \tag{2.8}$$

for all $(x-a) \in (0, R)$, *the coefficients* $a_n = 0$ *for* n *given by* $n\gamma + \lambda - \beta = 0$ *and either*

(a) $\lambda > \mu$, $\mu = \max(\beta + [\alpha], [\beta + \alpha])$

or

(b) $\lambda \leq \mu$, $a_k = 0$, *for* $k = 0, 1, \ldots, \left[\frac{\mu-\lambda}{\gamma}\right]$; *here* $[x]$ *denotes the greatest integer less than or equal to* x.

One important function used in this paper is the two-parameter Mittag-Leffler function $E_{\mu,\nu}(z)$ [6], which is defined in terms of the power series by

$$E_{\mu,\nu}(z) = \sum_{n=0}^{\infty} \frac{z^n}{\Gamma(\mu n + \nu)}, \qquad \mu > 0,\ \nu \in \mathbb{R},\ z \in \mathbb{C}. \tag{2.9}$$

In particular, the function $E_{\mu,\nu}(z)$ is entire of order $\rho = \frac{1}{\mu}$ and type $\sigma = 1$. The exponential, trigonometric and hyperbolic functions are expressed through (2.9) as follows (see [6]):

$$E_{1,1}(z) = e^z, \qquad E_{2,1}(-z^2) = \cos(z), \qquad E_{2,1}(z^2) = \cosh(z),$$
$$zE_{2,2}(-z^2) = \sin(z), \qquad zE_{2,2}(z^2) = \sinh(z).$$

Two important fractional integral and differential formulae involving the two-parametric Mittag-Leffler function are the following

$$I_{a+}^{\alpha}\left((x-a)^{\nu-1} E_{\mu,\nu}(k(x-a)^\mu)\right) = (x-a)^{\alpha+\nu-1} E_{\mu,\nu+\alpha}(k(x-a)^\mu) \tag{2.10}$$
$${}^C D_{a+}^{\alpha}\left((x-a)^{\nu-1} E_{\mu,\nu}(k(x-a)^\mu)\right) = (x-a)^{\nu-\alpha-1} E_{\mu,\nu-\alpha}(k(x-a)^\mu) \tag{2.11}$$

for all $\alpha > 0$, $\mu > 0$, $\nu \in \mathbb{R}$, $k \in \mathbb{C}$, $a > 0$, $x > a$.

Our approach leads to the resolution of a linear Abel integral equation of the second kind, which solution is given using the Mittag–Leffler function, accordingly with the next Theorem.

Theorem 2.6 ([6, Thm. 4.2]). *Let* $f \in L_1[a,b], \alpha > 0$ *and* $\lambda \in \mathbb{C}$. *Then the integral equation*

$$u(x) = f(x) + \frac{\lambda}{\Gamma(\alpha)} \int_a^x (x-t)^{\alpha-1} u(t)\, dt, \qquad x \in [a,b]$$

has a unique solution

$$u(x) = f(x) + \lambda \int_a^x (x-t)^{\alpha-1} E_{\alpha,\alpha}(\lambda(x-t)^\alpha) f(t)\, dt. \tag{2.12}$$

2.2. Clifford analysis

Let $\{e_1, \ldots, e_d\}$ be the standard basis of the Euclidean vector space in $\mathbb{R}^d$. The associated Clifford algebra $\mathbb{R}_{0,d}$ is the free algebra generated by $\mathbb{R}^d$ modulo $x^2 = -||x||^2\, e_0$, where $x \in \mathbb{R}^d$ and e_0 is the neutral element with respect to the multiplication operation in the Clifford algebra $\mathbb{R}_{0,d}$. The defining relation induces the multiplication rules

$$e_i e_j + e_j e_i = -2\delta_{ij}, \tag{2.13}$$

where δ_{ij} denotes Kronecker's delta. In particular, $e_i^2 = -1$ for all $i = 1, \ldots, d$. The standard basis vectors thus operate as imaginary units. A vector space basis for $\mathbb{R}_{0,d}$ is given by the set $\{e_A : A \subseteq \{1, \ldots, d\}\}$ with $e_A = e_{l_1} e_{l_2} \ldots e_{l_r}$, where $1 \leq l_1 < \cdots < l_r \leq d$, $0 \leq r \leq d$, $e_\emptyset := e_0 := 1$. Each $a \in \mathbb{R}_{0,d}$ can be written in the form $a = \sum_A a_A\, e_A$, with $a_A \in \mathbb{R}$. The conjugation in the Clifford algebra $\mathbb{R}_{0,d}$ is defined by $\overline{a} = \sum_A a_A\, \overline{e}_A$, where $\overline{e}_A = \overline{e}_{l_r}\, \overline{e}_{l_{r-1}} \ldots \overline{e}_{l_1}$, and $\overline{e}_j = -e_j$ for $j = 1, \ldots, d$, $\overline{e}_0 = e_0 = 1$. An important subspace of the real Clifford algebra $\mathbb{R}_{0,d}$ is the so-called space of paravectors $\mathbb{R}_1^d = \mathbb{R} \bigoplus \mathbb{R}^d$, being the sum of scalars and vectors. Each non-zero vector $a \in \mathbb{R}_1^d$ has a multiplicative inverse given by $\frac{\overline{a}}{||a||^2}$.

A $\mathbb{R}_{0,d}$-valued function f over $\Omega \subseteq \mathbb{R}_1^d$ has the representation $f = \sum_A e_A f_A$, with components $f_A : \Omega \to \mathbb{R}_{0,d}$. Properties such as continuity or differentiability have to be understood componentwise. Next, we recall the Euclidean Dirac operator $D = \sum_{j=1}^d e_j \partial_{x_j}$, which factorizes the d-dimensional Euclidean Laplace, i.e., $D^2 = -\Delta$. A $\mathbb{R}_{0,d}$-valued function f is called *left-monogenic* if it satisfies $Du = 0$ on Ω (resp. *right-monogenic* if it satisfies $uD = 0$ on Ω).

For more details about Clifford algebras and basic concepts of its associated function theory we refer the interested reader for example to [1].

3. Method of separation of variables

3.1. Eigenfunctions and fundamental solution of the fractional Laplace operator

We consider the eigenfunction problem for the fractional Laplace operator ${}^C\Delta_+^\alpha\, u(x) = \lambda\, u(x)$, i.e.,

$$\begin{aligned} \left({}^C D_{x_0^+}^{1+\alpha} u\right)(x,y,z) + \left({}^C D_{y_0^+}^{1+\beta} u\right)(x,y,z) \\ + \left({}^C D_{z_0^+}^{1+\gamma} u\right)(x,y,z) &= \lambda\, u(x,y,z), \end{aligned} \tag{3.1}$$

where $\lambda \in \mathbb{C}$, $(\alpha, \beta, \gamma) \in]0,1]^3$, $(x,y,z) \in \Omega = [x_0, X_0] \times [y_0, Y_0] \times [z_0, Z_0]$, $x_0, y_0, z_0 > 0$, $X_0, Y_0, Z_0 < \infty$, and $u(x,y,z)$ admits summable fractional derivatives ${}^C D_{x_0^+}^{1+\alpha}$, ${}^C D_{y_0^+}^{1+\beta}$ and ${}^C D_{z_0^+}^{1+\gamma}$. Taking the integral operator $I_{x_0^+}^{1+\alpha}$ from both sides of (3.1) and taking into account (2.4) we get

$$\begin{aligned} &u(x,y,z) - u(x_0,y,z) - (x-x_0)\, u'_x(x_0,y,z) \\ &\quad + \left(I^{1+\alpha}_{x_0^+}\, {}^C D^{1+\beta}_{y_0^+} u\right)(x,y,z) + \left(I^{1+\alpha}_{x_0^+}\, {}^C D^{1+\gamma}_{z_0^+} u\right)(x,y,z) \\ &\quad = \lambda \left(I^{1+\alpha}_{x_0^+} u\right)(x,y,z). \end{aligned} \tag{3.2}$$

Now, applying the operator $I^{1+\beta}_{y_0^+}$ to both sides of the previous expression and using Fubini's Theorem we get

$$\begin{aligned} &\left(I^{1+\beta}_{y_0^+} u\right)(x,y,z) - \left(I^{1+\beta}_{y_0^+} f_0\right)(y,z) - (x-x_0)\,\left(I^{1+\beta}_{y_0^+} f_1\right)(y,z) \\ &\quad + \left(I^{1+\alpha}_{x_0^+} u\right)(x,y,z) - \left(I^{1+\alpha}_{x_0^+} u\right)(x,y_0,z) \\ &\quad - (y-y_0)\,\left(I^{1+\alpha}_{x_0^+} u'_y\right)(x,y_0,z) + \left(I^{1+\alpha}_{x_0^+}\, I^{1+\beta}_{y_0^+}\, {}^C D^{1+\gamma}_{z_0^+} u\right)(x,y,z) \\ &= \lambda \left(I^{1+\alpha}_{x_0^+}\, I^{1+\beta}_{y_0^+} u\right)(x,y,z), \end{aligned} \tag{3.3}$$

where we denote Cauchy's fractional integral conditions by

$$f_0(y,z) = u(x_0,y,z), \qquad f_1(y,z) = u'_x(x_0,y,z). \tag{3.4}$$

Finally, applying the operator $I^{1+\gamma}_{z_0^+}$ to both sides of equation (3.3) and using Fubini's Theorem we get

$$\begin{aligned} &\left(I^{1+\beta}_{y_0^+}\, I^{1+\gamma}_{z_0^+} u\right)(x,y,z) - \left(I^{1+\beta}_{y_0^+}\, I^{1+\gamma}_{z_0^+} f_0\right)(y,z) - (x-x_0)\,\left(I^{1+\beta}_{y_0^+}\, I^{1+\gamma}_{z_0^+} f_1\right)(y,z) \\ &\quad + \left(I^{1+\alpha}_{x_0^+}\, I^{1+\gamma}_{z_0^+} u\right)(x,y,z) - \left(I^{1+\alpha}_{x_0^+}\, I^{1+\gamma}_{z_0^+} h_0\right)(x,z) \\ &\quad - (y-y_0)\,\left(I^{1+\alpha}_{x_0^+}\, I^{1+\gamma}_{z_0^+} h_1\right)(x,z) + \left(I^{1+\alpha}_{x_0^+}\, I^{1+\beta}_{y_0^+} u\right)(x,y,z) \\ &\quad - \left(I^{1+\alpha}_{x_0^+}\, I^{1+\beta}_{y_0^+} u\right)(x,y,z_0) - (z-z_0)\,\left(I^{1+\alpha}_{x_0^+}\, I^{1+\beta}_{y_0^+} u'_z\right)(x,y,z_0) \\ &= \lambda \left(I^{1+\alpha}_{x_0^+}\, I^{1+\beta}_{y_0^+}\, I^{1+\gamma}_{z_0^+} u\right)(x,y,z), \end{aligned}$$

which is equivalent to

$$\begin{aligned} &\left(I^{1+\beta}_{y_0^+}\, I^{1+\gamma}_{z_0^+} u\right)(x,y,z) + \left(I^{1+\alpha}_{x_0^+}\, I^{1+\gamma}_{z_0^+} u\right)(x,y,z) \\ &\quad + \left(I^{1+\alpha}_{x_0^+}\, I^{1+\beta}_{y_0^+} u\right)(x,y,z) - \lambda \left(I^{1+\alpha}_{x_0^+}\, I^{1+\beta}_{y_0^+}\, I^{1+\gamma}_{z_0^+} u\right)(x,y,z) \\ &\quad = \left(I^{1+\beta}_{y_0^+}\, I^{1+\gamma}_{z_0^+} f_0\right)(y,z) + (x-x_0)\,\left(I^{1+\beta}_{y_0^+}\, I^{1+\gamma}_{z_0^+} f_1\right)(y,z) \\ &\quad + \left(I^{1+\alpha}_{x_0^+}\, I^{1+\gamma}_{z_0^+} h_0\right)(x,z) + (y-y_0)\,\left(I^{1+\alpha}_{x_0^+}\, I^{1+\gamma}_{z_0^+} h_1\right)(x,z) \\ &\quad + \left(I^{1+\alpha}_{x_0^+}\, I^{1+\beta}_{y_0^+} g_0\right)(x,y) + (z-z_0)\,\left(I^{1+\alpha}_{x_0^+}\, I^{1+\beta}_{y_0^+} g_1\right)(x,y), \end{aligned} \tag{3.5}$$

where we denote Cauchy's fractional integral conditions by

$$h_0(x,z) = u(x,y_0,z), \qquad h_1(x,z) = u'_y(x,y_0,z), \tag{3.6}$$

$$g_0(x,y) = u(x,y,z_0), \qquad g_1(x,y) = u'_z(x,y,z_0). \tag{3.7}$$

Assume that the eigenfunctions are such that $u(x,y,z) = u_1(x)\,u_2(y)\,u_3(z)$. Substituting in (3.5) and taking into account the initial conditions (3.4), (3.6), and (3.7) we obtain

$$\begin{aligned} u_1(x)\ &\left(I^{1+\beta}_{y_0^+} u_2(y)\ I^{1+\gamma}_{z_0^+} u_3(z)\right) + u_2(y)\ \left(I^{1+\alpha}_{x_0^+} u_1(x)\ I^{1+\gamma}_{z_0^+} u_3(z)\right) \\ &+ u_3(z)\ \left(I^{1+\alpha}_{x_0^+} u_1(x)\ I^{1+\beta}_{y_0^+} u_2(y)\right) \\ &- \lambda \left(I^{1+\alpha}_{x_0^+} u_1\right)(x)\ \left(I^{1+\beta}_{y_0^+} u_2\right)(y)\ \left(I^{1+\gamma}_{z_0^+} u_3\right)(z) \\ &= a_1 \left(I^{1+\beta}_{y_0^+} u_2(y)\ I^{1+\gamma}_{z_0^+} u_3(z)\right) + a_2(x-x_0)\left(I^{1+\beta}_{y_0^+} u_2(y)\ I^{1+\gamma}_{z_0^+} u_3(z)\right) \\ &+ b_1 \left(I^{1+\alpha}_{x_0^+} u_1(x)\ I^{1+\gamma}_{z_0^+} u_3(z)\right) + b_2(y-y_0)\left(I^{1+\alpha}_{x_0^+} u_1(x)\ I^{1+\gamma}_{z_0^+} u_3(z)\right) \\ &+ c_1 \left(I^{1+\alpha}_{x_0^+} u_1(x)\ I^{1+\beta}_{y_0^+} u_2(y)\right) + c_2(z-z_0)\left(I^{1+\alpha}_{x_0^+} u_1(x)\ I^{1+\beta}_{y_0^+} u_2(y)\right), \end{aligned} \tag{3.8}$$

where

$$\begin{aligned} a_1 &= u_1(x_0), & a_2 &= u'_{1,x}(x_0), \\ b_1 &= u_2(y_0), & b_2 &= u'_{2,y}(y_0), \\ c_1 &= u_3(z_0), & c_2 &= u'_{3,z}(z_0), \end{aligned}$$

are constants defined by the initial conditions (3.4), (3.6), and (3.7). Supposing that

$$\left(I^{1+\alpha}_{x_0^+} u_1\right)(x)\left(I^{1+\beta}_{y_0^+} u_2\right)(y)\left(I^{1+\gamma}_{z_0^+} u_3\right)(z) \neq 0,$$

for $(x,y,z) \in \Omega$, we can divide (3.8) by this factor. Separating the variables we get the following three Abel type second kind integral equations:

$$u_1(x) - \mu\ \left(I^{1+\alpha}_{x_0^+} u_1\right)(x)\ = a_1 + a_2(x-x_0), \tag{3.9}$$

$$u_2(y) + \nu\ \left(I^{1+\beta}_{y_0^+} u_2\right)(y)\ = b_1 + b_2(y-y_0), \tag{3.10}$$

$$u_3(z) + (\mu - \lambda - \nu)\ \left(I^{1+\gamma}_{z_0^+} u_3\right)(z)\ = c_1 + c_2(z-z_0), \tag{3.11}$$

where $\lambda,\ \mu,\ \nu \in \mathbb{C}$ are constants. We observe that the equality

$$\left(I^{1+\alpha}_{x_0^+} u_1\right)(x)\left(I^{1+\beta}_{y_0^+} u_2\right)(y)\left(I^{1+\gamma}_{z_0^+} u_3\right)(z) = 0,$$

for at least one point (x^*, y^*, z^*) agrees with (3.8), (3.9), (3.10), and (3.11). Solving the latter equations using (2.12) in Theorem 2.6 and after straightforward computations we obtain the following family of eigenfunctions $u_{\lambda,\mu,\nu}(x,y,z) = u_1(x)\, u_2(y)\, u_3(z)$, with

$$u_1(x) = a_1\, E_{1+\alpha,1}\left(\mu(x-x_0)^{1+\alpha}\right) + a_2(x-x_0)\, E_{1+\alpha,2}\left(\mu(x-x_0)^{1+\alpha}\right), \quad (3.12)$$

$$u_2(y) = b_1\, E_{1+\beta,1}\left(-\nu(y-y_0)^{1+\beta}\right) + b_2(y-y_0)\, E_{1+\beta,2}\left(-\nu(y-y_0)^{1+\beta}\right), \quad (3.13)$$

$$u_3(z) = c_1\, E_{1+\gamma,1}\left((\mu-\lambda-\nu)(z-z_0)^{1+\gamma}\right) + c_2(z-z_0)\, E_{1+\gamma,2}\left((\mu-\lambda-\nu)(z-z_0)^{1+\gamma}\right). \quad (3.14)$$

For the particular case of $\lambda = 0$ (fundamental solution), $\mu = 2$, $\nu = 1$, $x_0 = y_0 = z_0 = 0$, $X_0 = Z_0 = 5$, $Y_0 = 15$, and $a_i = b_i = c_i = 1$, with $i = 1,2$, we show the graphical representation of the components u_1, u_2, u_3 for α, β, γ equal to $\frac{1}{4}$, $\frac{1}{2}$, $\frac{3}{4}$, 1. The plots were obtained using the software Mathematica 9 since this software is able to evaluate and to graphically represent functions involving the Mittag-Leffler functions.

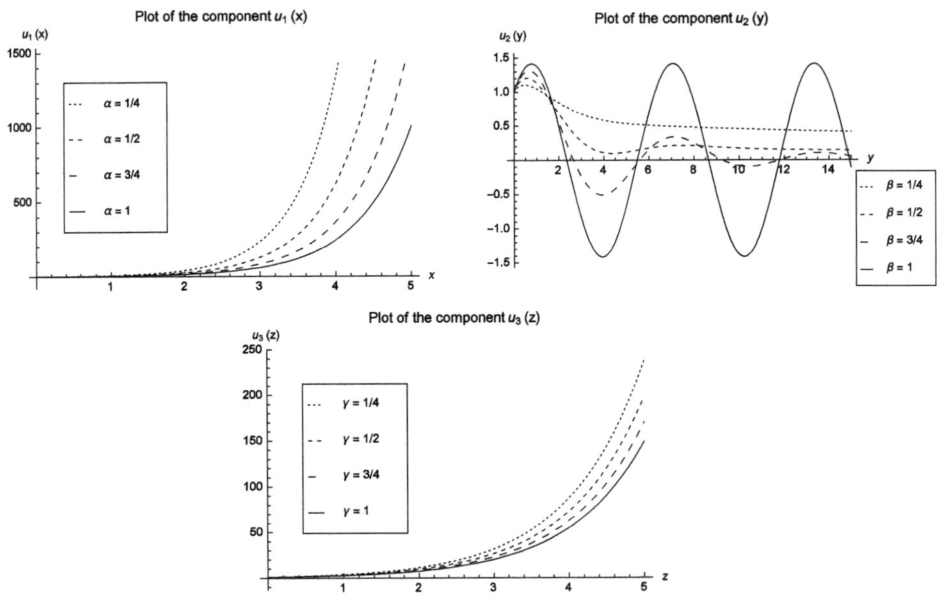

FIGURE 1. Plots of the components u_1, u_2 and u_3, when $\lambda = 0$, $\mu = 2$, $\nu = 1$, $x_0 = y_0 = z_0 = 0$, $X_0 = Z_0 = 5$, $Y_0 = 15$, and $a_i = b_i = c_i = 1$, and different values of α, β, and γ.

From the plots we observe that the components u_1 and u_3 are of exponential type and the increasing of the curve coincides with the decreasing of the parameters. For the component u_2 the sinusoidal behavior observed in the classical case $\beta = 1$ suffers a relaxation with the decreasing of the parameter β.

Remark 3.1. In the special case of $\alpha = \beta = \gamma = 1$ the functions u_1, u_2 and u_3 take the form

$$u_1(x) = a_1 \cosh\left(\sqrt{\mu}\,(x - x_0)\right) + \frac{a_2}{\sqrt{\mu}} \sinh\left(\sqrt{\mu}\,(x - x_0)\right), \tag{3.15}$$

$$u_2(y) = b_1 \cos\left(\sqrt{\nu}\,(y - y_0)\right) + \frac{b_2}{\sqrt{\nu}} \sin\left(\sqrt{\nu}\,(y - y_0)\right), \tag{3.16}$$

$$\begin{aligned} u_3(z) &= c_1 \cosh\left(\sqrt{\mu - \lambda - \nu}\,(z - z_0)\right) \\ &\quad + \frac{c_2}{\sqrt{\mu - \lambda - \nu}} \sinh\left(\sqrt{\mu - \lambda - \nu}\,(z - z_0)\right), \end{aligned} \tag{3.17}$$

which are the components of the fundamental solution of the Laplace operator in $\mathbb{R}^3$ obtained by the method of separation of variables.

3.2. Fundamental solution of the fractional Dirac operator

In this section we compute the fundamental solution for the three-dimensional fractional left Dirac operator defined via Caputo derivatives

$${}^C D_+^{(\alpha,\beta,\gamma)} := e_1\, {}^C D_{x_0^+}^{\frac{1+\alpha}{2}} + e_2\, {}^C D_{y_0^+}^{\frac{1+\beta}{2}} + e_3\, {}^C D_{z_0^+}^{\frac{1+\gamma}{2}}, \quad (\alpha,\beta,\gamma) \in]0,1]^3. \tag{3.18}$$

This operator factorizes the fractional Laplace operator ${}^C\Delta_+^{(\alpha,\beta,\gamma)}$ for Clifford-valued functions f given by

$$f(x,y,z) = \sum_A e_A f_A(x,y,z),$$

where $e_A \in \{1, e_1, e_2, e_3, e_1e_2, e_1e_3, e_2e_3, e_1e_2e_3\}$, and each real-valued function f_A satisfies one of the sufficient conditions presented in Theorems 2.3, 2.4 or 2.5. In fact, for such functions we can apply the semigroup property (2.5) to obtain

$$\begin{aligned} &{}^C D_{x_0^+}^{\frac{1+\gamma}{2}} \left({}^C D_{x_0^+}^{\frac{1+\gamma}{2}} f_A\right) = {}^C D_{x_0^+}^{1+\alpha} f_A, \qquad {}^C D_{y_0^+}^{\frac{1+\beta}{2}} \left({}^C D_{y_0^+}^{\frac{1+\beta}{2}} f_A\right) = {}^C D_{y_0^+}^{1+\beta} f_A, \\ &{}^C D_{z_0^+}^{\frac{1+\gamma}{2}} \left({}^C D_{z_0^+}^{\frac{1+\gamma}{2}} f_A\right) = {}^C D_{z_0^+}^{1+\gamma} f_A. \end{aligned} \tag{3.19}$$

Moreover, for the mixed fractional derivatives ${}^C D_{x_0^+}^{\frac{1+\alpha}{2}} \left({}^C D_{y_0^+}^{\frac{1+\beta}{2}} f_A\right)$, due to the Leibniz rule for the differentiation under integral sign, Fubini's Theorem and Schwarz' Theorem, we have

$${}^C D_{x_0^+}^{\frac{1+\alpha}{2}} \left({}^C D_{y_0^+}^{\frac{1+\beta}{2}} f_A\right) = {}^C D_{y_0^+}^{\frac{1+\beta}{2}} \left({}^C D_{x_0^+}^{\frac{1+\alpha}{2}} f_A\right), \tag{3.20}$$

$${}^C D_{x_0^+}^{\frac{1+\alpha}{2}} \left({}^C D_{z_0^+}^{\frac{1+\gamma}{2}} f_A\right) = {}^C D_{z_0^+}^{\frac{1+\gamma}{2}} \left({}^C D_{x_0^+}^{\frac{1+\alpha}{2}} f_A\right), \tag{3.21}$$

$${}^C D_{y_0^+}^{\frac{1+\beta}{2}} \left({}^C D_{z_0^+}^{\frac{1+\gamma}{2}} f_A\right) = {}^C D_{z_0^+}^{\frac{1+\gamma}{2}} \left({}^C D_{y_0^+}^{\frac{1+\beta}{2}} f_A\right). \tag{3.22}$$

From (3.19), (3.20), (3.21), (3.22) and the multiplication rules (2.13) of the Clifford algebra, we finally get

$$ {}^{C}D_{+}^{(\alpha,\beta,\gamma)}\left({}^{C}D_{+}^{(\alpha,\beta,\gamma)}f\right) = -\ {}^{C}\Delta_{+}^{(\alpha,\beta,\gamma)}f, \tag{3.23}$$

i.e., the fractional Dirac operator factorizes the fractional Laplace operator.
In order to get the fundamental solution of ${}^{C}D_{+}^{(\alpha,\beta,\gamma)}$ we apply this operator to the fundamental solution $u(x,y,z) = u_1(x)\,u_2(y)\,u_3(z)$, where u_i are given by (3.12), (3.13) and (3.14), respectively. To make the calculations we make use of the derivation rule (2.11) and the fractional analogous formula for differentiation of integrals depending on a parameter where the upper limit also depends on the same parameter (see [12, Section 2.7.4]). Hence,

$$\begin{aligned} U(x,y,z) &= \left({}^{C}D_{+}^{(\alpha,\beta,\gamma)}u\right)(x,y,z)\\ &= e_1\ u_2(y)\ u_3(z)\left({}^{C}D_{x_0^+}^{\frac{1+\gamma}{2}}u_1\right)(x) + e_2\ u_1(x)\ u_3(z)\left({}^{C}D_{y_0^+}^{\frac{1+\beta}{2}}u_2\right)(y)\\ &\quad + e_3\ u_1(x)\ u_2(y)\left({}^{C}D_{z_0^+}^{\frac{1+\gamma}{2}}u_3\right)(z), \end{aligned} \tag{3.24}$$

where u_1, u_2, u_3 are given respectively by (3.12), (3.13), (3.14) and

$$\begin{aligned} \left({}^{C}D_{x_0^+}^{\frac{1+\gamma}{2}}u_1\right)(x) &= a_1\ (x-x_0)^{-\frac{1+\alpha}{2}}\ E_{1+\alpha,\frac{1-\alpha}{2}}\left(\mu(x-x_0)^{1+\alpha}\right)\\ &\quad + a_2\ (x-x_0)^{\frac{1-\alpha}{2}}\ E_{1+\alpha,\frac{3-\alpha}{2}}\left(\mu(x-x_0)^{1+\alpha}\right), \end{aligned} \tag{3.25}$$

$$\begin{aligned} \left({}^{C}D_{y_0^+}^{\frac{1+\beta}{2}}u_2\right)(y) &= b_1\ (y-y_0)^{-\frac{1-\beta}{2}}\ E_{1+\beta,\frac{1-\beta}{2}}\left(-\nu(y-y_0)^{1+\beta}\right)\\ &\quad + b_2\ (y-y_0)^{\frac{1-\beta}{2}}\ E_{1+\beta,\frac{3-\beta}{2}}\left(-\nu(y-y_0)^{1+\beta}\right), \end{aligned} \tag{3.26}$$

$$\begin{aligned} \left({}^{C}D_{z_0^+}^{\frac{1+\gamma}{2}}u_3\right)(z) &= c_1\ (z-z_0)^{\frac{1-\gamma}{2}}\ E_{1+\gamma,\frac{1-\gamma}{2}}\left((-\mu+\lambda+\nu)(z-z_0)^{1+\gamma}\right)\\ &\quad + c_2\ (z-z_0)^{\frac{1-\gamma}{2}}\ E_{1+\gamma,\frac{3-\gamma}{2}}\left((-\mu+\lambda+\nu)(z-z_0)^{1+\gamma}\right). \end{aligned} \tag{3.27}$$

Remark 3.2. In the special case of $\alpha = \beta = \gamma = 1$, u_1, u_2 and u_3 take the form (3.15), (3.16) and (3.17), respectively, and expressions (3.25), (3.26), and (3.27) take the form

$$\begin{aligned} (D_x u_1)(x) &= a_1\sqrt{\mu}\ \sinh\left(\sqrt{\mu}\ (x-x_0)\right) + a_2\ \sinh\left(\sqrt{\mu}\ (x-x_0)\right),\\ (D_y u_2)(y) &= b_1\sqrt{\nu}\ \sin\left(\sqrt{\nu}\ (y-y_0)\right) + b_2\ \cos\left(\sqrt{\nu}\ (y-y_0)\right),\\ (D_z u_3)(z) &= c_1\sqrt{-\mu+\lambda+\nu}\ \sinh\left(\sqrt{\mu-\lambda-\nu}\ (z-z_0)\right)\\ &\quad + c_2\ \sinh\left(\sqrt{\mu-\lambda-\nu}\ (z-z_0)\right), \end{aligned}$$

which are the components of the fundamental solution of the Dirac operator in $\mathbb{R}^3$ obtained by the method of separation of variables.

Acknowledgment

The authors were supported by Portuguese funds through the CIDMA – Center for Research and Development in Mathematics and Applications, and the Portuguese Foundation for Science and Technology ("FCT – Fundação para a Ciência e a Tecnologia"), within project UID/MAT/ 0416/2013. N. Vieira was also supported by FCT via the FCT Researcher Program 2014 (Ref: IF/00271/2014).

References

[1] R. Delanghe, F. Sommen and V. Souček, *Clifford algebras and spinor-valued functions. A function theory for the Dirac operator*, Mathematics and its Applications-Vol. 53, Kluwer Academic Publishers, Dordrecht, 1992.

[2] K. Diethelm, *The analysis of fractional differential equations. An application-oriented exposition using differential operators of Caputo type*, Lecture Notes in Mathematics, Springer, Berlin, 2004.

[3] M. Ferreira and N. Vieira, *Eigenfunctions and fundamental solutions of the fractional Laplace and Dirac operators: the Riemman–Liouville case*, Complex Anal. Oper. Theory, **10** No. 5, (2016), 1081–1100.

[4] M. Ferreira and N. Vieira, *Eigenfunctions and fundamental solutions of the fractional Laplace and Dirac operators using Caputo derivatives*, submitted.

[5] M. Garg and P. Manohar, *Three-dimensional generalized differential transform method for space-time fractional diffusion equation in two space variables with variable coefficients*, Palest. J. Math., **4** No. 1, (2015), 127–135.

[6] R. Gorenflo, A.A. Kilbas, F. Mainardi and S. Rogosin, *Mittag-Leffler functions. Theory and applications*, Springer Monographs in Mathematics, Springer, Berlin, 2014.

[7] R. Gorenflo, Y. Luchko and M. Stojanovć, *Fundamental solution of a distributed order time-fractional diffusion-wave equation as probability density*, Fract. Calc. Appl. Anal., **16** No. 2, (2013), 297–316.

[8] R. Gorenflo and F. Mainardi, *Fractional calculus: integral and differential equations of fractional order*, In: Fractals and Fractional Calculus in Continuum Mechanics, eds.: A. Carpinteri and F. Mainardi, Springer-Verlag, Wien-New York, (1997), 223–276.

[9] B. Guo, X. Pu and F. Huang, *Fractional partial differential equations and their numerical solutions*, World Scientific, Hackensack, NJ, 2015.

[10] A. Kilbas, H.M. Srivastava and J.J. Trujillo, *Theory and applications of fractional differential equations*, North-Holland Mathematics Studies-Vol. 204, Elsevier, Amsterdam, 2006.

[11] S.K. Miller and B. Ross, *An introduction to the fractional calculus and fractional differential equations*, John Wiley & Sons Inco., New York, 1993.

[12] I. Podlubny, *Fractional differential equations. An introduction to fractional derivatives, fractional differential equations, to methods of their solution and some of their applications*, Mathematics in Science and Engineering-Vol. 198, Academic Press, San Diego, CA, 1999.

[13] S.G. Samko, A.A. Kilbas and O.I. Marichev, *Fractional integrals and derivatives: theory and aqpllications*, Gordon and Breach, New York, NY, 1993.

[14] J.L. Vázquez, *Recent progress in the theory of nonlinear diffusion with fractional Laplacian operators*, Discrete Contin. Dyn. Syst., **Ser .7** No. 4, (2014), 857–885.

[15] S. Yakubovich, *Eigenfunctions and fundamental solutions of the fractional two-parameter Laplacian*, Int. J. Math. Math. Sci., Article ID **541934**, (2010), 18pp.

[16] Y. Zhou, *Basic theory of fractional differential equations*, World Scientific, Hackensack, NJ, 2014.

Milton Ferreira
School of Technology and Management
Polytechnic Institute of Leiria
P-2411-901, Leiria, Portugal

and

Center for Research and Development
in Mathematics and Applications (CIDMA)
Department of Mathematics, University of Aveiro
Campus Universitário de Santiago
P-3810-193 Aveiro, Portugal
e-mail: milton.ferreira@ipleiria.pt
mferreira@ua.pt

Nelson Vieira
Center for Research and Development
in Mathematics and Applications (CIDMA)
Department of Mathematics, University of Aveiro
Campus Universitário de Santiago
P-3810-193 Aveiro, Portugal
e-mail: nloureirovieira@gmail.com

Modern Trends in Hypercomplex Analysis
Trends in Mathematics, 203–215

Three-dimensional Analogue of Kolosov–Muskhelishvili Formulae

Yuri Grigor'ev

Abstract. In the plane elasticity an effective method of using the holomorphic complex function theory is based on Kolosov–Muskhelishvili formulae. For a three-dimensional case monogenic Clifford functions or regular quaternion functions of a reduced quaternion variable are used. Such functions are solutions of the Moisil–Theodoresco system. In recent papers some variants of three-dimensional Kolosov–Muskhelishvili formulae are obtained but only for star-shaped regions. For applications it is very important to have these formulae for a wider class of domains. We propose the generalized Kolosov–Muskhelishvili formulae in arbitrary simply connected domains with a smooth boundary not only star-shaped, where a notion of harmonic primitive function is used. The method of proving is based on a new theorem about reconstruction of a regular function from a given scalar part.

Mathematics Subject Classification (2010). Primary 30G35; Secondary 74B05.

Keywords. Classical linear elasticity, functions of hypercomplex variables and generalized variables, regular quaternionic functions, Moisil–Theodoresco system, representation formulae.

1. Introduction

In two-dimensional problems of the theory of elasticity the methods of complex variable theory are effectively used. In plane problems the basis of this is the representation of the general solution of the equilibrium equations in terms of two arbitrary analytic functions called the Kolosov–Muskhelishvili formulae [34]. In axially symmetric problems different classes of generalized analytic functions of complex variable [1, 39], p- and (p, q)-analytic functions [38] are used. As a generalization of the method of complex functions in multidimensional problems the methods of hypercomplex functions are developed (see [9, 11, 26, 28, 29, 30], etc.). For three-dimensional problems of mathematical physics such an apparatus is the

This work was supported by RFBR grant 15-41-05081.

Moisil–Theodoresco system theory, which is developed as the theory of regular quaternion functions of reduced quaternion variable [15, 17, 20, 26, 22, 23, 24]. We note that this theory is partially covered by Clifford analysis. In [32] the first quaternion solution of the equilibrium equations of the theory of elasticity is obtained. In [16, 17, 18, 21, 22, 23, 24] some forms of a three-dimensional quaterniuonic analogue of the Kolosov–Muskhelishvili formulae for displacements and their applications are obtained in the star-shaped domains. Another variant of the spatial generalization of the Kolosov–Muskhelishvili formulae with expressions for stresses is obtained for the star-shaped domains by using a notion of monogenic primitive in [5, 6, 7]. By means of a quaternionic refinement of the classical harmonic Papkovich–Neuber solution in [37] a monogenic formulation of the three-dimensional elasticity problem is also obtained with some geometrical restrictions on a domain. In [8] an alternative Kolosov–Muskhelishvili formula for the displacement field by means of a (paravector-valued) monogenic, an anti-monogenic and a ψ-hyperholomorphic function is proposed. In [22, 37] one can find short reviews about some other results in using hypercomplex functions in the three-dimensional elasticity. In this paper we present the quaternionic Kolosov–Muskhelishvili formulae with the expressions for stresses in arbitrary simply connected domains with a smooth boundary not only star-shaped, where the notion of harmonic primitive function is used. The theorem about the reconstruction of the regular function from the given scalar part is proved.

2. Preliminaries and notations

Let $\mathbf{i}$, $\mathbf{j}$, $\mathbf{k}$ be the basic quaternions obeying the following rules of multiplication:

$$\mathbf{i}^2 = \mathbf{j}^2 = \mathbf{k}^2 = -1, \quad \mathbf{ij} = -\mathbf{ji} = \mathbf{k}, \quad \mathbf{jk} = -\mathbf{kj} = \mathbf{i}, \quad \mathbf{ki} = -\mathbf{ik} = \mathbf{j}.$$

An element q of the quaternion algebra $\mathbb{H}$ we write in the form $q = q_0 + \mathbf{i}q_x + \mathbf{j}q_y + \mathbf{k}q_z = q_0 + \mathbf{q}$, where q_0, q_x, q_y, q_z are the real numbers, q_0 is called the scalar part of the quaternion, $\mathbf{q} = \mathbf{i}q_x + \mathbf{j}q_y + \mathbf{k}q_z$ is called the vector part of the quaternion q. The quaternion conjugation is denoted as $\tilde{q} = q_0 - \mathbf{q}$.

Let x, y, z be the Cartesian coordinates in the Euclidean space $\mathbb{R}^3$. Let Ω be a domain of $\mathbb{R}^3$ with a piecewise smooth boundary. A quaternion-valued function or, briefly, $\mathbb{H}$-valued function f of a reduced quaternion variable $\mathbf{r} = \mathbf{i}x + \mathbf{j}y + \mathbf{k}z \in \mathbb{R}^3$ is a mapping

$$f : \Omega \longrightarrow \mathbb{H},$$

such that

$$f(\mathbf{r}) = f_0(\mathbf{r}) + \mathbf{f}(\mathbf{r}) = f_0(x, y, z) + \mathbf{i}f_x(x, y, z) + \mathbf{j}f_y(x, y, z) + \mathbf{k}f_z(x, y, z).$$

The functions f_0, f_x, f_y, f_z are real-valued defined in Ω. Continuity, differentiability or integrability of f are defined coordinate-wisely. The differential operator $\nabla = \mathbf{i}\partial_x + \mathbf{j}\partial_y + \mathbf{k}\partial_z$ is called the generalized Cauchy–Riemann operator.

According to [14, 33] a function f is called left-regular in Ω if

$$\nabla f = 0, \quad \mathbf{r} \in \Omega. \tag{2.1}$$

A similar definition can be given for right-regular functions. From now on we use only left-regular functions that, for simplicity, we call regular. With the vectorial notations the regularity condition is given as follows:

$$\nabla f(\mathbf{r}) = -\nabla \cdot \mathbf{f}(\mathbf{r}) + \nabla f_0(\mathbf{r}) + \nabla \times \mathbf{f}(\mathbf{r}) = 0, \tag{2.2}$$

where ∇f_0, $\nabla \cdot \mathbf{f}$, $\nabla \times \mathbf{f}$ are the usual gradient, divergence and curl, respectively. Thus, the coordinate-wise representations of the regularity condition are given as follows:

$$\begin{cases} f_{x,x} + f_{y,y} + f_{z,z} = 0, \\ f_{0,x} + f_{z,y} - f_{y,z} = 0, \\ f_{0,y} + f_{x,z} - f_{z,x} = 0, \\ f_{0,z} + f_{y,x} - f_{x,y} = 0. \end{cases} \tag{2.3}$$

The system (2.3) is called the Moisil–Theodoresco system (MTS) [33, 3, 9] and is a spatial generalization of the Cauchy–Riemann system (CRS). If we assume that f depends only on two variables, for example, x and y, then the MTS (2.3) splits into two CRS and the complex functions $f(\zeta) = f_x(x,y) - i f_y(x,y)$, $g(\zeta) = f_0(x,y) - i f_z(x,y)$ will be the analytic functions of complex variable $\zeta = x + iy$. If the MTS is written in the cylindrical coordinates ρ, φ, z, then in the case of axial symmetry the MTS splits into two generalized by Vekua [39] CRS and $f(\zeta) = f_0(z,\rho) - i f_\varphi(z,\rho)$, $g(\zeta) = f_z(z,\rho) - i f_\rho(z,\rho)$ will be the generalized analytic by Vekua functions of the complex variable $\zeta = z + i\rho$. Exactly these functions are used in axially symmetric problems [1].

A quaternion function F is called a primitive of a regular function f if $\nabla F = f$. Because a regular function f is a harmonic function then a primitive function F is also harmonic and the function F is also called as a harmonic primitive. In this approach the primitive function is also a solution of the inhomogeneous MTS and differs from that defined in [10], [27] because a concept of hyperderivative is not used and it is not a regular (monogenic) function. In the complex analysis solutions of inhomogeneous Cauchy–Riemann system $dF/d\bar{z} = f$ are expressed by the Theodoresco operator (transform). Therefore, the harmonic primitive can be considered as a variant of the generalized Theodoresco transform for the MTS. The generalized Theodoresco transform is used in a hypercomplex operator method [28].

The equation of elastic equilibrium is called the Lamé equation and has the next form

$$L\mathbf{u} \equiv (\lambda + 2\mu)\nabla(\nabla \cdot \mathbf{u}) - \mu \nabla \times (\nabla \times \mathbf{u}) = 0. \tag{2.4}$$

If we introduce the next scalar function f_0 and vector function $\mathbf{f}$

$$(\lambda + 2\mu)\nabla \cdot \mathbf{u} = f_0, \quad -\mu \nabla \times \mathbf{u} = \mathbf{f}, \tag{2.5}$$

then the Lamé equation (2.4) is transformed into the MTS

$$\nabla \cdot \mathbf{f} = 0, \quad \nabla f_0 + \nabla \times \mathbf{f} = 0, \tag{2.6}$$

thus, the quaternion function $f = f_0 + \mathbf{f}$ is regular. In the paper [32] it was indicated that the connection (2.5) between the Lamé equation and quaternion functions was first pointed by G. Moisil.

Hooke's Law in the Cartesian coordinates x_i $(i = 1, 2, 3)$ expresses connections between components σ_{ij} of a stress tensor and components ε_{ij} of a deformation tensor and for three-dimensional linear elasticity has the form

$$\begin{aligned} \sigma_{ij} &= (\lambda + 2\mu)(\nabla \cdot \mathbf{u})\delta_{ij} + 2\mu\varepsilon_{ij}, \\ \varepsilon_{ij} &= \frac{1}{2}\left(u_{i,j} + u_{j,i}\right), \quad (i, j = 1, 2, 3) \end{aligned} \tag{2.7}$$

hereinafter the index after the comma means the partial derivative of the appropriate variable, δ_{ij} is the Kronecker delta function.

We use the usual quaternionic notations in the MTS theory [15, 20, 22, 23, 24]. Some authors have used only matrix algebra methods for the MTS theory, for example, A. Bitsadze [3] in such a way defines three-dimensional analogues of the Cauchy type integral for the MTS. In Clifford analysis notations [5, 6, 7] in the MTS theory the reduced quaternionic variable has the form $x = x_0 + \mathbf{e}_1 x_1 + \mathbf{e}_2 x_2$, where $\mathbf{e}_1$ and $\mathbf{e}_2$ are Clifford algebra units. This approach provides a natural way for introducing notions of hypercomplex derivative and monogenic primitive [27]. However, in the usual quaternionic notations it is possible to define hypercomplex derivatives for the MTS [31].

We will also use the next notations:

- $\partial\Omega$ denotes the boundary of a domain $\Omega \in \mathbb{R}^3$;
- $\overline{\Omega} \equiv \Omega \cup \partial\Omega$ denotes the closure of Ω;
- $C^{l+\alpha}$ denotes the set of functions having continuous partial derivatives up to order l and whose lth-order partial derivatives satisfy a Holder condition with an exponent α;
- $\partial\Omega \in C^{l+\alpha}$ denotes the smooth boundary $\partial\Omega$ that can be covered by a finite number of spheres and intersections $\partial\Omega$ with these spheres can be described in local coordinates by means of functions from $C^{l+\alpha}$ (see [25]).

3. Reconstruction of regular function from given scalar part

All components of a regular function are harmonic functions. Let f_0 be a given scalar harmonic function. It is known then we can reconstruct a regular function $f = f_0 + \mathbf{f}$ with f_0 as a scalar part but only in the star-shaped domains [41, 23] up to a gradient of arbitrary harmonic function ψ_0:

$$f = f_0 + \mathbf{f} = f_0 + \mathbf{r} \times \int_0^1 t^\alpha \nabla f(\mathbf{r}t)dt + \nabla\psi_0.$$

In this section we will show that such the reconstruction is possible for any simply connected (not only star-shaped) arbitrary domain in $\mathbb{R}^3$ but with a smooth boundary. A domain $\Omega \in \mathbb{R}^3$ is said to be simply connected if every closed curve in Ω can be shrunk to a point without leaving Ω (equivalently, if every closed curve in Ω is the boundary of some surface contained completely within Ω). The procedure of such reconstruction is needed for the main result in Section 4.

We must use a solution of Dirichlet problem in the form of a double layer potential with certain smoothness of density. For this purpose it will be useful the next

Theorem 3.1. *Let $\Omega \in \mathbb{R}^3$ be a bounded domain with the boundary $\partial\Omega \in C^{l+\alpha}$, $l \geq 2$, $0 < \alpha < 1$; the double layer potential*

$$u(\mathbf{r}) = W[\mu](\mathbf{r}) \equiv \oint\limits_{\partial\Omega} \mu(\boldsymbol{\varrho}) \frac{\boldsymbol{\varrho} - \mathbf{r}}{|\boldsymbol{\varrho} - \mathbf{r}|^3} \cdot \mathbf{dS}_{\varrho} = - \oint\limits_{\partial\Omega} \mu(\boldsymbol{\varrho}) \frac{\partial}{\partial n} \frac{1}{|\boldsymbol{\varrho} - \mathbf{r}|} dS_{\varrho} \tag{3.1}$$

where $\partial/\partial n$ denotes a differentiation in the direction of the outward normal of $\partial\Omega$, is the solution of the next Dirichlet problem

$$\begin{aligned} \triangle u(\mathbf{r}) &= 0, & \mathbf{r} &\in \Omega, \\ u(\mathbf{r}) &= \varphi(\mathbf{r}) \in C^0(\partial\Omega), & \mathbf{r} &\in \partial\Omega. \end{aligned} \tag{3.2}$$

Then the function $u(\mathbf{r})$ belongs to the class $C^{l+\alpha}(\overline{\Omega})$ if and only if one of the next two conditions is fulfilled

$$\varphi \in C^{l+\alpha}(\partial\Omega), \tag{3.3}$$

$$\mu \in C^{l+\alpha}(\partial\Omega). \tag{3.4}$$

Proof. Let the conditions of the theorem be fulfilled and $u(\mathbf{r}) \in C^{l+\alpha}(\overline{\Omega})$. Then obviously the condition (3.3) is fulfilled and $\mu \in C^0(\partial\Omega)$. It is known [25] that for the boundary value of a double potential with continuous density and defined on the Lyapunov surface we have $W[\mu] \in C^{\beta}$ with an arbitrary value for $\beta \in (0,1)$. Let β_0 be subjected to the condition $\alpha < \beta_0 < 1$, thus $W[\mu] \in C^{\beta_0}$. It is known from the potential theory [25] that the function μ is a solution of the next integral equation of second type

$$\mu(\mathbf{r}) + \frac{1}{2\pi} W[\mu](\mathbf{r}) = \frac{1}{2\pi}\varphi(\mathbf{r}), \quad \mathbf{r} \in \partial\Omega, \tag{3.5}$$

where according to (3.3) $\varphi \in C^{l+\alpha}$. Thus, we have the inclusion $\mu \in C^{\beta_0}(\partial\Omega)$. Then according to [25] we have the inclusion $W[\mu] \in C^{1+\beta_1}(\partial\Omega)$, where $\beta_1 \in (\alpha, \beta_0)$. Again, referring to the equation (3.5), we see that $\mu \in C^{1+\beta_1}(\partial\Omega)$. Repeating these arguments as many times as necessary we get the inclusion $\mu \in C^{1+\gamma}(\partial\Omega)$, where $\gamma > \alpha$. Hence, we have the inclusion $\mu \in C^{1+\alpha}(\partial\Omega)$ and the necessity of the condition (3.4) is proved.

Now let $\varphi \in C^{l+\alpha}(\partial\Omega)$. Then according to Schauder's theorem [40] we have $u \in C^{l+\alpha}(\overline{\Omega})$ and the sufficiency of the condition (3.3) is proved. Let $\mu \in C^{l+\alpha}(\partial\Omega)$. Then we have the inclusion [25] $W[\mu] \in C^{l+1+\gamma}$, where $\gamma \in (0,\alpha)$ is arbitrary.

Because of $C^{l+1+\gamma}(\partial\Omega) \subset C^{l+\alpha}(\partial\Omega)$ by using (3.5) we get $\varphi \in C^{l+\alpha}(\partial\Omega)$ and the sufficiency condition (3.3) is fulfilled and also we obtain that $u(\mathbf{r}) \in C^{l+\alpha}(\overline{\Omega})$. □

Theorem 3.2. *Let $\Omega \in \mathbb{R}^3$ be a bounded simply connected arbitrary domain with the boundary $\partial\Omega \in C^{2+\alpha}$, $0 < \alpha < 1$; $\varphi \in C^{2+\alpha}(\overline{\Omega})$ be a given harmonic scalar function. Then in the domain Ω there exists a regular function f such that $f_0 = \varphi$ and the vector part of the function f is reconstructed in the next form*

$$\mathbf{f}(\mathbf{r}) = \nabla \times \mathbf{A} + \nabla\psi, \quad \mathbf{r} \in \Omega, \tag{3.6}$$

where

$$\mathbf{A}(\mathbf{r}) = -\frac{1}{4\pi}\int\limits_{\Omega} \frac{\nabla_{\varrho} h(\boldsymbol{\varrho})}{|\mathbf{r}-\boldsymbol{\varrho}|} dV_{\varrho}; \tag{3.7}$$

h is a scalar harmonic function in Ω and its boundary value on a $\partial\Omega$ coincides with the density of a double-layer potential in Ω for the function $4\pi\varphi$; ψ is an arbitrary harmonic function in Ω.

Proof. According to Theorem 3.1 the function $4\pi\varphi$ can be represented as a double layer potential:

$$\varphi(\mathbf{r}) = \frac{1}{4\pi} W[g](\mathbf{r}), \tag{3.8}$$

where a density function $g \in C^{2+\alpha}(\partial\Omega)$ is uniquely defined by φ and Ω. Let us consider the next Dirichlet problem:

$$\triangle h = 0, \text{ in } \Omega, \quad h = g \text{ on } \partial\Omega.$$

It is known that the solution of this problem is unique and $h \in C^{2+\alpha}(\overline{\Omega})$. For the Newtonian potential (3.7) we have $\mathbf{A} \in C^3(\Omega)$ [25]. Because of $\nabla h \in C^{1+\alpha}(\overline{\Omega})$ we have the Poisson equation

$$\triangle \mathbf{A}(\mathbf{r}) = \nabla h(\mathbf{r}), \quad \mathbf{r} \in \Omega. \tag{3.9}$$

Using the property of Newtonian potential to admit the differentiation under the integral sign and harmonicity of the function $h(\mathbf{r})$ we have

$$\nabla \cdot \mathbf{A} = \frac{1}{4\pi}\int\limits_{\Omega} \frac{\mathbf{r}-\boldsymbol{\varrho}}{|\mathbf{r}-\boldsymbol{\varrho}|^3} \cdot \nabla_{\varrho} h(\boldsymbol{\varrho}) dV_{\varrho}, \quad \mathbf{r} \in \Omega_K. \tag{3.10}$$

Using the harmonicity of the function $h(\mathbf{r})$ we have

$$\nabla_{\varrho} \cdot \left[\frac{\nabla_{\varrho} h(\boldsymbol{\varrho})}{|\mathbf{r}-\boldsymbol{\varrho}|}\right] = \frac{\mathbf{r}-\boldsymbol{\varrho}}{|\mathbf{r}-\boldsymbol{\varrho}|^3} \cdot \nabla_{\varrho} h(\boldsymbol{\varrho})$$

and we can use the Gauss theorem in (3.10)

$$\begin{aligned} \nabla \cdot \mathbf{A} &= \frac{1}{4\pi}\int\limits_{\Omega} \nabla_{\varrho} \cdot \left[\frac{\nabla_{\varrho} h(\boldsymbol{\varrho})}{|\mathbf{r}-\boldsymbol{\varrho}|}\right] dV_{\varrho} \\ &= \frac{1}{4\pi}\oint\limits_{\partial\Omega} \frac{\nabla_{\varrho} h(\boldsymbol{\varrho})}{|\mathbf{r}-\boldsymbol{\varrho}|} \cdot \mathbf{dS}_{\varrho} = \frac{1}{4\pi}\oint\limits_{\partial\Omega} \frac{1}{|\mathbf{r}-\boldsymbol{\varrho}|} \frac{\partial h(\boldsymbol{\varrho})}{\partial n} dS_{\varrho}, \quad \mathbf{r} \in \Omega_K. \end{aligned} \tag{3.11}$$

As is known, there exists the next integral representation for the harmonic function $h(\mathbf{r})$

$$h(\mathbf{r}) = \frac{1}{4\pi} \oint\limits_{\partial\Omega} \left[\frac{1}{|\mathbf{r}-\boldsymbol{\varrho}|} \frac{\partial h(\boldsymbol{\varrho})}{\partial n} - h(\boldsymbol{\varrho}) \frac{\partial}{\partial n} \frac{1}{|\mathbf{r}-\boldsymbol{\varrho}|} \right] dS_{\varrho}. \tag{3.12}$$

After substituting expressions from (3.11) and (3.8) with the equality $g = h$ on $\partial\Omega_K$ into the formula (3.12) we obtain

$$h(\mathbf{r}) = \nabla \cdot \mathbf{A} + \varphi(\mathbf{r})$$

and

$$\nabla[\nabla \cdot \mathbf{A}] = \nabla h(\mathbf{r}) - \nabla\varphi(\mathbf{r}), \quad \mathbf{r} \in \Omega_K. \tag{3.13}$$

Now let us introduce the vector function $\mathbf{f}_1(\mathbf{r}) \equiv \nabla \times \mathbf{A}(\mathbf{r})$. In view of (3.9), (3.13) and using the identity $\nabla \times (\nabla \times \mathbf{A}) \equiv \nabla(\nabla \cdot \mathbf{A}) - \triangle\mathbf{A}$ one can see that the function f_1 is a particular solution of the next system on K

$$\nabla \cdot \mathbf{f}_1 = 0, \quad \nabla \times \mathbf{f}_1 = -\nabla\varphi, \tag{3.14}$$

and the function $\varphi + \mathbf{f}_1$ is a regular function. Let $\mathbf{f}$ be an arbitrary solution of the system (3.14). Then for the difference $\mathbf{F} \equiv \mathbf{f}_1 - \mathbf{f}$ we have the homogenous system on K

$$\nabla \cdot \mathbf{F} = 0, \quad \nabla \times \mathbf{F} = 0. \tag{3.15}$$

As is well known [13], the general solution of this system in the simply connected domain is a function $\mathbf{F} = -\nabla\psi$, where ψ is an arbitrary harmonic function on Ω. Thus we have the formula (3.6) for the vector part $\mathbf{f}$ of the regular function on Ω. Conversely, if the function $\mathbf{f}$ is to be determined by the formula (3.6), where ψ is an arbitrary harmonic function, then by the straightforward differentiation one can check that the function $\mathbf{f}$ is a solution of the system (3.14). □

Remark 3.3. Some results of this section were announced in [19]. The requirement of simple connectivity for the domain Ω is necessary because the Riesz system (3.15) can be solvable as a gradient of a harmonic function only in such domains. In the case of multiply connected domain solutions of the Riesz system exist in other forms (see [2]). Some properties of the inhomogeneous Riesz system were investigated in [12] by means of a quaternionic treatment.

4. Three-dimensional analogue of Kolosov–Muskhelishvili Formulae

4.1. Representation for elastic displacement vector

For the case of star-shaped domains $\Omega^* \in \mathbb{R}^3$ the three-dimensional analogue of the Kolosov–Muskhelishvili formulae for displacements is proved in papers [17, 22] as the next theorem:

Theorem 4.1. *The general solution of the Lame equation* (2.4) *in* Ω^* *is expressed in terms of two regular in* Ω^* *functions* φ, ψ *in the form*

$$2\mu\mathbf{u}(\mathbf{r}) = \varkappa\Phi(\mathbf{r}) - \mathbf{r}\widetilde{\varphi(\mathbf{r})} - \widetilde{\psi(\mathbf{r})}, \;\; \varkappa = -\frac{3\lambda + 7\mu}{\lambda + \mu} = -7 + 8\nu, \tag{4.1}$$

where as Φ *one can take any primitive of function* φ, *having subordinated* ψ *to the condition* $\varkappa\Phi_0 = \mathbf{r}\cdot\boldsymbol{\varphi} + \psi_0$.

Here ν is Poisson's ratio. Thus, the vector of elastic displacement is represented in terms of two arbitrary regular functions φ, ψ and a harmonic primitive Φ of a function φ. It can be shown that from (4.1) the expressions for the divergence and circle have the next forms

$$\begin{aligned}(\lambda + 2\mu)\nabla\cdot\mathbf{u} &= 4(1-\nu)\varphi_0,\\ -\mu\nabla\times\mathbf{u} &= (5-4\nu)\boldsymbol{\varphi} + (\mathbf{r}\cdot\nabla)\boldsymbol{\varphi} - \mathbf{r}\times\nabla\varphi_0 + \nabla\psi_0.\end{aligned} \tag{4.2}$$

But in [17, 22] expressions for stresses were not established. In this section we show that the representation (4.1) remains faithful on any compact in the arbitrary simply connected domains not only star-shaped and expressions for components of a stress tensor are obtained.

Indeed, let a vector field $\mathbf{u}$ be in the form of (4.1) then by the straightforward differentiation one can check that we have a solution of the Lamé equation (2.4).

Now let $\Omega \in \mathbb{R}^3$ be a bounded simply connected arbitrary domain with the boundary $\partial\Omega \in C^{2+\alpha}$, $0 < \alpha < 1$; a vector function $\mathbf{u} \in C^{3+\alpha}(\overline{\Omega})$ be a solution of the Lamé equation (2.4) in the domain Ω. We will show that there exist such regular functions φ, ψ and a harmonic primitive Φ that the vector function $\mathbf{u}$ is represented in the form of (4.1) in the domain Ω. Let us introduce the harmonic scalar function φ_0

$$4(1-\nu)\varphi_0 = (\lambda + 2\mu)\nabla\cdot\mathbf{u}, \tag{4.3}$$

for this function we have $\varphi_0 \in C^{2+\alpha}(\overline{\Omega})$. Thus according to Theorem 3.2 we can introduce a regular function $\varphi = \varphi_0 + \boldsymbol{\varphi}$ in the domain Ω. Then let us introduce a vector function $\mathbf{A}$:

$$\mathbf{A} = -\mu\nabla\times\mathbf{u} - (5-4\nu)\boldsymbol{\varphi} - (\mathbf{r}\cdot\nabla)\boldsymbol{\varphi} + \mathbf{r}\times\nabla\varphi_0. \tag{4.4}$$

By the direct differentiation we have the Riesz system for $\mathbf{A}$: $\nabla\cdot\mathbf{A} = 0$, $\nabla\times\mathbf{A} = 0$. In such the case it is known [13] that in a simply connected domain there exists a harmonic function ψ_0 such that the vector function $\mathbf{A}$ can be represented as a gradient of this function: $\mathbf{A} = \nabla\psi_0$. Then according to Theorem 3.2 we can introduce a regular function $\psi = \psi_0 + \boldsymbol{\psi}$. And now let us introduce a function $\Phi = \Phi_0 + \boldsymbol{\Phi}$:

$$\begin{aligned}\varkappa\Phi_0 &= \mathbf{r}\cdot\boldsymbol{\varphi} + \psi_0,\\ \varkappa\boldsymbol{\Phi} &= 2\mu\mathbf{u} + \mathbf{r}\varphi_0 - \mathbf{r}\times\boldsymbol{\varphi} - \boldsymbol{\psi}.\end{aligned} \tag{4.5}$$

By the straightforward differentiation one can check that

$$\nabla\cdot\boldsymbol{\Phi} = -\varphi_0, \quad \nabla\Phi_0 + \nabla\times\boldsymbol{\Phi} = \boldsymbol{\varphi}, \tag{4.6}$$

thus $\nabla\Phi = \varphi$ and the function Φ is a harmonic primitive for the function φ and we proved Theorem 4.1 in the arbitrary simply connected domain with the smooth boundary.

4.2. Expressions for components of stress tensor

In this section for the quaternionic representation (4.1) using ideas similar [5, 6] we will construct the expressions for the stress tensor components in terms of two arbitrary regular functions. Let us write Hooke's Law (2.7) when we have elastic displacements in the form of (4.1):

$$\begin{aligned}
\sigma_{xx} &= \varkappa\Phi_{x,x} - (1-4\nu)\varphi_0 - x\varphi_{0,x} + y\varphi_{z,x} - z\varphi_{y,x} + \psi_{x,x},\\
\sigma_{yy} &= \varkappa\Phi_{y,y} - (1-4\nu)\varphi_0 - y\varphi_{0,y} + z\varphi_{x,y} - x\varphi_{z,y} + \psi_{y,y},\\
\sigma_{zz} &= \varkappa\Phi_{z,z} - (1-4\nu)\varphi_0 - x\varphi_{0,z} + x\varphi_{y,z} - y\varphi_{x,z} + \psi_{z,z},\\
2\sigma_{xy} &= \varkappa(\Phi_{x,y} + \Phi_{y,x}) - x\varphi_{0,y} - y\varphi_{0,x} + y\varphi_{z,y} - z\varphi_{y,y} + z\varphi_{x,x}\\
&\quad - x\varphi_{z,x} + \psi_{x,y} + \psi_{y,x},\\
2\sigma_{yz} &= \varkappa(\Phi_{y,z} + \Phi_{z,y}) - y\varphi_{0,z} - z\varphi_{0,y} + z\varphi_{x,z} - x\varphi_{z,z} + x\varphi_{y,y}\\
&\quad - y\varphi_{x,y} + \psi_{y,z} + \psi_{z,y},\\
2\sigma_{zx} &= \varkappa(\Phi_{z,x} + \Phi_{x,z}) - z\varphi_{0,x} - x\varphi_{0,z} + x\varphi_{y,x} - y\varphi_{x,x} + y\varphi_{z,z}\\
&\quad - z\varphi_{y,z} + \psi_{z,x} + \psi_{x,z}.
\end{aligned} \tag{4.7}$$

4.2.1. First Kolosov–Muskhelishvili formula for stresses. If we will follow the ideas of the complex case the first Kolosov–Muskhelishvili formula for the stresses includes only the normal stress components. From Hooke's Law (2.7) the next formula can be obtained:

$$\sigma_{xx} + \sigma_{yy} + \sigma_{zz} = 3\lambda\nabla\cdot\mathbf{u} + 2\mu\nabla\cdot\mathbf{u} = (3\lambda + 2\mu)\nabla\cdot\mathbf{u}.$$

Then using the formula (4.2) we obtain the first Kolosov–Muskhelishvili formula for the stresses:

$$\sigma_{xx} + \sigma_{yy} + \sigma_{zz} = 4(1-\nu)\varphi_0. \tag{4.8}$$

4.2.2. Second Kolosov–Muskhelishvili formula for stresses. Analogically to [5, 6] we start the construction by

$$\mathbf{i}(-\sigma_{xx} + \sigma_{yy} + \sigma_{zz}) - \mathbf{j}2\sigma_{xy} - \mathbf{k}2\sigma_{xz},$$

by substitution of expressions from (4.7) into the above formula and using the regularity of the function φ after some calculations we obtain the desired expression and in a similar way the other two:

$$\begin{aligned}
&\mathbf{i}(-\sigma_{xx} + \sigma_{yy} + \sigma_{zz}) - \mathbf{j}2\sigma_{xy} - \mathbf{k}2\sigma_{xz}\\
&\quad = \mathbf{i}6(1-2\nu)\varphi_0 - \nabla(\varkappa\Phi_x + \psi_x) - \varkappa\boldsymbol{\Phi}_{,x} - \boldsymbol{\psi}_{,x}\\
&\qquad + x\nabla\varphi_0 + \mathbf{r}\varphi_{0,x} - y\nabla\varphi_z + z\nabla\varphi_y + \boldsymbol{\varphi}_{,x} \times \mathbf{r},\\
&\mathbf{j}(\sigma_{xx} - \sigma_{yy} + \sigma_{zz}) - \mathbf{k}2\sigma_{yz} - \mathbf{i}2\sigma_{yx}\\
&\quad = \mathbf{j}6(1-2\nu)\varphi_0 - \nabla(\varkappa\Phi_y + \psi_y) - \varkappa\boldsymbol{\Phi}_{,y} - \boldsymbol{\psi}_{,y}\\
&\qquad + y\nabla\varphi_0 + \mathbf{r}\varphi_{0,y} - z\nabla\varphi_x + x\nabla\varphi_z + \boldsymbol{\varphi}_{,y} \times \mathbf{r},\\
&\mathbf{k}(\sigma_{xx} = \sigma_{yy} - \sigma_{zz}) - \mathbf{i}2\sigma_{zx} - \mathbf{j}2\sigma_{zy}
\end{aligned}$$

$$
\begin{aligned}
&= \mathbf{k}6(1-2\nu)\varphi_0 - \nabla(\varkappa\Phi_z + \psi_z) - \varkappa\boldsymbol{\Phi}_{,z} - \boldsymbol{\psi}_{,z} \\
&\quad + z\nabla\varphi_0 + \mathbf{r}\varphi_{0,z} - x\nabla\varphi_y + y\nabla\varphi_x + \boldsymbol{\varphi}_{,z} \times \mathbf{r}.
\end{aligned} \tag{4.9}
$$

The obtained three formulas (4.9) can be called as the second Kolosov–Muskhelishvili formula for the stresses.

4.2.3. Second Kolosov–Muskhelishvili formula for stresses in tensor notations. The second Kolosov–Muskhelishvili formula (4.9) for the stresses can written in a compact form by using the tensor notations for the Cartesian coordinates x_i $(i = 1, 2, 3)$. Let $\mathbf{e}_i$ be unit vectors of the Cartesian coordinates and $\mathbf{T}_i$ be a stress vector on the plane with normal $\mathbf{e}_i$ $(i = 1, 2, 3)$. It is known from the basis of elasticity theory that

$$
\mathbf{T}_i = \sum_{l=1}^{3} \mathbf{e}_i \sigma_{li}, \quad i = 1, 2, 3. \tag{4.10}
$$

By using (4.8) and (4.10) the left side expression of the first formula in (4.9) can be transformed into the next form

$$
\begin{aligned}
&\mathbf{i}(\sigma_{xx} + \sigma_{yy} + \sigma_{zz}) - \mathbf{i}2\sigma_{xx} - \mathbf{j}2\sigma_{xy} - \mathbf{k}2\sigma_{xz} \\
&\quad = \mathbf{i}4(1-\nu)\varphi_0 - 2\mathbf{T}_x,
\end{aligned}
$$

here $\mathbf{T}_x \equiv \mathbf{T}_1$. Thus, for $\mathbf{T}_x$ we have the formula

$$
\begin{aligned}
2\mathbf{T}_x &= \mathbf{i}2(1-4\nu)\varphi_0 + \nabla(\varkappa\Phi_x + \psi_x) + \varkappa\boldsymbol{\Phi}_{,x} + \boldsymbol{\psi}_{,x} - x\nabla\varphi_0 - \mathbf{r}\varphi_{0,x} \\
&\quad + y\nabla\varphi_z - z\nabla\varphi_y - \mathbf{r} \times \boldsymbol{\varphi}_{,x}.
\end{aligned}
$$

One can obtain in an analogical way formulas for $\mathbf{T}_y$ and $\mathbf{T}_y$. In the tensor notation these three formulas can be written as the next second Kolosov–Muskhelishvili formula for the stresses

$$
\begin{aligned}
2\mathbf{T}_i &= \mathbf{e}_i 2(1-4\nu)\varphi_0 + \nabla(\varkappa\Phi_i + \psi_i) + \varkappa\boldsymbol{\Phi}_{,i} + \boldsymbol{\psi}_{,i} - x_i\nabla\varphi_0 - \mathbf{r}\varphi_{0,x} \\
&\quad + \sum_{k,l=1}^{3} \varepsilon_{ikl} x_k \nabla\varphi_l + \mathbf{r} \times \boldsymbol{\varphi}_{,i}, \quad i = 1, 2, 3;
\end{aligned} \tag{4.11}
$$

here ε_{ikl} is the completely antisymmetric Levi-Civita tensor.

5. Conclusion

In this paper the theory of Moisil–Theodoresco system in terms of regular quaternionic functions of reduced quaternion variable is used. We obtain the generalized Kolosov–Muskhelishvili formulae in the three-dimensional linear elasticity where displacements and stresses are expressed in terms of two regular quaternion functions in arbitrary domains not only star-shaped but with some restrictions for the smoothness of the boundary and functions. For this result the theorem about the reconstruction of a regular function from the given scalar part is proved in such domains. The restrictions on the domain boundary and used functions are caused by the used methods of the potential theory.

References

[1] A.Ya. Aleksandrov and Yu.I. Solov'ev, *Three-Dimensional Problems of the Theory of Elasticity* [in Russian]. Nauka, Moscow, 1979.

[2] S. Bergman, *Integral operators in the theory of linear partial differential equations.* Springer-Verlag, Berlin–Göttingen–Heidelberg, 1961.

[3] A.V. Bitsadze, *Boundary value problems for second order elliptic equations.* North-Holland Publ.Co., Amsterdam, 1968.

[4] S. Bock and K. Gürlebeck, *On Hypercomplex Differential and Primitivation Operators with Applications to Representation Formulae of Linear Elastostatics.* AIP Conference Proceedings, **936** (2007), 717–720. DOI 10.1063/1.2790252

[5] S. Bock and K. Gürlebeck, *On a spatial generalization of the Kolosov–Muskhelishvili formulae.* Mathematical Methods in the Applied Sciences, **32**, Issue 2 (2009), 223–240. DOI: 10.1002/mma.1033.

[6] S. Bock and K. Gürlebeck, *On a Polynomial Basis Generated from the Generalized Kolosov–Muskhelishvili Formulae.* Advances in Applied Clifford Algebras, **19** (2009), 191–209. DOI. 10.1007/s00006-009-0156-5.

[7] S. Bock, *On Monogenic Series Expansions with Applications to Linear Elasticity.* Advances in Applied Clifford Algebras, **24** (2014), 931–943. DOI 10.1007/s00006-014-0490-0

[8] S. Bock, K. Gürlebeck, D. Legatiuk and H.M. Nguyen *ψ-Hyperholomorphic functions and a Kolosov–Muskhelishvili formula.* Mathematical Methods in the Applied Sciences, 2015. DOI: 10.1002/mma.3431

[9] F. Brackx, R. Delanghe, and F. Sommen, *Clifford Analysis*, Research Notes in Mathematics No. 76, Pitman, London, 1982.

[10] I. Cacao, K. Gürlebeck, *On monogenic primitives of monogenic functions.* Complex Variables and Elliptic Equations, **52**, no. 10-11 (2007), 1081–1100. DOI: 10.1080/17476930701399088.

[11] F. Colombo, I. Sabadini and D.C. Struppa, *Slice monogenic functions.* Israel J. of Math., **171**, No. 1 (2009), 385–403. DOI 10.1007/s11856-009-0055-4.

[12] F. Colombo, M.E. Luna-Elizarraras, I. Sabadini, M. Shapiro and D.C. Struppa, *A Quaternionic Treatment of the Inhomogeneous Div-Rot System.* Moscow Math. J., **12**, N. 1, January–March (2012), 37–48.

[13] G.M. Fikhtengolts, *The Fundamentals of Mathematical Analysis.* Pergamon Press, 1965.

[14] R. Fueter, *Über die analytischen Darstellungen der regulären Funktionen einer Quaternionenvariablen.* Com. Math. Helv., **4** (1932), 9–20.

[15] Yu.M. Grigor'ev and V.V. Naumov, *Approximation theorems for the Moisil–Theodorescu system.* Siberian Mathematical Journal September–October, **25**, Issue 5 (1984), 693–701. DOI 10.1007/BF00968681.

[16] Yu.M. Grigor'ev, *Solution of a problem for an elastic sphere in a closed form.* Dynamics of Continuous Medium [in Russian], No. 71 (1985), Inst. Gidrodin. Sib. Otd. Akad. Nauk SSSR, Novosibirsk, 50–54.

[17] Yu.M. Grigor'ev, *Solution of spatial problems of an elasticity theory by means of a quaternionic functions theory.* Candidat dissertation (PhD) [in Russian], Lavrent'ev Institute of Hydrodynamics, Novosibirsk (1985), 131 pp.

[18] Yu.M. Grigor'ev and V.V. Naumov, *Solution of third and fourth main problems of an equilibrium of an elastic sphere in a closed form.* Dynamics of Continuous Medium [in Russian], No. 87 (1988), Inst. Gidrodin. Sib. Otd. Akad. Nauk SSSR, Novosibirsk, 54–66.

[19] Yu.M. Grigor'ev and V.V. Naumov, *Erratum in: Approximation theorems for the Moisil–Theodorescu system.* Siberian Mathematical Journal, September–October, **25**, Issue 5 (1984), 693–701. Deposited in VINITI 11.05.89, N 5739-B89 [in Russian].

[20] Yu.M. Grigor'ev and V.V. Alekhin, *A quaternionic boundary element method.* Sib. jurn. industr. matem. [in Russian], Vol. 2, No. 1 (1999), Inst. Matem. Sib. Otd. Akad. Nauk SSSR, Novosibirsk, 47–52.

[21] Yu.M. Grigor'ev, *A spatial analogue of the integral equation of Mushelishvili.* Dynamics of Continuous Medium [in Russian], No. 114 (1999), Inst. Gidrodin. Sib. Otd. Akad. Nauk SSSR, Novosibirsk, 161–165.

[22] Yu. Grigor'ev, *Three-dimensional Quaternionic Analogue of the Kolosov–Muskhelishvili Formulae.* Hypercomplex Analysis: New perspectives and applications, Trends in Mathematics, (eds. S. Bernstein, U. Kaehler, I. Sabadini, F. Sommen), Birkhäuser, Basel (2014), 145–166. DOI 10.1007/978-3-319-08771-9-10.

[23] Yu. Grigoriev, *Radial integration method in quaternion function theory and its applications.* AIP Conference Proceedings, **1648**, 440003 (2015), DOI:http://dx.doi.org/10.1063/1.4912654

[24] Yu. Grigor'ev, *Regular quaternionic functions and their applications.* Proceedings of the 20th International Conference on the Applications of Computer Science and Mathematics in Architecture and Civil Engineering. Editors: K. Gürlebeck, T. Lahmer, Weimar, Germany, July 20–22, 2015. Pp. 45–50.

[25] N.M. Gunther, *Potential Theory and its Applications to Basic Problems of Mathematical Physics.* Frederick Ungar Publishing, New York, 1967.

[26] K. Gürlebeck, K. Habetha and W. Sprößig, *Quaternionic Calculus for Engineers and Physicists.* John Wiley & Sons, Cinchester, 1997.

[27] K. Gürlebeck and H.R. Malonek *A Hypercomplex Derivative of Monogenic Functions in R^{n+1} and its Applications.* Complex Variables, Theory and Applications, **39** (1999), 199–228.DOI 10.1080/17476939908815192.

[28] K. Gürlebeck and W. Sprößig, *Holomorphic functions in the plane and n-dimensional space.* Birkhäuser Verlag, 2008.

[29] V.V. Kravchenko, *Applied quaternionic analysis.* Research and Exposition in Mathematics **28** (2003), Lemgo, Heldermann Verlag.

[30] M. Ku, U. Kähler and D.S. Wang, *Riemann boundary value problems on the sphere in Clifford analysis.* Advances in Applied Clifford Algebras, **22**, Issue 2 (June 2012), 365–390. DOI 10.1007/s00006-011-0308-2.

[31] M.E. Luna-Elizarraras, M.A. Macas-Cedeño, M. Shapiro, *On the hyperderivatives of Moisil–Theodoresco hyperholomorphic functions.* Hypercomplex Analysis and Applications, Trends in Mathematics, (eds. I. Sabadini and F. Sommen), Springer Basel AG (2011), 181–193. DOI 10.1007/978-3-0346-0246-4-13

[32] M. Misicu, *Representarea ecuatilor echilibrului elastic prin functii monogene de cuaterninoni.* Bull. Stiint. Acad. RPR. Sect. st. mat.fiz., **9**, No. 2 (1957), 457–470.

[33] Gr.C. Moisil, N. Theodoresco, *Fonctions holomorphes dans l'espace.* Mathematica, **5**, (1931), 141–153.

[34] N.I. Muskhelishvili, *Some Basic Problems of the Mathematical Theory of Elasticity.* Springer, 1977.

[35] V.V. Naumov and Yu.M. Grigor'ev, *The Laurent series for the Moisil–Theodoresco system.* Dynamics of Continuous Medium [in Russian], No. 54 (1982), Inst. Gidrodin. Sib. Otd. Akad. Nauk SSSR, Novosibirsk, 115–126.

[36] V.V. Naumov, *Solution of two main problems of an equilibrium of an elastic sphere in a closed form.* Dynamics of Continuous Medium [in Russian], No. 54 (1986), Inst. Gidrodin. Sib. Otd. Akad. Nauk SSSR, Novosibirsk, 96–108.

[37] D. Weisz-Patrault, S. Bock and K. Gürlebeck, *Three-dimensional elasticity based on quaternion-valued potentials.* International Journal of Solids and Structures, **51** (2014), 3422–3430. DOI: 10.1016/j.ijsolstr.2014.06.002

[38] G. Polozij, *The Theory and Application of p-analytic and pq-Analytical Functions* [in Russian]. Naukova Dumka, Kiev, 1973.

[39] I. Vekua, *Generalized Analytic Functions.* Addison Wesley, Reading Mass., 1962.

[40] J. Schauder, *Über lineare elliptische Differentialgleichungen zweiter Ordnung.* Math. Z. **38** (1934), 257–282.

[41] A. Sudbery, *Quaternionic analysis.* Mathematical Proceedings of the Cambridge Philosophical Society, No. 85 (1979), 199–225. DOI:10.1017/S0305004100055638.

Yuri Grigor'ev
Belinsky Str. 58
North-Eastern Federal University
677000 Yakutsk, Russia
e-mail: grigyum@yandex.ru

Modern Trends in Hypercomplex Analysis
Trends in Mathematics, 217–226

On Some Properties of Pseudo-complex Polynomials

Klaus Gürlebeck and Dmitrii Legatiuk

Abstract. Monogenic functions are typically approximated by help of monogenic polynomials. Different systems of monogenic polynomials have been developed by several authors in the last years. One of available constructions are the so-called system of Pseudo-Complex Polynomials (PCP). PCP are 3D monogenic polynomials which have a structure similar to integer powers of one complex variable. In this paper we present an algorithm allowing us to find clear relations between different sets of PCP, which has not been studied before.

Mathematics Subject Classification (2010). Primary 30G35.

Keywords. Monogenic functions; polynomials; structural study; rotations.

1. Introduction

With the development of hypercomplex analysis monogenic functions become more and more popular in different fields of applications. Typically, to describe a monogenic function one uses its series expansion with respect to a system of monogenic polynomials. In recent years several different systems of monogenic polynomials were proposed by many authors, see for example [1, 2, 3, 10] and the references therein. To simplify applications of polynomial systems it is helpful to discover their properties.

In 1970 a special type of monogenic polynomials the so-called totally regular variables defined in $\mathbb{R}^n$ have been introduced in [7]. In this work the main intention was to introduce monogenic functions, such that their integer powers are also monogenic. To assure that, some specific sufficient conditions were formulated. The detailed study with necessary and sufficient conditions for totally regular variables with values in $\mathbb{R}^4$ was done in [9].

The research of the second author is supported by the German Research Foundation (DFG) and the Russian Foundation of Basic Research (RFBR).

Recently, another system of monogenic polynomials was proposed in [4, 5, 6], where the so-called Pseudo-Complex Polynomials (PCP) have been introduced. This system has a direct relation to works [7, 9]. PCP have a structure which is similar to integer powers of one complex variable, and they share some of very useful properties of complex powers. Due to this simple structure it is very attractive to use PCP in different fields of applications, e.g., interpolation of monogenic functions (Lagrange-type, Newton-type), etc., but to gain more benefit of the simple structure one has to study PCP in more detail.

In previous works the question of a construction of PCP has been studied in detail. But some related structural questions have not been discussed neither for PCP, nor for totally regular variables. Therefore in this paper we present basic results summarizing some interesting properties of PCP. We start with preliminaries from hypercomplex analysis followed by an introduction of PCP. After that we present a theorem stating that PCP can be obtained from a complex monomial by using an invertible multiplicative extension. As the main part of this paper we discuss in details a construction of one specific set of parameters defining PCP.

2. Preliminaries and notations

Let $1, \mathbf{e}_1, \mathbf{e}_2, \mathbf{e}_3$ be an orthonormal basis of the Euclidean vector space $\mathbb{R}^4$. As usual we identify the basis vector $\mathbf{e}_0$ with 1. We introduce an associative multiplication of the basis vectors subject to the multiplication rules:

$$\mathbf{e}_1^2 = \mathbf{e}_2^2 = \mathbf{e}_3^2 = -1, \quad \mathbf{e}_1\mathbf{e}_2 = -\mathbf{e}_2\mathbf{e}_1 = \mathbf{e}_3.$$

This non-commutative product generates the algebra of real quaternions denoted by $\mathbb{H}$. The real vector space $\mathbb{R}^4$ will be embedded in $\mathbb{H}$ by identifying the element $\mathbf{a} = (a_0, a_1, a_2, a_3) \in \mathbb{R}^4$ with the element

$$\mathbf{a} = a_0 + a_1\mathbf{e}_1 + a_2\mathbf{e}_2 + a_3\mathbf{e}_3 \in \mathbb{H}.$$

The real number $\mathbf{Sc}\,\mathbf{a} := a_0$ is called the scalar part of $\mathbf{a}$ and $\mathbf{Vec}\,\mathbf{a} := a_1\mathbf{e}_1 + a_2\mathbf{e}_2 + a_3\mathbf{e}_3$ is the vector part of $\mathbf{a}$. Analogous to the complex case, the conjugate of $\mathbf{a} := a_0 + a_1\mathbf{e}_1 + a_2\mathbf{e}_2 + a_3\mathbf{e}_3 \in \mathbb{H}$ is the quaternion $\bar{\mathbf{a}} := a_0 - a_1\mathbf{e}_1 - a_2\mathbf{e}_2 - a_3\mathbf{e}_3$. The norm of $\mathbf{a}$ is given by $|\mathbf{a}| = \sqrt{\mathbf{a}\bar{\mathbf{a}}}$ and coincides with the corresponding Euclidean norm of $\mathbf{a}$, as a vector in $\mathbb{R}^4$.

Additionally we introduce the subset

$$\mathcal{A} := \mathrm{span}_{\mathbb{R}}\{1, \mathbf{e}_1, \mathbf{e}_2\} \subset \mathbb{H}.$$

The real vector space $\mathbb{R}^3$ can be embedded in $\mathcal{A}$ by the identification of each element $\mathbf{x} = (x_0, x_1, x_2)^{\mathrm{T}} \in \mathbb{R}^3$ with the reduced quaternion or paravector

$$\mathbf{x} = x_0 + x_1\mathbf{e}_1 + x_2\mathbf{e}_2 \in \mathcal{A}.$$

As a consequence, we will often use the same symbol $\mathbf{x}$ to represent a point in $\mathbb{R}^3$ as well as to represent the corresponding reduced quaternion. For the coordinate axes x, y, z we will use the notation x_0, x_1, x_2.

Let Ω be an open subset of $\mathbb{R}^3$ with a piecewise smooth boundary. An $\mathbb{H}$-valued function is a mapping

$$f\colon \Omega \to \mathbb{H}$$

such that

$$f(\mathbf{x}) = \sum_{i=0}^{3} f^i(\mathbf{x})\mathbf{e}_i, \quad \mathbf{x} \in \Omega.$$

The coordinates f^i are real-valued functions defined in Ω, i.e.,

$$f^i\colon \Omega \to \mathbb{R}, \quad i = 0, 1, 2, 3.$$

Continuity, differentiability or integrability of f are defined coordinate-wisely. For continuously real-differentiable functions $f\colon \Omega \subset \mathbb{R}^3 \to \mathbb{H}$, which we will denote for simplicity by $f \in C^1(\Omega, \mathbb{H})$, the operator

$$\bar{\partial} = \frac{1}{2}\left(\frac{\partial}{\partial x_0} + \mathbf{e}_1\frac{\partial}{\partial x_1} + \mathbf{e}_2\frac{\partial}{\partial x_2}\right)$$

is called generalized Cauchy–Riemann operator. The corresponding conjugate generalized Cauchy–Riemann operator is defined as

$$\partial = \frac{1}{2}\left(\frac{\partial}{\partial x_0} - \mathbf{e}_1\frac{\partial}{\partial x_1} - \mathbf{e}_2\frac{\partial}{\partial x_2}\right).$$

We define and denote the Cauchy–Riemann operators analogous to the complex one-dimensional case. Here, we follow the notation used in [8], which is opposed to the commonly used notation in the Clifford analysis, but analogues to complex function theory.

Definition 2.1. A function $f \in C^1(\Omega, \mathbb{H})$ is called left (resp. right) monogenic in Ω if

$$\bar{\partial} f = 0 \quad \text{in} \quad \Omega \quad (\text{resp.}, f\bar{\partial} = 0 \quad \text{in} \quad \Omega).$$

Finally we introduce special reduced quaternions, usually called Fueter variables, defined by

$$z_k = x_k - x_0\mathbf{e}_k, \quad x_0, x_k \in \mathbb{R}, \quad k = 1, 2.$$

2.1. Pseudo-complex polynomials

In this section we are going to introduce a set of polynomials of the form

$$\mathcal{Z}_s^k(x) = (x_0 + y_s\mathbf{i}_s)^k, \tag{2.1}$$

where

$$y_s = \alpha_s x_1 + \beta_s x_2$$

and

$$\mathbf{i}_s = \alpha_s\mathbf{e}_1 + \beta_s\mathbf{e}_2,$$

with parameters α_s and β_s. It is easy to verify that if the parameters satisfy the condition

$$\alpha_s^2 + \beta_s^2 = 1,$$

then it follows immediately that $\mathbf{i}_s^2 = -1$, and therefore one can prove that the above polynomials are monogenic polynomials which are isomorphic to the complex powers with respect to a chosen parameter set. To underline this fact these polynomials are called pseudo-complex polynomials (PCP). We would like to remark that sometimes in literature one can find another name for PCP – Pseudo-Complex Powers, but we will not mix these names and we will follow only the name pseudo-complex polynomials.

The original idea of such polynomials is going back to 1970 where in [7] the so-called totally regular variables defined in $\mathbb{R}^n$ with values in Clifford algebra $\mathcal{C}\ell_{0,n}$ were introduced. Later, in 1982, a detailed study of totally regular variables in the case of quaternions has been done in [9]. These variables are defined as follows:

Definition 2.2. A linear hypercomplex holomorphic function of the form

$$g = x_0 d_0 + x_1 d_1 + x_2 d_2 + x_3 d_3,$$

whose integer powers are also holomorphic is called totally regular variable, where $d_k = a_{k0}\mathbf{e}_0 + a_{k1}\mathbf{e}_1 + a_{k2}\mathbf{e}_2 + a_{k3}\mathbf{e}_1\mathbf{e}_2 \in \mathbb{H}$, $a_{jk} \in \mathbb{R}$ $j,k = 0,1,2,3$.

Recently PCP have been considered by several authors in different contexts. In [2] it was proved that homogeneous monogenic polynomials of the form

$$H^k_{(a_i,b_i)} = (a_i z_1 + b_i z_2)^k, \quad a_i, b_i \in \mathbb{R}, i = 0,\dots,k$$

form a basis of the $\mathbb{H}$-linear space of homogeneous monogenic $\mathbb{H}$-valued polynomials of degree k if and only if $a_i^2 + b_i^2 = 1$ and $a_i b_j - a_j b_i \neq 0$, $i,j = 0,\dots,k$, $i \neq j$. These two conditions provide an explicit relation between these polynomials and PCP. In [3], a complete set of pseudo-complex polynomials, having prescribed properties was constructed. Later on, aspects of combinatorial nature were considered in [5], and some computational aspects related to the explicit constructions of PCP were already discussed in [4]. In [6] numerical properties of the implementation of PCP have been studied in details.

According to [5] the parameters α_s and β_s form the so-called parameter set $\mathbf{A} = \{(\alpha_s, \beta_s)\,,\ s = 0,\dots,k\}$ which can be associated with unit vectors in $\mathbb{R}^2$.

To finish the introduction of PCP we recall from [5] the following theorem:

Theorem 2.3. *The set of polynomials* $\left\{\mathcal{Z}_s^k\right\}_{s=0}^k$ *of the form* (2.1) *is a basis for the space of homogeneous monogenic paravector-valued polynomials of degree* k *in* $\mathbb{R}^3$*, provided that the* $k+1$ *unit vectors* $\gamma_s^k = (\alpha_s, \beta_s) \in \mathbb{R}^2$*, with* $s = 0,\dots,k$*, are pairwise noncollinear.*

The lower triangular part of Table 1 represents the basis generated by pseudo-complex polynomials (2.1). Sometimes it will be useful to identify unit vectors γ_s as complex numbers $\gamma_s = \alpha_s + i\beta_s$, and we will use the same notation. Moreover, we will typically work only with the PCP $\mathcal{Z}_s^1$ and to shorter the notations we omit the upper index 1.

	$s=0$	$s=1$	$s=2$	$s=3$	$\cdots$
$k=0$	1	1	1	1	$\cdots$
$k=1$	$\mathcal{Z}_0$	$\mathcal{Z}_1$	$\mathcal{Z}_2$	$\mathcal{Z}_3$	$\cdots$
$k=2$	$\mathcal{Z}_0^2$	$\mathcal{Z}_1^2$	$\mathcal{Z}_2^2$	$\mathcal{Z}_3^2$	$\cdots$
$k=3$	$\mathcal{Z}_0^3$	$\mathcal{Z}_1^3$	$\mathcal{Z}_2^3$	$\mathcal{Z}_3^3$	$\cdots$
$\vdots$	$\vdots$	$\vdots$	$\vdots$	$\vdots$	$\ddots$

TABLE 1. Pseudo-Complex Polynomials: columns correspond to different parameters α_s and β_s, rows correspond to the degree of a polynomial.

2.2. Representation of pseudo-complex polynomials

Theorem 2.4. *The pseudo-complex polynomial $\mathcal{Z}_s$ can be obtained from the complex mononial $z = x_0 + x_1\mathbf{e}_1$ by an action of a family of invertible multiplicative extensions as follows*

$$\mathcal{Z}_s = \mathcal{P}_s\, z = z\,\bar{\mathcal{P}}_s,$$

where the functions $\mathcal{P}_s$ and $\bar{\mathcal{P}}_s$ are given by

$$\mathcal{P}_s = \frac{x_0^2 + \alpha_s y_s x_1}{x_0^2 + x_1^2} + \frac{\alpha_s y_s x_0 - x_0 x_1}{x_0^2 + x_1^2}\mathbf{e}_1 + \frac{x_0 \beta_s y_s}{x_0^2 + x_1^2}\mathbf{e}_2 + \frac{x_1 \beta_s y_s}{x_0^2 + x_1^2}\mathbf{e}_3,$$

$$\bar{\mathcal{P}}_s = \frac{x_0^2 + \alpha_s y_s x_1}{x_0^2 + x_1^2} + \frac{\alpha_s y_s x_0 - x_0 x_1}{x_0^2 + x_1^2}\mathbf{e}_1 + \frac{x_0 \beta_s y_s}{x_0^2 + x_1^2}\mathbf{e}_2 - \frac{x_1 \beta_s y_s}{x_0^2 + x_1^2}\mathbf{e}_3,$$

with

$$y_s = \alpha_s x_1 + \beta_s x_2.$$

Proof. The theorem can be proved by straightforward calculations. □

Since the multiplicative extensions $\mathcal{P}_s$ and $\bar{\mathcal{P}}_s$ are invertible we have an obvious corollary:

Corollary 2.5. *The complex monomial $z = x_0 + x_1\mathbf{e}_1$ can be obtained from pseudo-complex polynomials by an action of a family of multiplicative restrictions as follows*

$$z = \mathcal{P}_s^{-1}\mathcal{Z}_s = \mathcal{Z}_s\bar{\mathcal{P}}_s^{-1},$$

where the functions $\mathcal{P}_s^{-1}$ and $\bar{\mathcal{P}}_s^{-1}$ are given by

$$\mathcal{P}_s^{-1} = \frac{x_0^2 + \alpha_s y_s x_1}{x_0^2 + y_s^2} - \frac{\alpha_s y_s x_0 - x_0 x_1}{x_0^2 + y_s^2}\mathbf{e}_1 - \frac{x_0 \beta_s y_s}{x_0^2 + y_s^2}\mathbf{e}_2 - \frac{x_1 \beta_s y_s}{x_0^2 + y_s^2}\mathbf{e}_3,$$

$$\bar{\mathcal{P}}_s^{-1} = \frac{x_0^2 + \alpha_s y_s x_1}{x_0^2 + y_s^2} - \frac{\alpha_s y_s x_0 - x_0 x_1}{x_0^2 + y_s^2}\mathbf{e}_1 - \frac{x_0 \beta_s y_s}{x_0^2 + y_s^2}\mathbf{e}_2 + \frac{x_1 \beta_s y_s}{x_0^2 + y_s^2}\mathbf{e}_3,$$

with

$$y_s = \alpha_s x_1 + \beta_s x_2.$$

Since multiplicative restrictions and extensions are invertible, we get
(i) $\mathcal{P}_s \mathcal{P}_s^{-1} = \mathcal{P}_s^{-1} \mathcal{P}_s = I$,
(ii) $\bar{\mathcal{P}}_s \bar{\mathcal{P}}_s^{-1} = \bar{\mathcal{P}}_s^{-1} \bar{\mathcal{P}}_s = I$,
where I is the identity.

Remark 2.6. The multiplicative extensions $\mathcal{P}_s$ and $\bar{\mathcal{P}}_s$ can also be written in a more compact form as follows

$$\mathcal{P}_s = 1 + \frac{\alpha_s y_s - x_1}{x_0^2 + x_1^2} \bar{z}\mathbf{e}_1 + \frac{\beta_s y_s}{x_0^2 + x_1^2} z\mathbf{e}_2,$$

$$\bar{\mathcal{P}}_s = 1 + \frac{\alpha_s y_s - x_1}{x_0^2 + x_1^2} \bar{z}\mathbf{e}_1 + \frac{\beta_s y_s}{x_0^2 + x_1^2} \bar{z}\mathbf{e}_2.$$

One may expect that by applying the same idea to the multiplicative restrictions $\mathcal{P}_s^{-1}$ and $\bar{\mathcal{P}}_s^{-1}$ one gets a similar more compact structure. However, this is not the case, and the multiplicative restrictions are written as follows

$$\mathcal{P}_s^{-1} = 1 + \frac{\alpha_s x_1 - y_s}{x_0^2 + y_s^2} (y_s - x_0\mathbf{e}_1) - \frac{\beta_s y_s}{x_0^2 + y_s^2} z\mathbf{e}_2,$$

$$\bar{\mathcal{P}}_s^{-1} = 1 + \frac{\alpha_s x_1 - y_s}{x_0^2 + y_s^2} (y_s - x_0\mathbf{e}_1) - \frac{\beta_s y_s}{x_0^2 + y_s^2} \bar{z}\mathbf{e}_2.$$

Remark 2.7. We would like to underline the fact, that we call PCP as monogenic polynomials which are *parameter-set isomorphic* to the complex powers. In the beginning, when PCP were introduced, they were called simply *isomorphic* to the complex powers. We would like to draw attention to it particularly in the context of Theorem 2.4. If we represent the complex monomial and PCP as sets, then it becomes clear that we cannot speak about a bijection between these sets. But if we fix the parameters α_s and β_s, then we get a unique bijection. In this case we get that each column in Table 1 is isomorphic to the complex powers. This fact is clearly underlined in Theorem 2.4. Thus, to avoid any misunderstanding it is better to refer to PCP as monogenic polynomials which are *parameter-set isomorphic* to the complex powers.

3. Study of a specific parameter set

In [5] it was discussed, that conditions which are formulated in Theorem 2.3 allow a wide choice of parameter-vectors, leading to different sets of polynomials $\left\{\mathcal{Z}_s^k\right\}_{s=0}^k$. Such freedom has positive and negative aspects:

- on one hand, from the theoretical point of view it is convenient to have such a flexible structure, because depending on a specific task one can construct a system of polynomials which is more suitable in a given situation;
- on the other hand, it is not completely satisfying to have different sets of PCP without clear relations between them.

The question of a relation between different PCP is of particular importance for the task of approximation of an arbitrary monogenic function by PCP. The most essential result in this case would be to find what kind of monogenic functions correspond to different columns in Table 1, or, at least, to find a possibility to express some columns in terms of the others. Therefore in this section we formalise a construction of one parameter set, which supports the idea of finding relations between different PCP.

We represent the parameters α_s and β_s in the following form

$$\gamma_s = \begin{pmatrix}\alpha_s & \beta_s\end{pmatrix} = \begin{pmatrix}\cos\varphi_s & \sin\varphi_s\end{pmatrix}, \tag{3.1}$$

where φ_s are real numbers describing the argument of the complex numbers γ_s taken to be equidistant on the unit circle. Figure 1 shows distribution of the first γ_s in terms of the arc length of $\widehat{\varphi_1\varphi_i}$. In this case the parameter set $\{\varphi_0, \varphi_1, \ldots, \varphi_k\}$ defines completely the set of unit vectors $\{\gamma_0, \gamma_1, \ldots, \gamma_k\}$, where k is the polynomial degree.

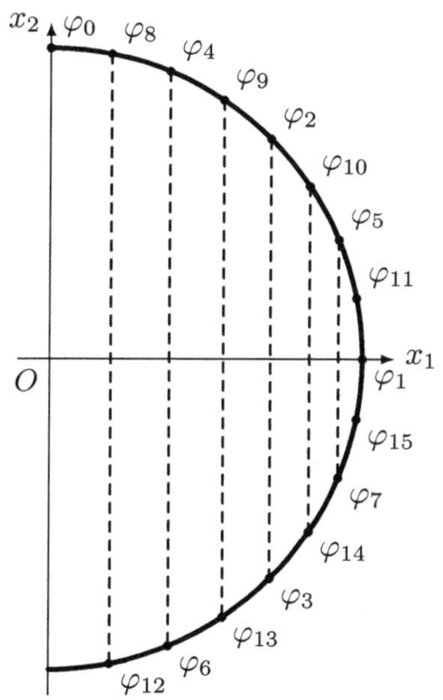

FIGURE 1. Distribution of γ_s on the unit disk

The construction of such a parameter set was discussed in [5], where in Tables I and II it was shown that such set of parameters satisfies the conditions of Theorem 2.3 and the corresponding basis functions are explicitly constructed. But the question of a relation between different PCP has not been studied there. Therefore in this paper we propose another approach to the construction of this parameter set which allows us finally to connect different PCP.

We construct the parameter set A as follows

$$A = \{A_1, A_2, A_3, \ldots\},$$

where $A_i, i = 0, 1, \dots$ are parameter sets for certain collections of PCP. According to that idea and taking into account Figure 1 we have the following subsets in the parameter set A

$$\begin{aligned} A_1 &:= \{\varphi_0\}, \\ A_2 &:= \{\varphi_1\}, \\ A_3 &:= \{\varphi_2, \varphi_3\}, \\ A_4 &:= \{\varphi_4, \varphi_5, \varphi_6, \varphi_7\} \\ A_5 &:= \{\varphi_8, \varphi_9, \varphi_{10}, \varphi_{11}, \varphi_{12}, \varphi_{13}, \varphi_{14}, \varphi_{15}\}, \\ &\vdots \end{aligned}$$

and it is easy to verify that the exact number of elements in each subset is given by

$$\mathrm{card}(A_i) = 2^{i-2}, \quad i = 2, 3, \dots .$$

From now on we consider the subsets $A_i, i = 0, 1, \dots$ as ordered subsets according to the construction described above. Thus, one obtains the following formulae for the elements of the subsets A_i, denoted by $A_i(j)$:

$$\begin{cases} A_i(j) = \dfrac{A_1(1) + \left[\frac{1}{2}\mathrm{card}(A_i) - j\right]\pi}{\mathrm{card}(A_i)}, \\ A_i\left(j + \dfrac{1}{2}\mathrm{card}(A_i)\right) = -A_i(j), \end{cases}$$

with $j = 1, 2, \dots, \frac{1}{2}\mathrm{card}(A_i)$, $i = 3, 4, \dots$, and $A_1 = \{\frac{\pi}{2}\}$, $A_2 = \{0\}$ are two basic sets. Finally, to construct a basis for the space of homogeneous monogenic paravector-valued polynomials of degree k in $\mathbb{R}^3$ one needs to take $k + 1$ elements of the parameter set $A_1 \cup A_2 \cup \cdots$.

Based on the proposed construction of the parameter set A and by using formula (3.1) we can introduce the following formula

$$\begin{aligned} \mathcal{Z}_s(x) = x_0 + &\left\{\left(\begin{bmatrix}\alpha_{s-1}\beta_{s-1}\end{bmatrix}\begin{bmatrix}\cos\vartheta_s & -\sin\vartheta_s \\ \sin\vartheta_s & \cos\vartheta_s\end{bmatrix}\right)\begin{bmatrix}x_1 \\ x_2\end{bmatrix}\right\} \\ \times &\left\{\left(\begin{bmatrix}\alpha_{s-1}\beta_{s-1}\end{bmatrix}\begin{bmatrix}\cos\vartheta_s & -\sin\vartheta_s \\ \sin\vartheta_s & \cos\vartheta_s\end{bmatrix}\right)\begin{bmatrix}\mathbf{e}_1 \\ \mathbf{e}_2\end{bmatrix}\right\}, \end{aligned} \tag{3.2}$$

with $s = 1, 2, \dots$, and the rotation angle ϑ_s is defined by

$$\vartheta_s = \begin{cases} \dfrac{A_1(1)}{\mathrm{card}(A_i)}, & \text{if } \varphi_s = A_i(1), \\ \dfrac{2\,A_1(1)}{\mathrm{card}(A_i)}, & \text{if } \varphi_s \in A_i, \varphi_s \neq A_i(1). \end{cases}$$

Formula (3.2) implies that next column in Table 1, i.e., PCP from that column, is obtained by a rotation of the vector $(\cos\varphi_s, \sin\varphi_s)$ of the previous PCP around x_0-axis. Let us illustrate it on first few PCP:

- to obtain the PCP $\mathcal{Z}_1$ corresponding to the subset A_2 we need to rotate the parameter set of $\mathcal{Z}_0$ by angle $\frac{\pi}{2}$;
- to obtain the PCP $\mathcal{Z}_2$ corresponding to the element $A_3(1)$ we need to rotate the parameter set of $\mathcal{Z}_1$ by angle $\frac{\pi}{4}$;
- to obtain the PCP $\mathcal{Z}_3$ corresponding to the element $A_3(2)$ we need to rotate the parameter set of $\mathcal{Z}_2$ by angle $\frac{\pi}{2}$.

Finally we see that by the proposed construction of the parameter set A we can easily calculate the rotation angle. Thus we obtained a clear relation between different PCP.

Remark 3.1. We would like to notice that PCP $\mathcal{Z}_{\varphi_0}$ and $\mathcal{Z}_{\varphi_1}$ can be obtained by rotations of the Fueter variables z_1 and z_2 as follows

$$\mathcal{Z}_{\varphi_0} = R_{x_1}\left(\frac{\pi}{2}\right) \cdot z_2, \quad \mathcal{Z}_{\varphi_1} = R_{x_2}\left(-\frac{\pi}{2}\right) \cdot z_1,$$

where $R_{x_1}(\vartheta)$ and $R_{x_2}(\vartheta)$ are elements of SO(3) representing rotations around axis x_1 and x_2, respectively.

We would like to notice that we have presented the detailed study of one specific parameter set which allows us finally to describe relations between different columns in the PCP basis. But of course, any other choice of parameters satisfying conditions $\alpha_s^2 + \beta_s^2 = 1$ is allowed, like for instance dyadic fractions $0, \frac{\pi}{2}, \pm\frac{2k+1}{2^m}\pi$ for integers $m \geq 2$ and $k \geq 0$ such that $2k + 1 < 2^{m-1}$. Therefore it is natural to ask a question if one choice of the parameter set is "better" than the others. But of course the meaning of "better" depends on a specific situation. For example, one can expect that different parameter sets would provide different approximation properties of the resulting system of PCP. This question has to be studied in future work.

4. Summary and outlook

In this paper we have presented a study of some interesting properties of pseudo-complex polynomials. Particularly, it was shown that by using multiplicative extensions and restrictions with chosen parameters α_s and β_s one can obtain PCP from the complex monomial and vice versa. By this construction it becomes clear that these polynomials are parameter-set isomorphic to the classical complex powers. Finally, by formalizing a construction of a specific parameter set we have obtained an explicit representation of PCP from each column in Table 1 in terms of the previous column. This construction is based on the idea of a rotation of PCP around x_0-axis. In future work the studied properties have to be checked in different applications.

Acknowledgement

We would like to express our deepest gratitude to Prof. Helmuth Malonek and to Prof. Maria Irene Falcão for a lot of useful discussions and hints. This work would not be possible without their inspiration and valuable support.

The research of the second author is supported by the German Research Foundation (DFG) via Research Training Group "Evaluation of Coupled Numerical Partial Models in Structural Engineering (GRK 1462)", and by the Russian Foundation of Basic Research via grant 16-31-60045 through the North-Eastern Federal University, which are gratefully acknowledged.

References

[1] S. Bock, K. Gürlebeck, *On a generalized Appell system and monogenic power series.* Mathematical Methods in the Applied Sciences, 33, 2010.

[2] I. Caçao, K. Gürlebeck, H. Malonek, *Special monogenic polynomials and L_2-approximation.* Advances in Applied Clifford Algebras, 11, 2001.

[3] I. Caçao, H. Malonek, *On complete sets of hypercomplex Appell polynomials.* AIP Conference Proceedings, vol. 1048, 2008.

[4] C. Cruz, M.I. Falcão, H.R. Malonek, *On pseudo-complex bases for monogenic polynomials.* AIP Conference Proceedings, vol. 1473, 2012.

[5] C. Cruz, M.I. Falcão, H.R. Malonek, *Monogenic pseudo-complex power functions and their applications.* Mathematical Methods in the Applied Sciences, 37, 2014.

[6] C. Cruz, M.I. Falcão, H.R. Malonek, *On Numerical Aspects of Pseudo-Complex Powers in* $\mathbb{R}^3$. Lecture Notes in Computer Science, Vol. 8579, 2014.

[7] R. Delanghe, *On regular-analytic functions with values in a Clifford algebra.* Mathematische Annalen 185, 1970.

[8] K. Gürlebeck, K. Habetha, W. Sprößig, *Holomorphic functions in the plane and n-dimensional spaces*, Birkhäuser, 2008.

[9] K. Gürlebeck, *Über Interpolation und Approximation verallgemeinert analytischer Funktionen*, volume 34 of Wissenschaftliche Informationen, Karl-Marx-Stadt, 1982.

[10] J. Morais and H.T. Le, *Orthogonal Appell systems of monogenic functions in the cylinder.* Mathematical Methods in the Applied Sciences, 34, 2011.

Klaus Gürlebeck
Chair of Applied Mathematics
Bauhaus-Universität Weimar
D-99423 Weimar, Germany
e-mail: klaus.guerlebeck@uni-weimar.de

Dmitrii Legatiuk
Research Training Group 1462
Bauhaus-Universität Weimar
D-99425 Weimar, Germany
e-mail: dmitrii.legatiuk@uni-weimar.de

Modern Trends in Hypercomplex Analysis
Trends in Mathematics, 227–245

Slice Regular Functions on Regular Quadratic Cones of Real Alternative Algebras

Guangbin Ren, Xieping Wang and Zhenghua Xu

Abstract. The theory of slice regular functions is a natural generalization of that of holomorphic functions of one complex variable to the setting of quaternions, octonions, paravectors in Clifford algebras, and more generally quadratic cones of real alternative algebras, in virtue of a slight modification of a well-known Fueter construction. In this paper, we focus on slice regular functions on the so-called regular quadratic cones, which are generally smaller than quadratic cones introduced by Ghiloni–Perotti and turn out to be the appropriate sets on which some nice properties of slice regular functions can be considered, including particularly the growth and distortion theorems for slice regular extensions of univalent holomorphic functions on the unit disc $\mathbb{D} \subset \mathbb{C}$, the Erdős–Lax inequality and the Turan inequality for a subclass of slice regular polynomials with all the coefficients in a same complex plane. It is noteworthy that the notion of regular quadratic cones also provides additionally an effective approach to unifying the theory of slice regular functions on quaternions, octonions, and paravectors in Clifford algebras.

Mathematics Subject Classification (2010). 30G35, 32A30.

Keywords. Slice regular functions; real alternative algebras; growth and distortion theorems; Erdős–Lax inequality; Turan inequality.

1. Introduction

A theory of slice regular functions on the so-called quadratic cones of real alternative algebras was recently initiated by Ghiloni and Perotti [24]. It contains and generalizes the theory of slice regular functions introduced initially by Gentili and Struppa in [20,21] for quaternions $\mathbb{H}$, and subsequently by Colombo, Sabadini and Struppa [7,8] in the real Clifford algebra $\mathbb{R}_n$, and later also by Gentili and Struppa in [22] for octonions $\mathbb{O}$. This new slice regular theory involves a notion of slice regularity, which goes back to a work of Cullen [10] and is significantly different from

This work was supported by Chinese NSF grant No. 11371337.

that of Cauchy–Fueter (see [2] for example). It also has elegant applications to the functional calculus for noncommutative operators [9], Schur analysis [1], and the construction and classification of orthogonal complex structures on dense open subsets of $\mathbb{R}^4$ [18].

The strategy proposed by Ghiloni and Perotti [24] is motivated by a well-known Fueter construction, which provides an effective way to generate quaternionic regular functions (in the sense of Cauchy–Fueter) starting from complex holomorphic functions (cf. [15,40]) and has been generalized by Sce [37], Qian [31], and Sommen [39] to the setting of Clifford algebras. Many variants have been given since then in Clifford analysis [6, 13, 30, 32] and Dunkl–Clifford analysis [14]. The approach introduced by Ghiloni and Perotti in [24] for an alternative algebra $\mathbb{A}$ over $\mathbb{R}$ makes use of the complexified algebra $\mathbb{A} \otimes_{\mathbb{R}} \mathbb{C}$, denoted by $\mathbb{A}_{\mathbb{C}}$. It turns out that for each holomorphic function F on an open set $D \subset \mathbb{C}$ invariant under the complex conjugate, there exists a unique slice regular function $f : [D] \longrightarrow \mathbb{A}$ on $[D] := \{\alpha + \beta J : \alpha + \beta i \in D, J \in \mathbb{S}_{\mathbb{A}}\} \subseteq \mathcal{Q}_{\mathbb{A}}$ such that the following diagram commutes for every $J \in \mathbb{S}_{\mathbb{A}}$:

$$\begin{array}{ccc} [D] & \xrightarrow{\ f\ } & \mathbb{A} \\ {\scriptstyle \Phi_J}\Big\uparrow & & \Big\uparrow{\scriptstyle \widetilde{\Phi}_J} \\ D & \xrightarrow{\ F\ } & \mathbb{A}_{\mathbb{C}} \end{array}$$

The construction above depends heavily on the so-called slice complex nature of $\mathcal{Q}_{\mathbb{A}}$, the so-called quadratic cone in $\mathbb{A}$ (see Section 2 below for the precise definition), i.e.,

$$\mathcal{Q}_{\mathbb{A}} = \bigcup_{J \in \mathbb{S}_{\mathbb{A}}} \mathbb{C}_J,$$

and

$$\mathbb{C}_I \cap \mathbb{C}_J = \mathbb{R}, \qquad \forall\, I, J \in \mathbb{S}_{\mathbb{A}} \text{ with } I \neq \pm J.$$

Here $\mathbb{S}_{\mathbb{A}}$ denotes the set of square roots of -1 in the algebra $\mathbb{A}$, defined by

$$\mathbb{S}_{\mathbb{A}} := \{J \in \mathcal{Q}_{\mathbb{A}} \mid J^2 = -1\},$$

and for every $J \in \mathbb{S}_{\mathbb{A}}$, $\Phi_J : \mathbb{C} \longrightarrow \mathcal{Q}_{\mathbb{A}}$ and $\widetilde{\Phi}_J : \mathbb{A}_{\mathbb{C}} \longrightarrow \mathbb{A}$ are two maps defined respectively by

$$\Phi_J(a + ib) = a + Jb, \qquad \forall\, a, b \in \mathbb{R},$$

and

$$\widetilde{\Phi}_J(\alpha + i\beta) = \alpha + J\beta, \qquad \forall\, \alpha, \beta \in \mathbb{A}.$$

As shown in [24], when $D \subset \mathbb{C}$ is a domain invariant under the complex conjugate and $\mathbb{A}$ is the quaternions $\mathbb{H}$, the set of slice regular functions f on $[D]$ obtained by the construction above from holomorphic functions F on D coincides exactly with the one studied in [3] on the symmetric slice domain $[D]$, a notion first introduced in [5] (as pointed out by the referee). As one easily sees, this is also the case when $\mathbb{A}$ is the octonions $\mathbb{O}$, even though the reference [22] dealt only with slice regular functions on the open balls centered at the origin, i.e., power series of one octonionic variable with octonionic coefficients on the right. Now slice regular functions on symmetric slice domains in $\mathbb{O}$ have been studied deeply in [42].

Now a rather natural question arises in the theory of slice regularity:

To what extent can one extend nice properties of holomorphic functions to the non-commutative setting for slice regular functions?

This paper attempts to answer this question. It turns out that the quadratic cones introduced by Ghiloni and Perotti is too big for the extensions in general and the suitable set should be the so-called *regular* quadratic cones $\mathcal{H} \subseteq \mathcal{Q}_{\mathbb{A}}$, which will be introduced in the next section, and guarantee the existence of an element $I \in \mathbb{S}_{\mathbb{A}}$ satisfying that

$$\mathcal{T}(IJ) \in \mathbb{R}, \qquad \forall\, J \in \mathbb{S}_{\mathbb{A}} \cap \mathcal{H}, \tag{1.1}$$

where $\mathcal{T}$ denotes a trace operator in the alternative algebra $\mathbb{A}$ (see Lemma 3.4). The importance of the trace condition (1.1) lies at its resulting fundamental identity for norms of slice functions. Condition (1.1) is known to be abided by each regular quadratic cone, while may fail for quadratic cone in general (see Remark 3.3 below for an explicit counterexample). This predicates that properties of stem functions may fail to extend to the biggest quadratic cone and also explains the reason why slice analysis should be resided in the regular quadratic cones. Moreover, it should be remarked that as already observed in the setting of quaternions for the study of slice Bergman spaces [4], for regular compositions [35] and for Bieberbach conjecture [16], a sufficient condition of preserving one slice is quite useful. Such a condition is even suspected to be necessary by Gal, González-Cervantes, and Sabadini for the Bieberbach conjecture in quaternions [16]. We also find that such a condition is best possible when generalizing the classical Erdős–Lax inequality to the setting of real alternative algebras.

The point of slice analysis on a regular quadratic cone is that it provides an effective approach to unifying the theory of slice regular functions on quaternions, octonions, and paravectors in Clifford algebras. It also fulfils the natural idea of extending nice properties of holomorphic functions to noncommutative case, at least for slice regular functions preserving one slice defined in the regular quadratic cones. We shall establish the sharp growth and distortion theorems, the Erdős–Lax inequality, and the Turan inequality for slice regular functions on regular quadratic cones as well as the structure theorem of the zero set for slice regular functions on the strong regular quadratic cones of real alternative algebras. In particular, we provide a systematic approach to the study of zeros of octonionic slice regular functions. For a series of studies on zeros of polynomials or power series over noncommutative algebra, we refer the interested readers to [23, 26, 28, 29, 44] and the references therein.

This paper is arranged as follows. In Section 2, we recall some basic definitions about real alternative algebras and introduce the notion of regular quadratic cones in real alternative algebras. We then establish the representation formula for the norm of slice functions on a real alternative algebra in Section 3, and extend the growth and distortion theorems, the Erdős–Lax inequality and the Turan inequality to the setting of real alternative algebras in Section 4. Finally, Section 5 comes the structure theorem of zeros of slice regular functions on strong regular quadratic cones of real alternative algebras.

2. Real alternative algebras and quadratic cones

In this section, we recall some definitions and results about real alternative algebras (see [12, 24, 25, 38]).

2.1. Real alternative algebras

Let $\mathbb{A}$ be a finite-dimensional real alternative algebra with a unity. We assume that $\mathbb{A}$ has dimension $d > 1$ as a real vector space, and identify the field of real numbers with the subalgebra of $\mathbb{A}$ generated by the unity. Recall that an real algebra $\mathbb{A}$ is alternative if the *associator* $(x, y, z) := (xy)z - x(yz)$ of three elements of $\mathbb{A}$ is an alternating function in its arguments. Artin's theorem (cf. [38]) asserts that *the subalgebra generated by two elements of* $\mathbb{A}$ *is associative.* Therefore there hold the following Moufang identities:

$$a(x(ay)) = (axa)y, \qquad ((xa)y)a = x(aya), \qquad (ax)(ya) = a(xy)a$$

for all $x, y, a \in \mathbb{A}$.

Assumption 2.1. *In what follows, we will assume that an anti-involution is fixed on the real alternative algebra* $\mathbb{A}$. *It is a linear map* $x \mapsto x^c$ *of* $\mathbb{A}$ *into itself satisfying the following properties:*

$$\begin{aligned} (x^c)^c &= x, & \forall\, x \in \mathbb{A}, \\ (xy)^c &= y^c x^c, & \forall\, x, y \in \mathbb{A}, \\ x^c &= x, & \forall\, x \in \mathbb{R}. \end{aligned}$$

2.2. The quadratic cone of a real alternative algebra

For each element x of $\mathbb{A}$, the *trace* of x is

$$\mathcal{T}(x) := x + x^c \in \mathbb{A}$$

and the (squared) *norm* of x is

$$\mathcal{N}(x) := xx^c \in \mathbb{A}.$$

Definition 2.2. (see [24]) The quadratic cone $\mathcal{Q}_{\mathbb{A}}$ of $\mathbb{A}$ is by definition a real cone given by

$$\mathcal{Q}_{\mathbb{A}} := \mathbb{R} \cup \Big\{ x \in \mathbb{A} \mid \mathcal{T}(x) \in \mathbb{R},\ \mathcal{N}(x) \in \mathbb{R},\ 4\mathcal{N}(x) > \mathcal{T}(x)^2 \Big\}.$$

For every $x \in \mathcal{Q}_{\mathbb{A}}$, we also write $\sqrt{\mathcal{N}(x)}$ as $|x|$.

The set of square roots of -1 in the algebra $\mathbb{A}$ is defined by

$$\mathbb{S}_{\mathbb{A}} := \{J \in \mathcal{Q}_{\mathbb{A}} \mid J^2 = -1\}.$$

For each $J \in \mathbb{S}_{\mathbb{A}}$, we will denote by

$$\mathbb{C}_J := \langle 1, J \rangle \cong \mathbb{C},$$

the subalgebra of $\mathbb{A}$ generated by J. Observe that for every $J \in \mathbb{S}_{\mathbb{A}}$ the restriction of the anti-involution to $\mathbb{C}_J$ becomes

$$x = \alpha + \beta J \mapsto x^c = \alpha - \beta J \qquad (\alpha, \beta \in \mathbb{R}).$$

Assumption 2.3. *We shall always assume that $\mathcal{Q}_{\mathbb{A}} \neq \mathbb{R}$, or equivalently $\mathbb{S}_{\mathbb{A}} \neq \emptyset$.*

Under the preceding assumption, the quadratic cone $\mathcal{Q}_{\mathbb{A}}$ has two fundamental properties [24]:

$$\mathcal{Q}_{\mathbb{A}} = \bigcup_{J \in \mathbb{S}_{\mathbb{A}}} \mathbb{C}_J;$$

$$\mathbb{C}_I \bigcap \mathbb{C}_J = \mathbb{R}, \qquad \forall\, I, J \in \mathbb{S}_{\mathbb{A}},\ I \neq \pm J.$$

Therefore, each non-zero element $x \in \mathcal{Q}_{\mathbb{A}} \setminus \{0\}$ has an multiplicative inverse $x^{-1} = \mathcal{N}(x)^{-1} x^c \in \mathcal{Q}_{\mathbb{A}} \setminus \{0\}$ and its nth power x^n also lies in $\mathcal{Q}_{\mathbb{A}}$ for each $n \in \mathbb{N}$. It is worth remarking here that in general the quadratic cone $\mathcal{Q}_{\mathbb{A}}$ is strictly contained in $\mathbb{A}$ and $\mathcal{Q}_{\mathbb{A}} = \mathbb{A}$ if and only if $\mathbb{A}$ is isomorphic to one of the division algebras $\mathbb{C}$, $\mathbb{H}$ and $\mathbb{O}$, in virtue of the well-known Hurwitz theorem.

2.3. Slice functions and slice regular functions

Given an open set D of $\mathbb{C}$, invariant under the complex conjugate. Let $[D]$ be its associated set, given by

$$[D] = \bigcup_{x \in D} [x],$$

where $[x] = \alpha + \beta \mathbb{S}_{\mathbb{A}}$ for any $x = \alpha + i\beta \in D$.

It is known that $[D]$ is a relatively open subset of $\mathcal{Q}_{\mathbb{A}}$. Such a set $[D]$ is a natural domains of definition for slice functions, which is called circular domains of $\mathbb{A}$ as it keeps invariant under the action of the square roots of -1.

Definition 2.4. (see [24]) A function $F : D \longrightarrow \mathbb{A} \otimes_{\mathbb{R}} \mathbb{C}$ on an open set $D \subseteq \mathbb{C}$ invariant under the complex conjugate is called a *stem function* if the $\mathbb{A}$-valued components F_1, F_2 of $F = F_1 + iF_2$ satisfy

$$F_1(\bar{z}) = F_1(z), \qquad F_2(\bar{z}) = -F_2(z), \qquad \forall\, z \in D.$$

Each stem function $F : D \longrightarrow \mathbb{A} \otimes_{\mathbb{R}} \mathbb{C}$ induces a (left) *slice function*

$$f = \mathcal{I}(F) : [D] \longrightarrow \mathbb{A}$$

via

$$f(x) := F_1(z) + JF_2(z), \qquad \forall\, x \in [z] \cap \mathbb{C}_J.$$

Each slice function f is induced by a unique stem function F since F_1 and F_2 can be obtained starting from f. We will denote the set of all such induced slice functions on $[D]$ by

$$\mathcal{S}([D]) := \Big\{ f = \mathcal{I}(F) : [D] \longrightarrow \mathbb{A} \mid \; F : D \longrightarrow \mathbb{A} \otimes_{\mathbb{R}} \mathbb{C} \text{ is a stem function} \Big\}.$$

Also, we will denote by

$$\mathcal{S}^1([D]) := \Big\{ f = \mathcal{I}(F) \in \mathcal{S}([D]) : F \in C^1(D) \Big\}$$

the set of slice functions with stem functions of class C^1. Let $f = \mathcal{I}(F) \in \mathcal{S}^1([D])$ and $z = \alpha + i\beta \in D$. Then the partial derivatives $\partial F/\partial \alpha$ and $i\partial F/\partial \beta$ are continuous stem functions on D. The same property holds for their linear combinations

$$\frac{\partial F}{\partial z} = \frac{1}{2}\left(\frac{\partial F}{\partial \alpha} - i\frac{\partial F}{\partial \beta}\right) \qquad \text{and} \qquad \frac{\partial F}{\partial \bar{z}} = \frac{1}{2}\left(\frac{\partial F}{\partial \alpha} + i\frac{\partial F}{\partial \beta}\right).$$

Definition 2.5. (see [24]) Let $f = \mathcal{I}(F) \in \mathcal{S}^1([D])$. We set

$$\frac{\partial f}{\partial x} := \mathcal{I}\left(\frac{\partial F}{\partial z}\right), \qquad \frac{\partial f}{\partial x^c} := \mathcal{I}\left(\frac{\partial F}{\partial \bar{z}}\right).$$

These functions are continuous slice functions on $[D]$.

Definition 2.6. (see [24]) A slice function $f = \mathcal{I}(F) \in \mathcal{S}^1([D])$ is *slice regular* if its stem function F is holomorphic. We will denote the vector space of slice regular functions on $[D]$ by

$$\mathcal{SR}([D]) := \Big\{ f = \mathcal{I}(F) \in \mathcal{S}^1([D]) : F \; \text{ is holomorphic} \Big\}.$$

For each slice regular function $f \in \mathcal{SR}([D])$, we define its *slice derivative* f' to be the slice regular function on $[D]$ given by

$$f' := \frac{\partial f}{\partial x} = \mathcal{I}\left(\frac{\partial F}{\partial z}\right).$$

In general, the pointwise product of two slice functions is not a slice function. However, the pointwise product in the algebra $\mathbb{A} \otimes_{\mathbb{R}} \mathbb{C}$ induces a natural product on slice functions.

Definition 2.7. (see [24]) Let $f = \mathcal{I}(F)$ and $g = \mathcal{I}(G)$ be two slice functions on $[D]$. The *slice product* of f and g is the slice function on $[D]$ given by

$$f * g := \mathcal{I}(FG).$$

If f, g are slice regular, then also $f * g$ is slice regular. In general, $(f * g)(x) \neq f(x)g(x)$. If the components F_1, F_2 of the first stem function F are real-valued, or if F and G are both $\mathbb{A}$-valued, then $(f * g)(x) = f(x)g(x)$ for every $x \in [D]$. In this case, we will use also the notation fg in place of $f * g$.

2.4. Regular Quadratic cones

Note that under Assumption 2.3, the quadratic cone $\mathcal{Q}_{\mathbb{A}}$ contains at least one 2-dimensional real subspace $\mathbb{C}_I \subseteq \mathbb{A}$, where I is an element of $\mathbb{S}_{\mathbb{A}}$. This trivial observation motivates us to introduce the following

Definition 2.8. A *regular quadratic cone* $\mathcal{H}$ is a subset of $\mathcal{Q}_{\mathbb{A}}$ for which there exist a real subspace $M \subset \mathbb{A}$ and an $I \in \mathbb{S}_{\mathbb{A}} \cap M$ such that $\mathcal{H} = \mathcal{Q}_{\mathbb{A}} \cap M$ and

$$I - J \in \mathcal{Q}_{\mathbb{A}}, \qquad \forall\, J \in \mathbb{S}_{\mathbb{A}} \cap M.$$

Notice that in a regular quadratic cone $\mathcal{H}$, we have

$$\mathbb{S}_{\mathbb{A}} \cap M = \mathbb{S}_{\mathbb{A}} \cap \mathcal{H}.$$

For clarity, we sometimes write $\mathcal{H}$ as $\mathcal{H}_I$ or $\mathcal{H}_{I,\mathbb{A}}$ with I an arbitrarily fixed element of $\mathbb{S}_{\mathbb{A}}$ satisfying the desired property in the definition of $\mathcal{H}$.

Example. The typical examples of the regular quadratic cone $\mathcal{H}_{I,\mathbb{A}}$ are given by

$$\mathcal{H} = \begin{cases} \mathbb{H}, & \mathbb{A} = \mathbb{H}, \qquad \mathcal{Q}_{\mathbb{A}} = \mathbb{H} = M, \\ \mathbb{O}, & \mathbb{A} = \mathbb{O}, \qquad \mathcal{Q}_{\mathbb{A}} = \mathbb{O} = M, \\ \mathbb{R}^{n+1}, & \mathbb{A} = Cl_{0,n}, \quad \mathcal{Q}_{\mathbb{A}} \supset \mathbb{R}^{n+1} = M. \end{cases}$$

This enables us to unify the theory of slice regular functions on quaternions, octonions, and paravectors in Clifford algebras. Moreover, the above typical examples are also strong regular quadratic cones (see Definition 5.1 below).

3. Representation formula for the norm of slice functions

In this section, a representation formula for the norm of slice functions is established, which states that the norm of every slice function is uniquely determined by its value on a plane $\mathbb{C}_I$. The corresponding result in the setting of quaternions or Clifford algebras was obtained in [33] in order to achieve the sharp growth and distortion theorems for slice regular extensions of univalent holomorphic functions on the open unit disc $\mathbb{D} \subset \mathbb{C}$. It has found many applications in the theory of slice regular functions (see [34–36, 42]).

As usual, our starting point is still the following fundamental representation formula for slice functions.

Lemma 3.1. (see [24]) *Let $f \in \mathcal{S}([D])$ and $I \in \mathbb{S}_{\mathbb{A}}$. Then the following formula holds true:*

$$f(x) = \frac{1}{2}\Big(f(\alpha+\beta I) + f(\alpha-\beta I)\Big) + \frac{J}{2}\Big(I(f(\alpha-\beta I) - f(\alpha+\beta I))\Big) \tag{3.1}$$

for every $J \in \mathbb{S}_{\mathbb{A}}$ and every $x = \alpha + \beta J \in D_J := [D] \cap \mathbb{C}_J$.

Theorem 3.2. *Let $f \in \mathcal{S}([D])$ be such that $f(D_I) \subseteq \mathbb{C}_I$ for some $I \in \mathbb{S}_{\mathbb{A}}$. Then there holds the identity:*

$$\mathcal{N}(f(\alpha+\beta J)) = \frac{2-\mathcal{T}(JI)}{4}\mathcal{N}(f(\alpha+\beta I)) + \frac{2+\mathcal{T}(JI)}{4}\mathcal{N}(f(\alpha-\beta I))$$

for every $\alpha+\beta J \in [D]$.

Proof. Denote $x = \alpha+\beta J$, $z = \alpha+\beta I$ and $\bar{z} = \alpha-\beta I$. By the representation formula (Lemma 3.1), we have

$$2f(x) = a+b$$

with

$$a := f(z)+f(\bar{z}), \qquad b := J\Big(I\big(f(\bar{z})-f(z)\big)\Big).$$

Noticing that

$$\mathcal{N}(a+b) = (a+b)(a+b)^c = \mathcal{N}(a)+\mathcal{N}(b)+\mathcal{T}(ab^c),$$

we can rewrite the norm $\mathcal{N}(f)$ as

$$4\mathcal{N}(f(x)) = \mathcal{N}(a)+\mathcal{N}(b)+\mathcal{T}(ab^c). \tag{3.2}$$

Since $f(D_I) \subseteq \mathbb{C}_I$, we have

$$\mathcal{N}(a) = \mathcal{N}(f(z)+f(\bar{z})) \in \mathbb{R}$$

and

$$\mathcal{N}(f(\bar{z})-f(z)), \quad \mathcal{N}(f(z)), \quad \mathcal{N}(f(\bar{z})) \in \mathbb{R}.$$

By Moufang identities in Section 2, we thus have

$$\begin{aligned}
\mathcal{N}(b) &= \Big(JI\big(f(\bar{z})-f(z)\big)\Big)\Big(\big(f(\bar{z})-f(z)\big)^c IJ\Big)\\
&= J\Big(I\big(f(\bar{z})-f(z)\big)\big(f(\bar{z})-f(z)\big)^c I\Big)J\\
&= J\Big(I\mathcal{N}\big(f(\bar{z})-f(z)\big)I\Big)J\\
&= \mathcal{N}\big(f(\bar{z})-f(z)\big)
\end{aligned}$$

and

$$\begin{aligned}
\mathcal{T}(ab^c) &= \mathcal{T}\Big(\big(f(z)+f(\bar{z})\big)\big(f(\bar{z})-f(z)\big)^c IJ\Big)\\
&= \Big(\mathcal{N}\big(f(\bar{z})\big)-\mathcal{N}\big(f(z)\big)\Big)\mathcal{T}(IJ).
\end{aligned}$$

Inserting the above two formulas into (3.2) yields that

$$\mathcal{N}(f(\alpha+\beta J)) = \frac{1}{4}(2-\mathcal{T}(JI))\mathcal{N}(f(\alpha+\beta I)) + \frac{1}{4}(2+\mathcal{T}(JI))\mathcal{N}(f(\alpha-\beta I)). \qquad \square$$

Remark 3.3. We point out that in the Clifford algebra $\mathbb{R}_n$ it may fail that

$$\mathcal{T}(IJ) \in \mathbb{R}$$

for all $I, J \in \mathbb{S}_{\mathbb{A}}$. In contrast, if $\mathbb{A}$ is one of the division algebras $\mathbb{C}$, $\mathbb{H}$ or $\mathbb{O}$, then $\mathcal{T}(IJ) \in \mathbb{R}$ for all $I, J \in \mathbb{S}_{\mathbb{A}}$.

For simplicity we consider the case of $\mathbb{R}_3$. An element $x \in \mathbb{R}_3$ can be represented uniquely as a sum

$$x = x_0 + \sum_{i=1}^{3} x_i e_i + \sum_{1 \leq j < k \leq 3} x_{jk} e_j e_k + x_{123} e_1 e_2 e_3$$

with real coefficients $x_0, x_i, x_{jk}, x_{123}$. It is known [24] that the quadratic cone in $\mathbb{R}_3$ is the six-dimensional real algebraic set

$$\mathcal{Q}_{\mathbb{R}_3} = \Big\{ x \in \mathbb{R}_3 \mid x_{123} = 0, \ x_1 x_{23} - x_2 x_{13} + x_3 x_{12} = 0 \Big\}$$

and $\mathbb{S}_{\mathbb{R}_3}$ is a four-dimensional sphere in $\mathbb{R}_3$:

$$\mathbb{S}_{\mathbb{R}_3} = \Big\{ x \in \mathcal{Q}_{\mathbb{R}_3} \mid x_0 = 0, \ \sum_i x_i^2 + \sum_{j<k} x_{jk}^2 = 1 \Big\}.$$

Now we take

$$I = x_1 e_1 + x_2 e_2 + x_3 e_3 + x_{12} e_{12} + x_{13} e_{13} + x_{23} e_{23} \in \mathbb{S}_{\mathbb{R}_3},$$
$$J = y_1 e_1 + y_2 e_2 + y_3 e_3 + y_{12} e_{12} + y_{13} e_{13} + y_{23} e_{23} \in \mathbb{S}_{\mathbb{R}_3}.$$

By a direct calculation, we have

$$\mathcal{T}(IJ) = -2\langle I, J\rangle + 2\big(x_1 y_{23} - x_2 y_{13} + x_3 y_{12} + y_1 x_{23} - y_2 x_{13} + y_3 x_{12}\big) e_{123}.$$

It is evident that $\mathcal{T}(IJ) \notin \mathbb{R}$ in general. This means that it may fail that

$$\mathcal{T}(IJ) \in \mathbb{R}$$

for all $I, J \in \mathcal{Q}_{\mathbb{R}_3}$.

Consequently, $\mathbb{Q}_{\mathbb{R}_3}$ is a quadratic cone but not a regular quadratic cone as shown by the following lemma.

Lemma 3.4. *In each regular quadratic cone $\mathcal{H}_I \subseteq \mathbb{A}$, we always have*

$$\mathcal{T}(IJ) \in \mathbb{R} \qquad \text{and} \qquad |\mathcal{T}(IJ)| \leq 2$$

for all $J \in \mathbb{S}_{\mathbb{A}} \cap \mathcal{H}_I$.

Proof. By definition, we have $I - J \in \mathcal{Q}_{\mathbb{A}}$ so that $\mathcal{N}(I - J) \in \mathbb{R}$. On the other hand, we have

$$\mathcal{N}(I - J) = (I - J)(I - J)^c = 2 + \mathcal{T}(IJ)$$

so that $\mathcal{T}(IJ) \in \mathbb{R}$. Since $\mathcal{T}(IJ) = IJ + JI$, it implies that $|\mathcal{T}(IJ)| \leq 2$. □

With this lemma, the identity in Theorem 3.2 thus becomes exactly a convex combination identity in each regular quadratic cone $\mathcal{H}_I$:

Theorem 3.5. *Let $\mathcal{H}_I \subseteq \mathcal{Q}_{\mathbb{A}}$ be a regular quadratic cone and $f \in \mathcal{S}([D] \cap \mathcal{H}_I)$ be such that $f(D_I) \subseteq \mathbb{C}_I$. Then for each $J \in \mathbb{S}_{\mathbb{A}} \cap \mathcal{H}_I$ there exists a real number $\lambda_{I,J} \in [0,1]$ such that*

$$\mathcal{N}\big(f(\alpha+\beta J)\big) = \lambda_{IJ}\mathcal{N}\big(f(\alpha+\beta I)\big) + (1-\lambda_{IJ})\mathcal{N}\big(f(\alpha-\beta I)\big).$$

4. Some consequences of Theorem 3.5

Throughout this section, we denote by $\mathcal{H}_I$ a regular quadratic cone in a real alternative algebra $\mathbb{A}$, where I is an arbitrarily fixed element of $\mathbb{S}_{\mathbb{A}}$ satisfying the desired property in the definition of $\mathcal{H}_I$ (see Definition 2.8). Set

$$\mathbb{B} = \{x \in \mathcal{H}_I : |x| < 1\}, \qquad \mathbb{B}_I = \mathbb{B} \cap \mathbb{C}_I.$$

As a direct consequence of Theorem 3.5, we conclude that the maximum as well as the minimum modulus of f is actually attained on the preserved slice.

Theorem 4.1. *Let $f \in \mathcal{S}([D] \cap \mathcal{H}_I)$ be such that $f(D_I) \subseteq \mathbb{C}_I$. Then for each $\alpha + \beta i \in D$, we have the following equalities:*

$$\max_{J \in \mathbb{S}_{\mathbb{A}} \cap \mathcal{H}_I} \mathcal{N}\big(f(\alpha+\beta J)\big) = \max\Big\{\big|f(\alpha+\beta I)\big|^2, \big|f(\alpha-\beta I)\big|^2\Big\},$$

and

$$\min_{J \in \mathbb{S}_{\mathbb{A}} \cap \mathcal{H}_I} \mathcal{N}\big(f(\alpha+\beta J)\big) = \min\Big\{\big|f(\alpha+\beta I)\big|^2, \big|f(\alpha-\beta I)\big|^2\Big\}.$$

The preceding theorem in turn results in the growth and distortion theorems for slice regular functions on a regular quadratic cone in an alternative algebra $\mathbb{A}$, which were first proved in [33] in the settings of quaternions and Clifford algebras.

Theorem 4.2 (Growth and Distortion Theorems). *Let f be a slice regular function on $\mathcal{H}_I$ such that its restriction f_I to $\mathbb{B}_I$ is injective and $f(\mathbb{B}_I) \subseteq \mathbb{C}_I$. If $f(0) = 0$ and $f'(0) = 1$, then for all $x \in \mathbb{B}$, the following inequalities hold:*

$$\frac{|x|}{(1+|x|)^2} \leq \mathcal{N}\big(f(x)\big)^{1/2} \leq \frac{|x|}{(1-|x|)^2}; \tag{4.1}$$

$$\frac{1-|x|}{(1+|x|)^3} \leq \mathcal{N}\big(f'(x)\big)^{1/2} \leq \frac{1+|x|}{(1-|x|)^3}; \tag{4.2}$$

$$\frac{1-|x|}{1+|x|} \leq \mathcal{N}\big(xf'(x) * f^{-*}(x)\big)^{1/2} \leq \frac{1+|x|}{1-|x|}, \tag{4.3}$$

where $f^{-*} : \mathbb{B}\backslash\{0\} \longrightarrow \mathbb{A}$ is the slice regular extension to $\mathbb{B}\backslash\{0\}$ of the holomorphic function $1/f_I : \mathbb{B}_I \setminus \{0\} \longrightarrow \mathbb{C}_I$. Moreover, equality holds for one of these six inequalities at some point $x_0 \neq 0$ if and only if f is of the form

$$f(x) = x(1 - xe^{I\theta})^{-*2}, \quad \forall\, x \in \mathbb{B},$$

for some $\theta \in \mathbb{R}$.

Proof. The result easily follows from Theorem 4.1 and a similar argument as that in the proof of [33, Theorem 3.5], together with the classical growth and distortion theorems (cf. [11]) for univalent holomorphic functions on the open unit disc of $\mathbb{C}$. □

The same argument can be used to proved the Erdős–Lax and the Turan inequalities for a subclass of slice regular polynomials on $\mathcal{H}_I$. Recall first that the classical Erdős–Lax inequality (cf. [27]) states that

$$\max_{|z|\leq 1} |p'(z)| \leq \frac{n}{2} \max_{|z|\leq 1} |p(z)| \tag{4.4}$$

for those complex polynomials p of degree n that have no zeros in the open unit disk of $\mathbb{C}$.

As pointed out in [17], this result fails in the quaternionic setting in general, but holds true for a subclass of slice regular quaternionic polynomials of degree n. Following the idea in [43], we now extend this result to slice regular polynomials on $\mathcal{H}_I \subseteq \mathbb{A}$.

Theorem 4.3. *Let $p : \mathcal{H}_I \longrightarrow \mathbb{A}$ be a slice regular polynomial of degree n with $p(x) = \sum_{j=0}^{n} x^j a_j$ and all coefficients $a_j \in \mathbb{C}_I$. If the restriction p_I has no zeros on $\mathbb{B}_I$, then*

$$\max_{x\in\mathbb{B}} \mathcal{N}(p'(x)) \leq \frac{n^2}{4} \max_{x\in\mathbb{B}} \mathcal{N}(p(x)).$$

Proof. The classical result applied to p_I yields that

$$\max_{|z|\leq 1} |p_I'(z)| \leq \frac{n}{2} \max_{|z|\leq 1} |p_I(z)|.$$

Note that $\mathcal{T}(IJ) \in \mathbb{R}$ for all $J \in \mathbb{S}_{\mathbb{A}}$, in view of Lemma 3.4. By the convex combination identity in Theorem 3.5 we have, for all $\alpha + \beta J \in \mathbb{B}_J$,

$$\begin{aligned}\mathcal{N}(p'(\alpha+\beta J)) &= \frac{1}{4}(2-\mathcal{T}(JI))|p'(\alpha+\beta I)|^2 + \frac{1}{4}(2+\mathcal{T}(JI))|p'(\alpha-\beta I)|^2 \\ &\leq \frac{n^2}{4} \max_{|z|\leq 1} |p_I(z)|^2,\end{aligned}$$

i.e.,

$$\mathcal{N}(p'(\alpha+\beta J)) \leq \frac{n^2}{4} \max_{x\in\mathbb{B}} \mathcal{N}(p(x)), \quad \forall \alpha+\beta J \in \mathbb{B}_J, \quad J \in \mathbb{S}_{\mathbb{A}} \cap \mathcal{H}_I,$$

as desired. □

Example. The result in the preceding theorem may fail after removing the restriction of preserving one slice: A counterexample is provided in [17]. Let $p(q) = q^2 - q(i+j) + ij$ be a slice regular quaternionic polynomial. The only zero of this polynomial is $q = i$. However,

$$\max_{|z|\leq 1} |p'(z)| > (6+4\sqrt{2})^{\frac{1}{2}} > (4+4\sqrt{2})^{\frac{1}{2}} \geq \frac{2}{2} \max_{|z|\leq 1} |p(z)|.$$

An inverse inequality analogous to (4.4) was proved by Turan [41], which states that

$$\max_{|z|\leq 1} |p'(z)| \geq \frac{n}{2} \max_{|z|\leq 1} |p(z)| \tag{4.5}$$

for complex polynomials of degree n with all zeros in $\{z \in \mathbb{C} : |z| \leq 1\}$.

Theorem 4.4. *Let $p : \mathcal{H}_I \longrightarrow \mathbb{A}$ be a slice regular polynomial of degree n with $p(x) = \sum_{j=0}^{n} x^j a_j$ and all coefficients $a_j \in \mathbb{C}_I$. If the restriction p_I has all its zeros in $\overline{\mathbb{B}}_I$, then*

$$\max_{x\in\mathbb{B}} \mathcal{N}(p'(x)) \geq \frac{n^2}{4} \max_{x\in\mathbb{B}} \mathcal{N}(p(x)).$$

Proof. The proof is completely similar as that of Theorem 4.3. □

5. Structure of zeros

In this section, we discuss the properties of the zeros of slice regular functions on the so-called strong regular quadratic cones in real alternative algebras $\mathbb{A}$.

5.1. Strong regular quadratic cones

Definition 5.1. We say a regular quadratic cone $\mathcal{H}_I$ in $\mathbb{A}$ is *strong* if for each $J \in \mathbb{S}_\mathbb{A} \cap \mathcal{H}_I$, $\frac{I+J}{2} \in \mathbb{S}_\mathbb{A}$ if and only if $J = I$.

It is easy to see that quaternions $\mathbb{H}$ and octonions $\mathbb{O}$ are respectively the strong regular quadratic cones contained in themselves, and the space of paravectors $\mathbb{R}^{n+1}$ is a strong regular quadratic cone in the Clifford algebras $\mathbb{R}_n$.

Lemma 5.2. *Let $\mathcal{H}_I$ be a strong regular quadratic cone in $\mathbb{A}$. Then for every $J \in \mathbb{S}_\mathbb{A} \cap \mathcal{H}_I$ with $J \neq I$, $I - J$ is always invertible.*

Proof. Let J be an element of $\mathbb{S}_\mathbb{A} \cap \mathcal{H}_I$ such that $J \neq I$. Then

$$(I - J)^2 = -2 - \mathcal{T}(IJ) =: \mu$$

is a real number, in view of Lemma 3.4. We claim that $\mu \neq 0$, so that $I - J$ admits an inverse

$$(I - J)^{-1} = \frac{1}{\mu}(I - J).$$

Indeed, if $\mu = 0$, then $\mathcal{T}(IJ) = -2$ so that

$$\left(\frac{I+J}{2}\right)^2 = \frac{-2 + \mathcal{T}(IJ)}{4} = -1.$$

This means that $(I + J)/2 \in \mathbb{S}_\mathbb{A} \cap \mathcal{H}_I$ and hence $J = I$, contradicting the choice of J. □

5.2. Representation formula and splitting lemma for slice regular functions

In what follows, we always denote by $\mathcal{H} = \mathcal{H}_I$ a *strong regular quadratic cone* in a real alternative algebra $\mathbb{A}$ and by $\mathbb{S}_\mathcal{H}$ the intersection $\mathbb{S}_\mathbb{A} \cap \mathcal{H}$.

Definition 5.3. A domain Ω in $\mathcal{H}$ is called an (*axially*) *symmetric slice domain* if there exists a domain $D \subseteq \mathbb{C}$ intersected with $\mathbb{R}$ and invariant under the complex conjugate such that $\Omega = [D]$.

For each symmetric slice domain $\Omega \subseteq \mathcal{H}$ and each $J \in \mathbb{S}_\mathcal{H}$, we denote by Ω_J the intersection $\Omega \cap \mathbb{C}_J$.

Theorem 5.4 (Representation formula). *Let $f : \Omega \longrightarrow \mathbb{A}$ be a slice regular function on a symmetric slice domain $\Omega \subseteq \mathcal{H}_I$ and let $\alpha + \beta I \in \Omega$. Then the following equality*

$$f(\alpha+\beta J) = (J-K)\big((I-K)^{-1}f(\alpha+\beta I)\big) + (I-J)\big((I-K)^{-1}f(\alpha+\beta K)\big).$$

holds for all $J \in \mathbb{S}_\mathcal{H} \setminus \{I\}$ and all $K \in \mathbb{S}_\mathcal{H}$.

Proof. Lemma 5.2 guarantees that for each $K \in \mathbb{S}_\mathcal{H}\setminus\{I\}$, $(I-K)^{-1}$ is well defined. The result follows immediately from the fact that the restriction of f to the sphere $\alpha+\beta\mathbb{S}_\mathcal{H} \in \Omega$ is affine in the imaginary unit; see [24, Proposition 6] for details. □

As a direct consequence of the preceding theorem, we have the following properties of zeros of slice regular functions.

Proposition 5.5. *Let $f : \Omega \longrightarrow \mathbb{A}$ be a slice regular function on a symmetric slice domain $\Omega \subseteq \mathcal{H}_I$. If f vanishes at a point $\alpha+\beta I$, then either $f : \Omega \longrightarrow \mathbb{A}$ vanishes identically in $\alpha+\beta\mathbb{S}_{\mathcal{H}_I}$ or f does not have other zero in $\alpha+\beta\mathbb{S}_{\mathcal{H}_I}$.*

We next digress to the splitting lemma for slice regular functions on symmetric slice domains $\Omega \subseteq \mathcal{H}$. Recall that each real alternative algebra $\mathbb{A}$ has a complex structure induced by the left multiplication of an element from $\mathbb{S}_\mathbb{A}$ so that its real dimension is even. We denote

$$h := \frac{1}{2}\dim_\mathbb{R} \mathbb{A} - 1 \in \mathbb{N} \cup \{0\}.$$

The following splitting lemma clarifies the relation between slice regularity and complex holomorphy (cf. [25, Lemma 2.4]).

Lemma 5.6 (Splitting lemma). *Let $f : \Omega \longrightarrow \mathbb{A}$ be a slice regular function on a symmetric slice domain $\Omega \subseteq \mathcal{H}$. Then for each $J \in \mathbb{S}_\mathcal{H}$, there exist $K_1, \dots, K_h \in \mathbb{S}_\mathbb{A}$ and holomorphic functions $F_0, F_1, \dots, F_h : \Omega_J \longrightarrow \mathbb{C}_J$ such that*

$$\Big\{K_0 := 1, K_1, \dots, K_h, J, JK_1, \dots, JK_h\Big\}$$

forms a real vector basis for $\mathbb{A}$ and

$$f_J(z) = \sum_{j=0}^{h} F_j(z)K_j, \qquad \forall\, z \in \Omega_J.$$

5.3. Auxiliary functions associated with slice regular functions

Suppose that $\Omega = [D] \subseteq \mathcal{H}$ is a symmetric slice domain, where D is a domain of $\mathbb{C}$ intersected with $\mathbb{R}$ and invariant under the complex conjugate. The preceding splitting lemma enables us to introduce for each slice regular function f on Ω an auxiliary function f^σ, which is crucial to study the structure of zeros of slice regular functions. For each $J \in \mathbb{S}_{\mathcal{H}}$, we split the restriction f_J into

$$f_J = \sum_{j=0}^{h} F_j K_j$$

with $K_1, \ldots, K_h \in \mathbb{S}_{\mathbb{A}}$ and holomorphic functions $F_0, F_1, \ldots, F_h : \Omega_J \longrightarrow \mathbb{C}_J$ as described in Lemma 5.6. We then define the holomorphic function $f_J^\sigma : \Omega_J \longrightarrow \mathbb{C}_J$ by

$$f_J^\sigma(z) = \sum_{j=0}^{h} F_j(z)\overline{F_j(\bar{z})}.$$

It induces a slice regular function f^σ defined on Ω via equality (3.1). The function f^σ has two main properties:

(i) f^σ vanishes on the zeros of f;

(ii) f^σ preserves each slice, i.e., $f^\sigma(\Omega_J) \subset \mathbb{C}_J$ for all $J \in \mathbb{S}_{\mathcal{H}}$.

From now on, we denote by Z_f the zeros of the function f.

Lemma 5.7. *Let $f : \Omega \longrightarrow \mathbb{A}$ be a non-identically vanishing slice regular function on a symmetric slice domain $\Omega \subset \mathcal{H}$. Then for each $J \in \mathbb{S}_{\mathcal{H}}$, the intersection $Z_f \cap \mathbb{C}_J$ is closed and discrete in Ω_J.*

Proof. For each $J \in \mathbb{S}_{\mathcal{H}}$, the closeness of $Z_f \cap \mathbb{C}_J$ in Ω_J is obvious and the discreteness can be proved as follows. By Lemma 5.6, for each $J \in \mathbb{S}_{\mathcal{H}}$, there exist $K_1, \ldots, K_h \in \mathbb{S}_{\mathbb{A}}$ and holomorphic functions $F_0, F_1, \ldots, F_h : \Omega_J \longrightarrow \mathbb{C}_J$ such that $\{K_0 := 1, K_1, \ldots, K_h, J, JK_1, \ldots, JK_h\}$ forms a real vector basis for $\mathbb{A}$ and

$$f_J(z) = \sum_{j=0}^{h} F_j(z) K_j, \qquad \forall\, z \in \Omega_J.$$

Since $f \not\equiv 0$, it follows from the representation formula (Lemma 3.1) that the restriction f_J does not vanish identically, and so does some F_{j_0} with $0 \le j_0 \le h$. Therefore, $Z_f \cap \mathbb{C}_J \subseteq Z_{F_{j_0}} \cap \mathbb{C}_J$ is discrete in Ω_J. □

Lemma 5.8. *Let $f : \Omega \longrightarrow \mathbb{A}$ be a slice regular function on a symmetric slice domain $\Omega \subset \mathcal{H}$. Then f vanishes identically on Ω if and only if so does f^σ.*

Proof. We only need to prove the sufficiency. As above, we write the restriction f_J as

$$f_J(z) = \sum_{j=0}^{h} F_j(z) K_j\ ,$$

where $F_j : \Omega_J \longrightarrow \mathbb{C}_J$ are holomorphic functions. Take $y_0 \in \Omega_J \cap \mathbb{R}$ and consider the series expansion

$$F_j(z) = \sum_{m=0}^{+\infty} (z - y_0)^m a_{j,m}, \quad a_{j,m} \in \mathbb{C}_J,$$

which holds in a suitable disc $\Delta(y_0, R) \subseteq \Omega_J$ of radius R and centered at $y_0 \in \mathbb{R}$. Then, on $\Delta(y_0, R)$, we have

$$\overline{F_j(\overline{z})} = \sum_{m=0}^{+\infty} (z - y_0)^m \overline{a}_{j,m}, \quad a_{j,m} \in \mathbb{C}_J.$$

Moreover on $\Delta(y_0, R)$ we can write

$$f_J^\sigma(z) = \sum_{j=0}^{h} F_j(z)\overline{F_j(\overline{z})} = \sum_{m=0}^{+\infty} (z - y_0)^m \sum_{j=0}^{h} c_{j,m},$$

where

$$c_{j,m} = \sum_{i=0}^{m} a_{j,i}\overline{a}_{j,m-i}.$$

By assumption, we have $f^\sigma \equiv 0$ so that $f_J^\sigma \equiv 0$ in the disc $\Delta(y_0, R)$. Hence,

$$\sum_{j=0}^{h} c_{j,0} = \sum_{j=0}^{h} |a_{j,0}|^2 = 0$$

so $a_{j,0} = 0$ for all multi-indices j. Now, by induction, assume that

$$a_{j,i} = 0, \qquad i = 0, 1, \ldots, k-1, \quad k \geq 1, \quad j = 0, 1, \ldots, h.$$

Consider the coefficient

$$\sum_{j=0}^{h} c_{j,2k} = \sum_{j=0}^{h} \sum_{i=0}^{2k} a_{j,i}\overline{a}_{j,2k-i}$$

which is zero because $f_J^\sigma \equiv 0$. Thus,

$$\sum_{j=0}^{h} c_{j,2k} = \sum_{j=0}^{h} |a_{jk}|^2$$

is zero if and only if $a_{jk} = 0$ for all multi-indices j. We conclude that $f_J^\sigma \equiv 0$ in the disc $\Delta(y_0, R)$ implies that all the coefficients a_{ji} vanish, thus also f_J vanishes identically on Ω_J. By the representation formula in Lemma 3.1, f vanishes identically. □

The zeros of slice regular functions f^σ is described in the following result:

Lemma 5.9. *Let $f : \Omega \longrightarrow \mathbb{A}$ be a non-identically vanishing slice regular function on a symmetric slice domain $\Omega \subset \mathcal{H}$. If there exists one point $x \in \Omega$ for which $f^\sigma(x) = 0$, then $f^\sigma(y) = 0$ for all $y \in [x] \cap \mathcal{H}$. Moreover, the zero set of f^σ consists of isolated spheres (which might reduce to points on the real axis).*

Proof. The first part of the statement is trivial. The second part follows from an argument by contradiction: if the spheres of zeros were not isolated, on each plane we would get accumulation points of zeros and thus f^σ would be identically zero by the identity principle for holomorphic functions and thus it follows from Lemma 5.8 that $f \equiv 0$, giving a contradiction. □

Lemma 5.10. *Let $\Omega \subset \mathcal{H}$ be a symmetric slice domain and $f : \Omega \longrightarrow \mathbb{A}$ a slice regular function. Then every zero of f is also a zero of f^σ.*

Proof. Assume that $f(\alpha + \beta J) = 0$. By the splitting lemma,

$$f_J(z) = \sum_{j=0}^{h} F_j(z) K_j.$$

Therefore, $F_j(\alpha + \beta J) = 0$ for all j. By definition,

$$f_J^\sigma(z) = \sum_{j=0}^{h} F_j(z)\overline{F_j(\overline{z})},$$

we thus have $f_J^\sigma(\alpha + \beta J) = 0$ and the result follows. □

5.4. Structure of zeros

We now state the topological property of the zeros of slice regular functions; see [19, Theorem 3.12] for the quaternionic case and [9, Theorem 2.5.14] for the Clifford algebra case.

Please replace it by the following sentences:

Theorem 5.11 (Structure of zeros). *Let $\mathcal{H} \subseteq \mathbb{A}$ be a strong regular quadratic cone such that for each pair of distinct elements $I,\ J \in \mathbb{S}_\mathcal{H}$, $I - J$ is invertible, and $f : \Omega \longrightarrow \mathbb{A}$ a non-identically vanishing slice regular function on a symmetric slice domain $\Omega \subset \mathcal{H}$. Then the zero set Z_f of f consists of isolated points or isolated spheres of the form $\alpha + \beta\mathbb{S}_\mathcal{H}$ with $\alpha, \beta \in \mathbb{R}$.*

Proof. Note that under our assumption, we can deduce from Theorem 5.4 that each sphere of the form $\alpha + \beta\mathbb{S}_\mathcal{H}$ with $\alpha, \beta \in \mathbb{R}$, either is contained in Z_f or contains only one point of Z_f. We proceed with an argument by contradiction and suppose that the conclusion asserted in the theorem were incorrect. Then there would be a sequence $\{x_n\} \subseteq Z_f$ satisfying that for any $j, k \in \mathbb{N}$ with $j \neq k$, $[x_j] \cap [x_k] = \emptyset$, and converging to some point $x_\infty \in \alpha + \beta\mathbb{S}_\mathcal{H}$ with $\alpha, \beta \in \mathbb{R}$. This together with Lemma 5.10 implies that the zero set of f^σ consists of non-isolated spheres. However, this is impossible in view of Lemmas 5.8 and 5.9. □

If we weaken the assumption in the preceding theorem, we can only obtain the following

Theorem 5.12. *Let $\mathcal{H} \subseteq \mathbb{A}$ be a strong regular quadratic cone and $f : \Omega \longrightarrow \mathbb{A}$ a non-identically vanishing slice regular function on a symmetric slice domain $\Omega \subset \mathcal{H}$. Then if some sphere $\alpha + \beta\mathbb{S}_\mathcal{H}$ with $\alpha, \beta \in \mathbb{R}$ is contained Z_f, it must be isolated in Z_f.*

Acknowledgment

The authors would like to thank the anonymous referee for the valuable suggestions and comments.

References

[1] D. Alpay, F. Colombo, I. Sabadini, *Slice Hyperholomorphic Schur Analysis*, Oper. Theory Adv. Appl., vol. 256, Springer Basel, 2017.

[2] F. Brackx, R. Delanghe, F. Sommen, *Clifford Analysis*, Research Notes in Mathematics, 76. Pitman, Boston, MA, 1982.

[3] F. Colombo, G. Gentili, I. Sabadini, D.C. Struppa, *Extension results for slice regular functions of a quaternionic variable.* Adv. Math. **222** (2009), 1793–1808.

[4] F. Colombo, J.O. González-Cervantes, I. Sabadini, *On slice biregular functions and isomorphisms of Bergman spaces.* Complex Var. Elliptic Equ. **57** (2012), 825–839.

[5] F. Colombo, I. Sabadini, *A structure formula for slice monogenic functions and some of its consequences*, in Hypercomplex Analysis, Trends in Mathematics, Birkhäuser Basel, 2009, 101–114.

[6] F. Colombo, I. Sabadini, F. Sommen, *The Fueter mapping theorem in integral form and the $\mathcal{F}$-functional calculus.* Math. Methods Appl. Sci. **33** (2010), 2050–2066.

[7] F. Colombo, I. Sabadini, D.C. Struppa, *Slice monogenic functions*, Israel J. Math. **171** (2009), 385–403.

[8] F. Colombo, I. Sabadini, D.C. Struppa, *An extension theorem for slice monogenic functions and some of its consequences*, Israel J. Math. **177** (2010), 369–389.

[9] F. Colombo, I. Sabadini, D.C. Struppa, *Noncommutative functional calculus. Theory and applications of slice hyperholomorphic functions.* Progress in Mathematics, vol. 289. Birkhäuser/Springer, Basel, 2011.

[10] C.G. Cullen, *An integral theorem for analytic intrinsic functions on quaternions.* Duke Math. J. **32** (1965), 139–148.

[11] P.L. Duren, *Univalent Functions*, Grundlehren der mathematischen Wissenschaften 259, New York: Springer-Verlag, 1983.

[12] H.D. Ebbinghaus, H. Hermes, F. Hirzebruch, M. Koecher, K. Mainzer, J. Neukirch, A. Prestel, R. Remmert, *Numbers*, Grad. Texts in Math., vol. 123, Springer-Verlag, New York, 1990.

[13] D. Eelbode, *The biaxial Fueter theorem.* Israel J. Math. **201** (2014), 233–245.

[14] M. Fei, P. Cerejeiras, U. Kähler, *Fueter's theorem and its generalizations in Dunkl–Clifford analysis.* J. Phys. A **42** (2009), no. 39, 395209, 15 pp.

[15] R. Fueter, *Die Funktionentheorie der Differentialgleichungen $\Delta u = 0$ und $\Delta\Delta u = 0$ mit vier reellen Variablen*, Comment. Math. Helv. **7** (1934), 307–330.

[16] S.G. Gal, J.O. González-Cervantes, I. Sabadini, *Univalence results for slice regular functions of a quaternion variable.* Complex Var. Elliptic Equ. **60** (2015), 1431–1455.

[17] S.G. Gal, I. Sabadini, *On Bernstein and Erdős–Lax inequalities for quaternionic polynomials*, C. R. Math. Acad. Sci. Paris. **353** (2015), 5–9.

[18] G. Gentili, S. Salamon, C. Stoppato, *Twistor transforms of quaternionic functions and orthogonal complex structures.* J. Eur. Math. Soc. **16** (2014), 2323–2353.

[19] G. Gentili, C. Stoppato, D.C. Struppa, *Regular functions of a quaternionic variable.* Springer Monographs in Mathematics, Springer, Berlin-Heidelberg, 2013.

[20] G. Gentili, D.C. Struppa, *A new approach to Cullen-regular functions of a quaternionic variable*, C. R. Math. Acad. Sci. Paris **342** (2006), 741–744.

[21] G. Gentili, D.C. Struppa, *A new theory of regular functions of a quaternionic variable*, Adv. Math. **216** (2007), 279–301.

[22] G. Gentili, D.C. Struppa, *Regular functions on the space of Cayley numbers*, Rocky Mountain J. Math. **40** (2010), 225–241.

[23] R. Ghiloni, A. Perotti, *Zeros of regular functions of quaternionic and octonionic variable: a division lemma and the cam-shaft effect.* Ann. Mat. Pura Appl. **190** (2011), 539–551.

[24] R. Ghiloni, A. Perotti, *Slice regular functions on real alternative algebras*, Adv. Math. **226** (2011), 1662–1691.

[25] R. Ghiloni, A. Perotti, *Power and spherical series over real alternative *-algebras.* Indiana Univ. Math. J. **63** (2014), 495–532.

[26] R. Ghiloni, A. Perotti, C. Stoppato, *The algebra of slice functions*, Trans. Amer. Math. Soc., in press.

[27] P. D. Lax, *Proof of a conjecture of P. Erdős on the derivative of a polynomial*, Bull. Amer. Math. Soc. **50** (1944), 509–513.

[28] I. Niven, *Equations in quaternions.* Amer. Math. Monthly **48** (1941), 654–661.

[29] I. Niven, *The roots of a quaternion.* Amer. Math. Monthly **49** (1942), 386–388.

[30] D. Peña Peña, F. Sommen, *Fueter's theorem: the saga continues*, J. Math. Anal. Appl. **365** (2010), 29–35.

[31] T. Qian, *Generalization of Fueter's result to* $\mathbb{R}^{n+1}$, Atti Accad. Naz. Lincei Cl. Sci. Fis. Mat. Natur. Rend. Lincei (9) Mat. Appl. **8** (1997), 111–117.

[32] T. Qian, F. Sommen, *Deriving harmonic functions in higher dimensional spaces.* Z. Anal. Anwendungen **22** (2003), 275–288.

[33] G.B. Ren, X.P. Wang, *The growth and distortion theorems for slice monogenic functions*, arXiv: 1410.4369v2.

[34] G.B. Ren, X.P. Wang, *Julia Theory for slice regular functions*, Trans. Amer. Math. Soc., in press.

[35] G.B. Ren, X.P. Wang, *Slice regular composition operators*, Complex Var. Elliptic Equ. **61** (2016), 682–711.

[36] G.B. Ren, Z.H. Xu, *Slice Lebesgue measure of quaternions*, Adv. Appl. Clifford Algebr. **26** (2016), 399–416.

[37] M. Sce, *Osservazioni sulle serie di potenze nei moduli quadratici*, Atti Accad. Naz. Lincei Rend. Cl. Sci. Fis. Mat. Natur. **23** (1957), 220–225.

[38] R.D. Schafer, *An introduction to nonassociative algebras*, Pure and Applied Mathematics, Vol. 22 Academic Press, New York, 1966.

[39] F. Sommen, *On a generalization of Fueter's theorem*, Z. Anal. Anwend. **19** (2000), 899–902.

[40] A. Sudbery, *Quaternionic analysis*, Math. Proc. Cambridge Philos. Soc. **85** (1979), 199–224.

[41] P. Turan, *Über die Ableitung von Polynomen*, Compositio Math. **7** (1939), 89–95.

[42] X.P. Wang, *On geometric aspects of quaternionic and octonionic slice regular functions.* submitted.

[43] Z.H. Xu, *The Bernstein inequality for slice regular polynomials*, arXiv:1602.08545, 2016.

[44] Y. Yang, T. Qian, *Zeroes of slice monogenic functions.* Math. Methods Appl. Sci. **34** (2011), 1398–1405.

Guangbin Ren (corresponding author),
Xieping Wang and Zhenghua Xu
School of Mathematical Sciences
University of Science and
Technology of China
Hefei 230026, China
e-mail: rengb@ustc.edu.cn
pwx@mail.ustc.edu.cn
xzhengh@mail.ustc.edu.cn

Modern Trends in Hypercomplex Analysis
Trends in Mathematics, 247–263

Differential Forms and Clifford Analysis

Irene Sabadini and Franciscus Sommen

Abstract. In this paper we use a calculus of differential forms which is defined using an axiomatic approach. We then define integration of differential forms over chains in a new way and we present a short proof of Stokes' formula using distributional techniques. We also consider differential forms in Clifford analysis, vector differentials and their powers. This framework enables an easy proof for a Cauchy formula on a k-surface. Finally, we discuss how to compute winding numbers in terms of the monogenic Cauchy kernel and the vector differentials with a new approach which does not involve cohomology of differential forms.

Mathematics Subject Classification (2010). MSC: 58A10, 55M25, 30G35.

Keywords. Differential forms, Clifford algebras, monogenic functions, winding numbers.

1. Introduction

This paper is a continuation of our former papers [9, 10, 11, 12] in which the calculus of differential forms has been combined with the Clifford algebra. Using Clifford analysis techniques, and monogenic functions in particular, we were able to establish a Cauchy-type formula for the Dirac operator on surfaces (see [10]), a theory of monogenic differential forms allowing a cohomology theory (see [9, 12]) and a formula for the winding number of a k-cycle and a $(m-k-1)$-cycle in $\mathbb{R}^m$ (see [9]). This extends the work of Hodge [7] in which the homology of a domain is measured in terms of integrals over cycles of harmonic differential forms. To understand these ideas, one has to recall that the theory of monogenic functions in Clifford analysis deals with nullsolutions of the Dirac operator $\partial_{\underline{x}}$ in $\mathbb{R}^m$, which is a higher-dimensional generalization of the theory of holomorphic functions in the plane. Consider a point p in the plane (or a number of points) and a closed Jordan curve (a 1-cycle) $\Gamma \subset \mathbb{C} \setminus \{p\}$; then the winding number of Γ around p is given by the Cauchy integral

$$\frac{1}{2\pi i} \int_{\Gamma} \frac{dz}{z-p}$$

which is a special case of the residue formula

$$\frac{1}{2\pi i}\int_\Gamma \frac{f(z)\,dz}{z-p}.$$

The analog Cauchy formula for monogenic functions has the form (see [1])

$$f(\underline{x}) = \int_{\partial C} E(\underline{u}-\underline{x})\,\sigma(d\underline{u})\,f(\underline{u})$$

where C is an open bounded set in $\mathbb{R}^m$, $\underline{x}\in C$, $E(\underline{u}-\underline{x})$ is the Cauchy kernel and $\sigma(d\underline{u})$ is a suitable $(m-1)$-form with values in a Clifford algebra that represents the oriented surface measure. Using this Cauchy formula in special cases, one can establish a formula for the winding number of an $(m-1)$-cycle around one or several points.

However, in $\mathbb{R}^m$ one can also consider k-cycles C_k and $(m-k-1)$-cycles C_{m-k-1} in $\mathbb{R}^m \setminus C_k$ for which there is a winding number that can be defined in terms of the intersection number; it cannot be measured in terms of monogenic functions right away. This makes it necessary to combine a calculus of differential forms with the theory of monogenic functions, as we do in this work.

The paper consists of 5 sections, besides this introduction. In Section 2, we define the calculus of differential forms from scratch using an axiomatic approach which is inspired by the use of differential forms in analysis. In Section 3 we define integration of differential forms over chains in a novel way which also includes partial integration operators that are anti-commuting. In Section 4 we present a short proof of Stokes' formula using distributional techniques. Section 5 is devoted to differential forms in Clifford analysis, starting with a short introduction to Clifford algebras and monogenic functions. Then we introduce the vector differential $d\underline{x}=\sum_{j=1}^m e_j\,dx_j$, that generalizes the complex differential $dz = dx + idy$, and its powers $d\underline{x}^k$ represent the oriented k-dimensional surface measure. This enables an easy proof for a Cauchy formula on a k-surface. The final Section 6 is devoted to the calculation of the winding number in terms of the monogenic Cauchy kernel and the vector differentials $d\underline{x}$, $d\underline{u}$, etc. The formulas thus obtained are easier to present and understand than the ones presented in [9], moreover the approach is new and does not involve cohomology of differential forms.

2. Differential forms

Let $\Omega \subseteq \mathbb{R}^n$ be an open set, and let $\mathcal{C}^\infty(\Omega)$ be the ring of real (or complex)-valued smooth functions on Ω. We begin by defining the algebra of differential forms:

Definition 2.1. The algebra $\Lambda(C^\infty(\Omega))$ of smooth differential forms on Ω is defined as the smallest associative algebra over $\mathcal{C}^\infty(\Omega)$ satisfying the following axioms:

(A_{-1}) $\mathcal{C}^\infty(\Omega)\subset\Lambda(C^\infty(\Omega))$;

and there is a map $d:\ \Lambda(C^\infty(\Omega))\to\Lambda(C^\infty(\Omega))$ such that

(A_0) $d1=0$;

(A_1) for $\varphi \in \mathcal{C}^\infty(\Omega)$, $F \in \Lambda(C^\infty(\Omega))$

$$d(\varphi F) = d\varphi\, F + \varphi\, dF;$$

(A_2) for $\varphi \in \mathcal{C}^\infty(\Omega)$, $F \in \Lambda(C^\infty(\Omega))$

$$d(d\varphi\, F) = -d\varphi\, dF.$$

Let $\mathcal{P} = \text{Alg}\{x_1, \ldots, x_m\}$ be the algebra of polynomials in $x_1, \ldots, x_m$ with real (or complex) coefficients. Then the generators $x_1, \ldots, x_m$, interpreted as coordinate functions, give rise to the differential $dx_1, \ldots, dx_m$. We can then give the following:

Definition 2.2. The subalgebra $\Lambda(\mathcal{P})$ of $\Lambda(C^\infty(\Omega))$ is generated by $\mathcal{P}$ and satisfies, for any $F \in \Lambda(\mathcal{P})$, the axioms

(A_1') for $F \in \Lambda(\mathcal{P})$ $\qquad d(x_j F) = dx_j\, F + x_j dF;$

(A_2') for $F \in \Lambda(\mathcal{P})$ $\qquad d(dx_j\, F) = -dx_j\, dF.$

Proposition 2.3. *The following properties hold:*

(i) $d(x_k\, dx_j) = dx_k\, dx_j$;
(ii) $d(dx_j\, x_k) = -dx_j dx_k$;
(iii) $dx_j\, dx_k = -dx_j\, dx_k$.

Proof. Property (i) follows from

$$d(x_k\, dx_j) = dx_k\, dx_j + x_k d^2 x_j = dx_k\, dx_j,$$

since, by (A_0) and (A_2)

$$d^2\varphi = d(d\varphi\, 1) = -d\varphi\, d1 = 0$$

for $\varphi \in \mathcal{C}^\infty(\Omega)$.

As a special case of (A_2'), we also have

$$d(dx_j\, x_k) = -dx_j dx_k,$$

so (ii) follows. As a consequence of (i) and (ii) we obtain $dx_j dx_k = -dx_j dx_k$. □

Remark 2.4. The previous result implies that $dx_1, \ldots, dx_m$ generate a Grassmann algebra of dimension 2^m.

From the definition of $\Lambda(\mathcal{P})$ it follows that every $F \in \Lambda(\mathcal{P})$ has the form

$$F = \sum_{A \subset M} F_A(\underline{x}) dx_A, \qquad F_A(\underline{x}) \in \mathcal{P},$$

where $M = \{1, \ldots, m\}$, $dx_A = dx_{\alpha_1} \ldots dx_{\alpha_k}$ for $A = \{\alpha_1, \ldots, \alpha_k\}$ and with $\alpha_1 < \cdots < \alpha_k$. It follows that

$$dF = \sum_{A \subset M} dF_A(\underline{x}) dx_A,$$

so it suffices to calculate $d\varphi$ for $\varphi \in \mathcal{P}$. By using iteratively the axiom (A_1') one can prove by induction on the degree of $\varphi \in \mathcal{P}$ that

$$d\varphi = \sum_{j=1}^{m} dx_j\, \partial_{x_j} \varphi.$$

Now note that $\mathcal{P}$ is dense in $\mathcal{C}^\infty(\Omega)$ and $\Lambda(\mathcal{P})$ is dense in $\Lambda(C^\infty(\Omega))$. So, it follows that every $F \in \Lambda(C^\infty(\Omega))$ is of the form

$$F = \sum_{A\subset M} F_A(\underline{x})\, dx_A, \qquad F_A \in \mathcal{C}^\infty(\Omega),$$

and, in general,

$$dF = \sum_{j=1}^{m} dx_j \sum_{A\subset M} \partial_{x_j} F_A(\underline{x})\, dx_A = \sum_{j=1}^{m} dx_j\, \partial_{x_j} F.$$

However, the definition of $\Lambda(\mathcal{C}^\infty(\Omega))$ and of d are independent of any coordinate system. Hence, if $(y_1,\dots,y_m)$ is another $\mathcal{C}^\infty$-coordinate system on Ω, then we have that

$$d = \sum_{j=1}^{m} dx_j\, \partial_{x_j} = \sum_{j=1}^{m} dy_j\, \partial_{y_j},$$

so that we also have the chain rule

$$dx_j = \sum_{\ell=1}^{m} \frac{\partial x_j}{\partial y_\ell}\, dy_\ell$$

and for $A = \{\alpha_1,\dots,\alpha_k\} \subseteq M$ with $\alpha_1 < \dots < \alpha_k$ we have

$$\begin{aligned} dx_A &= \sum_{\ell_1\dots\ell_k} \frac{\partial x_{\alpha_1}}{\partial y_{\ell_1}} \cdots \frac{\partial x_{\alpha_k}}{\partial y_{\ell_k}}\, dy_{\ell_1}\dots dy_{\ell_k} \\ &= \sum_{|B|=k} \left(\sum_{\pi\in\mathrm{Sym}(k)} \mathrm{sgn}\pi \frac{\partial x_{\alpha_1}}{\partial y_{\beta_{\pi(1)}}} \cdots \frac{\partial x_{\alpha_k}}{\partial y_{\beta_{\pi(k)}}} \right) dy_B \\ &= \sum_{|B|=k} J_{AB}\, dy_B, \qquad B = \{\beta_1,\dots,\beta_k\},\ \ \beta_1 < \dots < \beta_k, \end{aligned}$$

where

$$J_{AB} = \sum_{\pi\in\mathrm{Sym}(k)} \mathrm{sgn}\pi \frac{\partial x_{\alpha_1}}{\partial y_{\beta_{\pi(1)}}} \cdots \frac{\partial x_{\alpha_k}}{\partial y_{\beta_{\pi(k)}}}$$

are the generalized Jacobians. So, in the coordinates $(y_1,\dots,y_m)$ we have

$$F = \sum_{A\subseteq M} F_A(\underline{x})\, dx_A = \sum_{B\subseteq M} \left(\sum_{|A|=|B|} F_A(\underline{x}(y)) J_{AB} \right) dy_B.$$

Hence, the chain rule and Jacobians are an automatic consequence of the axioms.

3. Integration of differential forms

We extend the notion of differential form to the case where the components $F_A(\underline{x})$ are generalized functions or distributions in Ω. Let

$$F = F_M(x_1, \ldots, x_m)\, dx_1 \ldots dx_m$$

be a distributional form of maximum degree with $\operatorname{supp}(F_M) = K \subset \Omega$ compact. Then, the integral

$$\int_\Omega F = \int_{-\infty}^{\infty} \cdots \int_{-\infty}^{\infty} F_M(\underline{x}) dx_1 \ldots dx_m$$

is well defined (note that this is a formal way of writing: we are using the density of $\mathcal{D}(\Omega)$ in $\mathcal{E}'(\Omega)$ and thus the integrals are meant in the sense of functionals, see, e.g., [5]). Denote by $\Lambda_k(\mathcal{C}^\infty(\Omega))$ the subspace of k-forms, namely of elements $F = \sum_{|A|=k} F_A(\underline{x})\, dx_A$, where $F_A \in \mathcal{C}^\infty(\Omega)$ and denote by $\overline{\Lambda}_k(\mathcal{C}^\infty(\Omega))$ its closure in the distributions, namely the subspace of the k-forms $F = \sum_{|A|=k} F_A(\underline{x})\, dx_A$, with $F_A \in \mathcal{D}'(\Omega)$. Let Σ be an infinitely differentiable k-surface in $\mathbb{R}^m$ defined as the image of a $\mathcal{C}^\infty$-map:

$$\underline{x}(\cdot) : \ (u_1, \ldots, u_k) \to \underline{x}(u_1, \ldots, u_k),$$

where $\underline{u} = (u_1, \ldots, u_k) \in \Omega' \subset \mathbb{R}^k$, i.e., $\Sigma = \underline{x}(\Omega')$. Next, let $F \in \overline{\Lambda}_k(\mathcal{C}^\infty(\mathbb{R}^m))$ with $\operatorname{supp}(F) \cap \Sigma$ compact. Then we can define

$$\int_\Sigma F := \int_{\Omega'} \sum_A F_A(\underline{x}(u_1, \ldots, u_k)) J_A(\underline{u})\, du_1 \ldots du_k$$

where

$$J_A(\underline{u}) = \sum_\pi \operatorname{sgn}\pi \frac{\partial x_{\alpha_1}}{\partial u_{\pi(1)}} \cdots \frac{\partial x_{\alpha_k}}{\partial u_{\pi(k)}}$$

is the Jacobian that appears from the chain rule. This also implies that the above definition will not depend on the coordinate system in use. Indeed, if we use another coordinate system $(y_1, \ldots, y_k)$ that locally has the same orientation as $(u_1, \ldots, u_k)$, then for any $\varphi \in \mathcal{C}^\infty(\Omega')$

$$\begin{aligned} \int_{\Omega'} \varphi(\underline{u})\, du_1 \ldots du_k &= \int_{\Omega''} \varphi(\underline{u}(\underline{y})) \left| \frac{\partial^k u_1 \ldots u_k}{\partial y_1 \ldots \partial y_k} \right| dy_1 \ldots dy_k \\ &= \int_{\Omega''} \varphi(\underline{u}(\underline{y})) \frac{\partial^k u_1 \ldots u_k}{\partial y_1 \ldots \partial y_k}\, dy_1 \ldots dy_k, \end{aligned}$$

but we also have that

$$J_A(\underline{u}(\underline{y})) = J_A(\underline{u}) \cdot \frac{\partial^k u_1 \ldots u_k}{\partial y_1 \ldots \partial y_k}.$$

In other words, the calculus with differential forms automatically keeps track of Jacobians.

In the sequel we also need partial integration of differential forms. For a form $F(\underline{y})\,dy_1\ldots dy_\ell$ with compact support, this is defined as the operator

$$\int_{y_j} F(\underline{y})\,dy_1\ldots dy_\ell := (-1)^{j-1}\left(\int_{-\infty}^{+\infty} F(\underline{y})\,dy_j\right) dy_1\ldots dy_{j-1}dy_{j+1}dy_\ell$$

that transforms differential forms into differential forms. From this definition, it is clear that, as operators:

$$\int_{y_j}\int_{y_\ell} \cdot = -\int_{y_\ell}\int_{y_j} \cdot$$

and also

$$dy_j \int_{y_\ell} \cdot = -\int_{y_\ell} \cdot\, dy_j,$$

while the integral of k-forms may now be defined as

$$\begin{aligned}\int_\Sigma F &= \int_{\Omega'} \sum_A F_A(\underline{x}(\underline{u}))J_A(\underline{u})\,du_1\ldots du_k\\ &= \int_{u_k}\cdots\left(\int_{u_1}\sum_A F_A(\underline{x}(\underline{u}))J_A(\underline{u})\,du_1\right)\ldots du_k.\end{aligned}$$

In other words, variables of integration have to be moved to the left side of a differential form. It is important to note that the above definition of integral automatically keeps track of the orientation on Σ: it is determined by the order of the coordinates $u_1,\ldots,u_k$.

4. Stokes' formula

Let $F\in\overline{\Lambda}_{k-1}(\mathcal{C}^\infty(\Omega))$ with $\mathrm{supp}F\cap\Sigma$ compact and Σ is as above. Then we have that (where $\widehat{x}$ means that x is suppressed):

$$\begin{aligned}\int_\Sigma dF &= \int_{\mathbb{R}^k}\sum_{j=1}^k du_j\,\partial_{u_j}\sum_A F_A(\underline{x}(\underline{u}))\frac{\partial^{k-1}x_{\alpha_1}\ldots x_{\alpha_{k-1}}}{\partial u_1\ldots\widehat{\partial u_j}\ldots\partial u_k}\,du_1\ldots\widehat{du_j}\ldots du_m\\ &= \int_{\mathbb{R}^k}\sum_{j=1}^k \partial_{u_j} g_j(\underline{u})du_1\ldots du_k = 0,\end{aligned}$$

since $\mathrm{supp}F\cap\Sigma$ is compact, with

$$g_j(\underline{u}) = \sum_A F_A(\underline{x}(\underline{u}))\frac{\partial^{k-1}x_{\alpha_1}\ldots x_{\alpha_{k-1}}}{\partial u_1\ldots\widehat{\partial u_j}\ldots\partial u_k}.$$

Let C be a compact set in $\mathbb{R}^m$ with nonempty interior and $\mathcal{C}^\infty$ boundary. Let $\varphi\in\mathcal{C}^\infty(\mathbb{R}^m)$ be a defining function for C, i.e., $\varphi<0$ in $\mathrm{int}(C)$, $\varphi>0$ in $\mathbb{R}^m\setminus C$ and $\varphi=0$, $\nabla\varphi\neq 0$ on ∂C. Then, if Y denotes the Heaviside function, we have that $Y(-\varphi) = \chi_C$ where χ_C is the characteristic function of C. Moreover, for $F \in \Lambda_{k-1}(\mathcal{C}^\infty(\Omega))$ with $C \subset \Omega$ we would have that

$$\int_\Sigma d(\chi_C(x)F) = 0,$$

where

$$d(\chi_C F) = dY(-\varphi)F + \chi_C dF, \qquad dY(-\varphi) = -\delta(\varphi)\, d\varphi,$$

where δ is the Dirac distribution on the real line. This leads to

Theorem 4.1 (Stokes' formula). *With the above notations, the following formula holds:*

$$\int_\Sigma \delta(\varphi)\, d\varphi F = \int_\Sigma Y(-\varphi)\, dF.$$

The formula can be also written in the more familiar form

$$\int_{\partial C \cap \Sigma} F = \int_{C \cap \Sigma} dF.$$

Here one has to choose local coordinates $(v_1, \ldots, v_{k-1})$ on $\partial C \cap \Sigma$ such that the orientation of the system of coordinates $(\varphi, v_1, \ldots, v_{k-1})$ is the same as the orientation of $(u_1, \ldots, u_k)$.

Indeed, we have that

$$\int_\Sigma \delta(\varphi)\, d\varphi\, F = \int_{v_{k-1}} \cdots \int_{v_1} \int_\varphi \delta(\varphi)\, d\varphi\, F = \int_{v_{k-1}} \cdots \int_{v_1} F_{|\varphi=0} = \int_{\partial C \cap \Sigma} F.$$

5. Clifford differential forms

The complex Clifford algebra $\mathbb{C}_m$ is the complex associative algebra with generators $e_1, \ldots, e_m$ together with the defining relations $e_j e_k + e_k e_j = -2\delta_{jk}$. Every element $a \in \mathbb{C}_m$ can be written in the form

$$a = \sum_{A \subset M} a_A e_A, \qquad a_A \in \mathbb{C},$$

where, as before, $M = \{1, \ldots, m\}$ and for any multi-index $A = \{\alpha_1, \ldots, \alpha_k\} \subseteq M$, with $\alpha_1 < \cdots < \alpha_k$ we put $e_A = e_{\alpha_1} \cdots e_{\alpha_k}$.

Every $a \in \mathbb{C}_m$ admits a multivector decomposition

$$a = \sum_{k=0}^m [a]_k, \quad \text{where} \quad [a]_k = \sum_{|A|=k} a_A e_A,$$

so $[\cdot]_k : \mathbb{C}_m \to \mathbb{C}_m^k$ denotes the canonical projection of $\mathbb{C}_m$ onto the space $\mathbb{C}_m^k$ of k-vectors. Note that $\mathbb{C}_m^0 = \mathbb{C}$, the set of scalars while $\mathbb{C}_m^1$ is the space of 1-vectors $\underline{v} = \sum_{j=1}^m v_j e_j$. So the map

$$(v_1, \ldots, v_m) \to \underline{v} = \sum_{j=1}^m v_j e_j$$

leads to the identification of $\mathbb{C}^m$ with $\mathbb{C}_m^1$. For any $\underline{v}, \underline{w} \in \mathbb{C}_m^1$ we have

$$\underline{v}\,\underline{w} = \underline{v} \cdot \underline{w} + \underline{v} \wedge \underline{w},$$

$$\underline{v} \cdot \underline{w} = -\langle \underline{v}, \underline{w} \rangle = -\sum_{j=1}^{m} v_j w_j,$$

$$\underline{v} \wedge \underline{w} = \sum_{j<\ell} e_{j\ell}(v_j w_\ell - v_\ell w_j) \in \mathbb{C}_m^2.$$

More in general, for $\underline{v}_1, \ldots, \underline{v}_k \in \mathbb{C}_m^1$ we define the wedge (or Grassmann) product in terms of the Clifford product by

$$\underline{v}_1 \wedge \cdots \wedge \underline{v}_k = \frac{1}{k!} \sum_{\pi \in \mathrm{Sym}(k)} \mathrm{sgn}\pi \ \underline{v}_{\pi(1)} \cdots \underline{v}_{\pi(k)} \in \mathbb{C}_m^k.$$

We also call $\underline{v}_1 \wedge \cdots \wedge \underline{v}_k$ a *k-blade*. The k-blades span $\mathbb{C}_m^k$, but not every element in $\mathbb{C}_m^k$ is a k-blade.

For $\underline{v} \in \mathbb{C}_m^1$ and $a \in \mathbb{C}_m^k$ we set

$$\underline{v}a = [\underline{v}a]_{k-1} + [\underline{v}a]_{k+1} = \underline{v} \cdot a + \underline{v} \wedge a$$

where

$$\underline{v} \cdot a = \frac{1}{2}(\underline{v}a + (-1)^{k-1} a\underline{v}), \quad \underline{v} \wedge a = \frac{1}{2}(\underline{v}a + (-1)^{k} a\underline{v}).$$

More in general, for $a \in \mathbb{C}_m^k$, $b \in \mathbb{C}_m^\ell$, $k \geq \ell$ we have

$$ab = [ab]_{k-\ell} + [ab]_{k-\ell+2} + \cdots + [ab]_{k+\ell}$$

and we define the wedge product as

$$[ab]_{k+\ell} = a \wedge b.$$

So we have the Grassmann product in terms of the Clifford product. The variable $(x_1, \ldots, x_m) \in \mathbb{R}^m$ is identified with the vector variable $\underline{x} = \sum_{j=1}^m x_j e_j$ and $\mathbb{C}_m$-valued functions in $\mathbb{R}^m$ are denoted by $f(\underline{x}) = \sum_A f_A(\underline{x}) e_A$, f_A are $\mathbb{C}$-valued functions.

Definition 5.1. A function $f : \Omega \subseteq \mathbb{R}^m \to \mathbb{C}_m$ real differentiable will be called left monogenic in Ω if it satisfies $\partial_{\underline{x}} f(\underline{x}) = 0$ for $\underline{x} \in \Omega$, where $\partial_{\underline{x}} = \sum_{j=1}^m e_j \partial_{x_j}$ is the Dirac operator (or vector derivative).

We have the following formulas

$$\underline{x}\,\partial_{\underline{x}} = \underline{x} \cdot \partial_{\underline{x}} + \underline{x} \wedge \partial_{\underline{x}} = -E_{\underline{x}} - \Gamma_{\underline{x}}$$

where $E_{\underline{x}} = -\underline{x} \cdot \partial_{\underline{x}} = \sum_{j=1}^m x_j \partial_{x_j}$ is the Euler operator and $\Gamma_{\underline{x}} = -\underline{x} \wedge \partial_{\underline{x}} = -\sum_{j<k}^m e_{jk} L_{jk}$, $L_{jk} = x_j \partial_{x_k} - x_k \partial_{x_j}$, are the angular momentum operators. Moreover we have the overdot notation introduced by Hestenes

$$\partial_{\underline{x}}(\underline{x}\, f) = -mf + \dot{\partial}_{\underline{x}}(\underline{x}\dot{f})$$

where

$$\dot{\partial}_{\underline{x}}(\underline{x}\dot{f}) = -\underline{x}\partial_{\underline{x}} f - 2E_{\underline{x}} f.$$

Remark 5.2. The elements $dx_1, \dots, dx_m$ generate a Grassmann algebra and also $e_1, \dots, e_m$ form a Grassmann algebra with respect to the wedge product. Yet, we do not identify dx_j with e_j as some authors do. The elements e_j are imaginary units and so symbolic constants, while the elements dx_j are the differentials of the real coordinates $x_1, \dots, x_m$. The wedge notation will be used only for Clifford numbers, not for the differential forms $dx_1, \dots, dx_m$. However, we may use it for vector differentials (see below and the last section).

The vector variable $\underline{x} = \sum_{j=1}^m x_j e_j$ can be seen as a $\mathbb{R}_m^1$-valued function. Its differential, called *vector differential* is given by $d\underline{x} = \sum_{j=1}^m e_j dx_j$. Combining the Clifford product and the differential form product, we have that

$$(d\underline{x})^2 = \sum_{j,\ell}^m dx_j e_j dx_\ell e_\ell = 2\sum_{j<\ell}^m dx_j dx_\ell e_j e_\ell = d\underline{x} \wedge d\underline{x} = [d\underline{x}^2]_2,$$

and, more in general,

$$(d\underline{x})^k = k! \sum_{|A|=k} dx_A e_A = d\underline{x} \wedge \cdots \wedge d\underline{x} = [d\underline{x}^k]_k.$$

In particular

$$\frac{d\underline{x}^m}{m!} = dx_1 \dots dx_m e_1 \dots e_m = V(d\underline{x}) e_M,$$

$$\frac{d\underline{x}^{m-1}}{(m-1)!} = \sum_{j=1}^m dx_{M\setminus\{j\}} e_{M\setminus\{j\}} = -\sum_{j=1}^m e_j (-1)^{j-1} dx_{M\setminus\{j\}} e_M = -\sigma(d\underline{x}) e_M$$

where $V(d\underline{x})$ denotes the Euclidean volume form and

$$\sigma(d\underline{x}) = \sum_{j=1}^m (-1)^{j-1} e_j \, dx_1 \dots \widehat{dx_j} \dots dx_m$$

is called σ-form.

Let $f, g : \Omega \to \mathbb{C}_m$, then

$$\begin{aligned} d(f\,\sigma\,g) &= \sum_{j=1}^m \partial_{x_j} (f\, e_j\, g)\, dx_1 \dots dx_m \\ &= (\dot{f}\dot{\partial}_{\underline{x}} g + f \dot{\partial}_{\underline{x}} \dot{g})\, V(d\underline{x}). \end{aligned}$$

Hence, for a compact subset $C \subset \Omega$ with nonempty interior and with smooth boundary, we have (see [1, 10]):

Theorem 5.3 (Cauchy–Borel–Pompeiu). *Let $\Omega \subseteq \mathbb{R}^m$ be an open set and $f, g : \Omega \to \mathbb{C}_m$. Let $C \subset \Omega$ with nonempty interior and with smooth boundary. Then*

$$\int_{\partial C} d(f\,\sigma\,g) = \int_C (\dot{f}\dot{\partial}_{\underline{x}} g + f \dot{\partial}_{\underline{x}} \dot{g})\, V(d\underline{x}).$$

We are now going to generalize this result to smooth k-surfaces $C \cap \Sigma$ where, as before, Σ is the infinitely differentiable image of a map $\underline{u} = (u_1, \ldots, u_m) \to \underline{x}(\underline{u}) \in \Sigma$.

First of all, we have that for $\underline{x} \in \Sigma$:

$$d\underline{x} = \sum_{j=1}^{k} du_j\, \partial_{u_j}(\underline{x}) = \partial_{u_j}\underline{x}\, du_j,$$

$$d\underline{x}^2 = \sum_{j<\ell} (\partial_{u_j}\underline{x}\, \partial_{u_\ell}\underline{x} - \partial_{u_\ell}\underline{x}\, \partial_{u_j}\underline{x}) du_j\, du_\ell$$

$$= 2\sum_{j<\ell} \frac{\partial \underline{x}}{\partial u_j} \wedge \frac{\partial \underline{x}}{\partial u_\ell} du_j du_\ell$$

$$\frac{d\underline{x}^\ell}{\ell!} = \sum_{|A|=\ell} \frac{\partial \underline{x}}{\partial u_{j_1}} \wedge \cdots \wedge \frac{\partial \underline{x}}{\partial u_{j_\ell}} du_{j_1} \ldots du_{j_\ell} \tag{5.1}$$

and eventually

$$\frac{d\underline{x}^k}{k!} = \frac{\partial \underline{x}}{\partial u_1} \wedge \cdots \wedge \frac{\partial \underline{x}}{\partial u_k} du_1 \ldots du_k \tag{5.2}$$

is the oriented k-vector-valued surface element on Σ. All these surface forms are coordinate-independent.

We now prove the following crucial result, see also [10]:

Lemma 5.4. *We have the formal identity*

$$d\frac{d\underline{x}^{k-1}}{(k-1)!} = -\partial_{\underline{x}} \cdot \frac{d\underline{x}^k}{k!}.$$

Proof. We have the following chain of equalities

$$\begin{aligned} -\partial_{\underline{x}} \cdot \frac{d\underline{x}^k}{k!} &= -\frac{1}{2}\left(\partial_{\underline{x}} \frac{d\underline{x}^k}{k!} + (-1)^{k-1} \frac{d\underline{x}^k}{k!} \partial_{\underline{x}}\right) \\ &= -\frac{1}{2}\Big(\{\partial_{\underline{x}}, d\underline{x}\} \frac{d\underline{x}^{k-1}}{k!} - d\underline{x}\{\partial_{\underline{x}}, d\underline{x}\}\frac{d\underline{x}^{k-2}}{k!} + \cdots \\ &\qquad \cdots + (-1)^{k-1} \frac{d\underline{x}^{k-1}}{k!}\{\partial_{\underline{x}}, d\underline{x}\}\ \Big) \\ &= -\frac{1}{2}\{\partial_{\underline{x}}, d\underline{x}\} \frac{d\underline{x}^{k-1}}{(k-1)!} \end{aligned}$$

and clearly $\{\partial_{\underline{x}}, d\underline{x}\} = -2d$. □

Theorem 5.5 (Stokes). *Let Σ be a smooth k-surface, let C be a compact set with non empty interior whose boundary ∂C is a smooth $(n-1)$-surface, (so $C \cap \Sigma$ is a compact set). Let f, g be $\mathbb{C}_m$-valued smooth functions on Σ. Then*

$$\int_{\partial C \cap \Sigma} f \frac{d\underline{x}^{k-1}}{(k-1)!} g = -\int_{C\cap\Sigma} \left(\dot{f}\partial_{\underline{x}} \cdot \frac{d\underline{x}^k}{k!} g + f\partial_{\underline{x}} \cdot \frac{d\underline{x}^k}{k!}\dot{g}\right).$$

Now we have to characterize the restriction to Σ of $\partial_{\underline{x}} \cdot \frac{d\underline{x}^k}{k!}$. We already know, see (5.2), that

$$\frac{d\underline{x}^k}{k!} = \mathbb{T}(\underline{x}) V(du_1, \ldots, du_k),$$

where $\mathbb{T}(\underline{x})$ is the unit k-blade tangent to Σ at the point $\underline{x}$ and $V(du_1, \ldots, du_k)$ is the Euclidean volume form. Let $p \in \Sigma$ and consider an orthonormal basis $\underline{\varepsilon}_1, \ldots, \underline{\varepsilon}_k$ of k-planes tangent to Σ at the point p. Assume that the orthonormal basis has the same orientation as the coordinate frame $u_1, \ldots, u_k$. Then

$$\mathbb{T} = \underline{\varepsilon}_1 \ldots \underline{\varepsilon}_k = \omega \frac{\partial \underline{x}}{\partial u_1} \wedge \ldots \wedge \frac{\partial \underline{x}}{\partial u_k}$$

where ω is a positive weight, namely a function with strictly positive real values. Let $\underline{\nu}_1, \ldots, \underline{\nu}_{m-k}$ be the remaining $(m-k)$ unit vectors such that

$$(\underline{\varepsilon}_1, \ldots, \underline{\varepsilon}_k; \underline{\nu}_1, \ldots, \underline{\nu}_{m-k})$$

is an orthonormal basis of $\mathbb{R}^m$. Then

$$\partial_{\underline{x}} = \partial_{\underline{x}\|} + \partial_{\underline{x}\perp}$$

$$\partial_{\underline{x}\|} = \sum_{j=1}^{k} \underline{\varepsilon}_j \langle \underline{\varepsilon}_j, \partial_{\underline{x}} \rangle$$

$$\partial_{\underline{x}\perp} = \sum_{j=1}^{m-k} \underline{\nu}_j \langle \underline{\nu}_j, \partial_{\underline{x}} \rangle$$

so that after restriction to Σ we have

$$\partial_{\underline{x}} \cdot \frac{d\underline{x}^k}{k!} = \partial_{\underline{x}\|} \frac{d\underline{x}^k}{k!} = (-1)^{k-1} \frac{d\underline{x}^k}{k!} \partial_{\underline{x}\|}.$$

We then have (compare with [10]):

Theorem 5.6 (Cauchy). *Let Σ be a smooth k-surface, let C be a compact set with non empty interior whose boundary ∂C is a smooth $(n-1)$-surface. Let f, g be $\mathbb{C}_m$-valued smooth functions on Σ. Then*

$$\int_{\partial C \cap \Sigma} f \frac{d\underline{x}^{k-1}}{(k-1)!} g = -\int_{C \cap \Sigma} (f \partial_{\underline{x}\|}) \frac{d\underline{x}^k}{k!} g + (-1)^k \int_{C \cap \Sigma} f \frac{d\underline{x}^k}{k!} (\partial_{\underline{x}\|} g).$$

6. Winding numbers from monogenic functions

Let us recall that the Cauchy kernel for monogenic functions is

$$E(\underline{x}) = -\frac{1}{A_m} \frac{\underline{x}}{|\underline{x}|^m}, \qquad A_m = \frac{2\pi^{n/2}}{\Gamma(n/2)}$$

The function $E(\underline{x})$ is both left and right monogenic in $\mathbb{R}^m \setminus \{0\}$ and takes values in the space of 1-vectors $\mathbb{R}^1_m$. We also have the validity of the following equalities

$$\partial_{\underline{x}} E(\underline{x}) = E(\underline{x}) \partial_{\underline{x}} = \delta(\underline{x}) = \delta(x_1) \ldots \delta(x_m)$$

which hold in the distributional sense, see also [3]. Hence, we also have that, in view of Lemma 5.4:

$$\begin{aligned} d\left(E(\underline{x})\frac{d\underline{x}^{m-1}}{(m-1)!}\right) &= -\dot{E}(\underline{x})\dot{\partial}_{\underline{x}}\cdot\frac{d\underline{x}^m}{m!} \\ &= -(E(\underline{x})\partial_{\underline{x}})\frac{d\underline{x}^m}{m!} \\ &= -\delta(\underline{x})\frac{d\underline{x}^m}{m!}, \end{aligned}$$

whereby as before $\frac{d\underline{x}^m}{m!} = dx_1 \dots dx_m e_M$. We are now going to consider two sets of coordinates $x_1, \dots, x_m$, and $u_1, \dots, u_m$ and the corresponding differentials. Then, by translation, we have that

$$\begin{aligned} \partial_{\underline{x}} E(\underline{x}-\underline{u}) &= E(\underline{x}-\underline{u})\partial_{\underline{x}} = \delta(\underline{x}-\underline{u}) \\ &= \delta(x_1-u_1)\dots\delta(x_m-u_m) \\ &= -\partial_{\underline{u}} E(\underline{x}-\underline{u}) = -E(\underline{x}-\underline{u})\partial_{\underline{u}}. \end{aligned}$$

Now, by replacing also the vector differential

$$d\underline{x} \to d\underline{y} = d(\underline{x}-\underline{u}) = d\underline{x} - d\underline{u}, \qquad \text{where } \underline{y} = \underline{x}-\underline{u},$$

we still have that

$$d_{\underline{y}} E(\underline{y})\frac{d\underline{y}^{m-1}}{(m-1)!} = -\delta(\underline{y})\frac{d\underline{y}^m}{m!},$$

where $d_{\underline{y}} = \sum_{j=1}^m dy_j\partial_{y_j} = \sum_{j=1}^m (dx_j - du_j)\partial_{y_j}$ and also

$$\partial_{y_j} E(\underline{y}) = \partial_{x_j} E(\underline{x}-\underline{u}) = -\partial_{u_j} E(\underline{x}-\underline{u}).$$

Hence $d_{\underline{y}} = d_{\underline{x}} + d_{\underline{u}}$ and the above identity may be rewritten as

$$(d_{\underline{x}} + d_{\underline{u}})E(\underline{x}-\underline{u})\frac{(d\underline{x}-d\underline{u})^{m-1}}{(m-1)!} = -\delta(\underline{x}-\underline{u})\frac{(d\underline{x}-d\underline{u})^m}{m!} \tag{6.1}$$

where

$$\frac{(d\underline{x}-d\underline{u})^m}{m!} = (dx_1-du_1)\dots(dx_m-du_m)e_M = \sum_{k=0}^m V_k(d\underline{x}, d\underline{u})e_M$$

and (see also [11])

$$\begin{aligned} V_k(d\underline{x}, d\underline{u}) &= (-1)^k \sum_{|A|=k} \operatorname{sgn} A\, du_A\, dx_{M\setminus A}, \\ du_A &= du_{\alpha_1}\dots du_{\alpha_k}, & A &= \{\alpha_1\dots\alpha_k\},\ \alpha_1 < \dots < \alpha_k \\ dx_{M\setminus A} &= dx_{\beta_1}\dots dx_{\beta_{m-k}}, & M\setminus A &= \{\beta_1\dots\beta_{m-k}\},\ \beta_1 < \dots < \beta_{m-k} \end{aligned}$$

and sgn$A$ denotes the signature of the permutation $(\alpha_1, \ldots, \alpha_k, \beta_1, \ldots, \beta_{m-k})$ with respect to $(1, \ldots, m)$. This can also be obtained as follows:

$$d\underline{x} \cdot d\underline{u} = [d\underline{x}\, d\underline{u}]_0 = -\sum_{j=1}^{m} dx_j\, du_j$$

$$d\underline{x} \wedge d\underline{u} = [d\underline{x}\, d\underline{u}]_2 = \frac{1}{2}(d\underline{x}\, d\underline{u} + d\underline{u}\, d\underline{x}) = d\underline{u} \wedge d\underline{x}.$$

So, in general, we have that

$$\begin{aligned}\frac{(d\underline{x} - d\underline{u})^k}{k!} &= \frac{[(d\underline{x} - d\underline{u})^k]_k}{k!} \\ &= \frac{(d\underline{x} - d\underline{u}) \wedge \cdots \wedge (d\underline{x} - d\underline{u})}{k!} \\ &= \sum_{\ell=0}^{k} (-1)^\ell \binom{k}{\ell} \frac{[d\underline{u}^\ell\, d\underline{x}^{k-\ell}]_k}{k!} \\ &= \sum_{\ell=0}^{k} (-1)^\ell \frac{d\underline{u}^\ell}{\ell!} \wedge \frac{d\underline{x}^{k-\ell}}{(k-\ell)!}\end{aligned}$$

is a k-vector. In particular:

$$V_k(d\underline{x}, d\underline{u}) e_M = (-1)^k \frac{d\underline{u}^k}{k!} \wedge \frac{d\underline{x}^{m-k}}{(m-k)!}$$

while also

$$\frac{(d\underline{x} - d\underline{u})^{m-1}}{(m-1)!} = -\sigma(d\underline{x} - d\underline{u}) e_M = -\sum_{k=0}^{m-1} \sigma_k(d\underline{x}, d\underline{u})$$

with

$$\sigma_k(d\underline{x}, d\underline{u}) = (-1)^{k+1} \frac{d\underline{u}^k}{k!} \wedge \frac{d\underline{x}^{m-k-1}}{(m-k-1)!}.$$

Thus we arrive at the fundamental identity contained in the following result:

Theorem 6.1. *For every $k = 0, \ldots, m$ the following identity holds:*

$$\begin{aligned}&d_{\underline{x}} \left[E(\underline{x} - \underline{u}) \frac{d\underline{u}^k}{k!} \wedge \frac{d\underline{x}^{m-k-1}}{(m-k-1)!} \right] \\ &= d_{\underline{u}} \left[E(\underline{x} - \underline{u}) \frac{d\underline{u}^{k-1}}{(k-1)!} \wedge \frac{d\underline{x}^{m-k}}{(m-k)!} \right] - \delta(\underline{x} - \underline{u}) \frac{d\underline{u}^k}{k!} \wedge \frac{d\underline{x}^{m-k}}{(m-k)!}.\end{aligned}$$

Proof. The result follows by identifying differential forms with same degree in $du_1, \ldots, du_m$ within the formula (6.1). □

Let C_k be a k-chain in $\mathbb{R}^m$ that can be realized as a compact subset C_k with nonempty interior of an oriented infinitely differentiable surface of dimension k (with respect to the relative topology).

Definition 6.2. The indicatrix $I(C_k)(\underline{x})$ of the k-chain C_k in $\mathbb{R}^m$ is the $(m-k-1)$-form

$$I(C_k)(\underline{x}) = \int_{\underline{u}\in C_k} E(\underline{x}-\underline{u})\frac{d\underline{u}^k}{k!} \wedge \frac{d\underline{x}^{m-k-1}}{(m-k-1)!}.$$

The indicatrix is an $(m-k-1)$-form with left monogenic component. The above theorem clearly leads to the following result (compare with the result in [9]):

Theorem 6.3. *Let ∂C_k be the boundary of C_k with proper orientation, then*

$$(-1)^k dI(C_k)(\underline{x}) = I(\partial C_k)(\underline{x}) - \int_{\underline{u}\in C_k} \delta(\underline{x}-\underline{u})\frac{d\underline{u}^k}{k!} \wedge \frac{d\underline{x}^{m-k}}{(m-k)!}.$$

Proof. The result follows from Theorem 6.1 by integrating with respect to

$$\int_{\underline{u}\in C_k} \cdot$$

The factor $(-1)^k$ in front arises because of the anti-commutativity:

$$\int_{v_j} \cdot\, dx_j = -dx_j \int_{v_j} \cdot$$

between integral operators and differentials. □

Remark 6.4. The second term is a distributional $(m-k)$-form given by

$$-\left[\Delta(C_k)(\underline{x})\frac{d\underline{x}^{m-k}}{(m-k)!}\right]_m = (-1)^{k(m-k)+1}\left[\frac{d\underline{x}^{m-k}}{(m-k)!}\Delta(C_k)(\underline{x})\right]_m$$

where $\Delta(C_k)(\underline{x})$ denotes the distribution supported by C_k defined as

$$\Delta(C_k)(\underline{x}) = \int_{\underline{u}\in C_k} \delta(\underline{x}-\underline{u})\frac{d\underline{u}^k}{k!}.$$

Corollary 6.5. *Let $\partial C_k = 0$, i.e., let C_k be a k-cycle. Then*

$$(-1)^k dI(C_k)(\underline{x}) = -(-1)^{k(m-k)}\frac{d\underline{x}^{m-k}}{(m-k)!} \wedge \Delta(C_k)(\underline{x}),$$

which vanishes in $\mathbb{R}^m \setminus C_k$.

Now consider a k-cycle C_k and let C_{m-k} be an infinitely differentiable $(m-k)$-chain with infinitely differentiable boundary $\partial C_{m-k} \subset \mathbb{R}^m \setminus C_k$. We choose C_{m-k} such that it intersect generically C_k in finitely many points. Then, in view of Stokes' formula and using the previous corollary, we have

$$\begin{aligned}\int_{\partial C_{m-k}} I(C_k)(\underline{x}) &= \int_{C_{m-k}} dI(C_k)(\underline{x})\\ &= -(-1)^{k(m-k-1)}\int_{C_{m-k}} \frac{d\underline{x}^{m-k}}{(m-k)!} \wedge \Delta(C_k)(\underline{x}).\end{aligned}$$

Theorem 6.6. *Under the above assumptions*

$$\int_{C_{m-k}} I(C_k)(\underline{x}) = -(-1)^k \mathrm{Int}(C_k, C_{m-k})e_M,$$

where $\mathrm{Int}(C_k, C_{m-k})$ *is the intersection number of* C_k *with respect to* C_{m-k} *inside* $\mathbb{R}^m$.

We will show the result just in one case. The general result follows from homological arguments. First we note that all the above results extend to the case where C_k is an unbounded chain or cycle, as long as all the needed integrals converge. Let us consider the case where $C_k = W_k$, W_k being the oriented k-space with coordinates $u_1, \ldots, u_k$. In this case we have

$$\begin{aligned} I(W_k) &= \int_{u_m \in \mathbb{R}} \cdots \int_{u_1 \in \mathbb{R}} E(\underline{x} - \underline{u}) du_i \ldots du_k \left(e_1 \ldots e_k \wedge \frac{d\underline{x}^{m-k-1}}{(m-k-1)!} \right) \\ &= -\frac{1}{A_m} \frac{\underline{x}_\perp}{|\underline{x}_\perp|^{m-k}} \left(e_1 \ldots e_k \wedge \frac{d\underline{x}_\perp^{m-k-1}}{(m-k-1)!} \right) \end{aligned}$$

where $\underline{x}_\perp = \sum_{j=k+1}^m x_j e_j$, $d\underline{x}_\perp = \sum_{j=k+1}^m dx_j e_j$, and therefore by Lemma 5.4

$$\begin{aligned} dI(W_k)(\underline{x}) &= \frac{1}{A_{m-k}} \frac{\dot{\underline{x}}_\perp}{|\underline{x}_\perp|^{m-k}} e_1 \ldots e_k \left(\dot{\partial}_{\underline{x}_\perp} dx_{k+1} \ldots dx_m e_{k+1} \ldots e_m \right) \\ &= -(-1)^k \delta(\underline{x}_\perp) dx_{k+1} \ldots dx_m e_M. \end{aligned}$$

So if $C_{m-k} = B(1) \cap W_{m-k}$, W_{m-k} being the $(m-k)$-space with coordinates $x_{k+1}, \ldots, x_m$, we obtain that $\partial C_{m-k} = \mathbb{S}^{m-k-1}$, the unit sphere in W_{m-k} and

$$\begin{aligned} \int_{\mathbb{S}^{m-k-1}} I(W_k) &= \int_{|\underline{x}_\perp|<1} -(-1)^k \delta(\underline{x}_\perp) dx_{k+1} \ldots dx_m e_M \\ &= -(-1)^k e_M \\ &= -(-1)^k \mathrm{Int}(W_k, W_{m-k}) e_M. \end{aligned}$$

In general, we have the following:

Definition 6.7. The intersection number $\mathrm{Int}(C_k, C_{m-k})$ is defined as

$$\sum_{p \in C_k \cap C_{m-k}} \mathrm{Int}(T_p C_k, T_p C_{m-k}),$$

where $T_p C_k$, $T_p C_{m-k}$ denote the oriented tangent spaces to C_k and C_{m-k} at the point p, respectively, and where

$$\mathrm{Int}(T_p C_k, T_p C_{m-k}) = \mathrm{sgn} \det G,$$

where $G \in \mathrm{GL}(m, \mathbb{R})$ is the matrix of a linear transformation mapping $W_k \to T_p C_k$ and $W_{m-k} \to T_p C_{m-k}$.

Definition 6.8. The intersection number $\mathrm{Int}(T_p C_k, T_p C_{m-k})$ is also called the winding number of ∂C_{m-k} around C_k.

Remark 6.9. At a given point p, the number $\mathrm{Int}(T_pC_k, T_pC_{m-k})$ gives the signature of the orientation. The sum $\sum_{p\in C_k\cap C_{m-k}} \mathrm{Int}(T_pC_k, T_pC_{m-k})$ is equal to the total intersection number between C_k and C_{m-k}; it is also equal to the number of times that ∂C_{m-k} rotates around C_k.

Proof of the theorem. As $I(C_k)$ is closed in $\mathbb{R}^m \setminus C_k$, we have that whenever C'_{m-k} is an $(m-k)$-chain for which $\partial C'_{m-k}$ is homologous to ∂C_{m-k} in $\mathbb{R}^m \setminus C_k$, then

$$\int_{\partial C'_{m-k}} I(C_k)(\underline{x}) = \int_{\partial C_{m-k}} I(C_k)(\underline{x}).$$

Moreover, C'_{m-k} may be chosen to be a sum of unit discs. Due to the symmetry, one may also change C_k to a homologous cycle C'_k inside $\mathbb{R}^m \setminus \partial C'_{m-k}$ and choose C'_k to be a sum of spheres (or even oriented k-spaces). This reduces to the general case to a k-space W_k and a disc $B(1) \cap W_{m-k}$ for which we have the result. □

Remark 6.10. For the indicatrix of C_k we have the expressions

$$I(C_k)(\underline{x}) = \int_{\underline{u}\in C_k} E(\underline{x}-\underline{u})\frac{d\underline{u}^k}{k!} \wedge \frac{d\underline{x}^{m-k-1}}{(m-k-1)!}$$

from which we get

$$\begin{aligned}\int_{\underline{x}\in\partial C_{m-k}} I(C_k)(\underline{x}) &= \left[\int_{\underline{x}\in\partial C_{m-k}}\int_{\underline{u}\in C_k} E(\underline{x}-\underline{u})\frac{d\underline{u}^k}{k!}\frac{d\underline{x}^{m-k-1}}{(m-k-1)!}\right]_m \\ &= (-1)^{(k+1)(m-k-1)}\left[\int_{\underline{x}\in\partial C_{m-k}} \frac{d\underline{x}^{m-k-1}}{(m-k-1)!}\int_{\underline{u}\in C_k} E(\underline{x}-\underline{u})\frac{d\underline{u}^k}{k!}\right]_m \\ &= \pm\left[\int_{\partial C_{m-k}} \frac{d\underline{x}^{m-k-1}}{(m-k-1)!} M(C_k)(\underline{x})\right]_m\end{aligned}$$

whereby $M(C_k)$ is a left monogenic function in $\mathbb{R}^m \setminus C_k$ given by

$$M(C_k)(\underline{x}) = \int_{\underline{u}\in C_k} E(\underline{x}-\underline{u})\frac{d\underline{u}^k}{k!} = [M(C_k)]_{k+1} + [M(C_k)]_{k-1}$$

and in fact

$$\int_{\underline{x}\in\partial C_{m-k}} I(C_k)(\underline{x}) = (-1)^{(k+1)(m-k-1)}\int_{\underline{x}\in C_{m-k}} \frac{d\underline{x}^{m-k-1}}{(m-k-1)!} \wedge [M(C_k)]_{k+1}.$$

References

[1] F. Brackx, R. Delanghe, F. Sommen, *Clifford Analysis*, Pitman Res. Notes in Math., 76, 1982.

[2] F. Colombo, I. Sabadini, F. Sommen, D.C. Struppa, *Analysis of Dirac Systems and Computational Algebra*, Progress in Mathematical Physics, Vol. 39, Birkhäuser, Boston, 2004.

[3] F. Colombo, I. Sabadini, F. Sommen, D.C. Struppa, *Twisted plane wave expansions using hypercomplex methods*, Publ. Res. Inst. Math. Sci., **50** (2014), 1–18.

[4] R. Delanghe, F. Sommen, V. Souček, *Clifford Algebra and Spinor-valued Functions*, Mathematics and Its Applications 53, Kluwer Academic Publishers, 1992.

[5] I.M. Gelfand, G.E. Shilov, *Generalized functions. Vol. 2. Spaces of fundamental and generalized functions*, Translated from the Russian, Academic Press, New York-London, 1968.

[6] D. Hestenes, G. Sobczyk, *Clifford algebra to geometric calculus*, Reidel, Dordrecht, 1984.

[7] W. Hodge, *The theory and applications of harmonic integrals*, Cambridge Univ. Press, 1959.

[8] J.M. Lee, *Introduction to smooth manifolds*, Graduate Text in Mathematics, 218, Springer New York, 2003.

[9] F. Sommen, *Monogenic differential calculus*, Trans. Amer. Math. Soc., **326** (1991), 613–631.

[10] F. Sommen, *Cauchy-type theorems on surfaces*, AIP Conf. Proc. **1479** (2012), 336–339.

[11] F. Sommen, *Clifford Superanalysis*, AIP Conf. Proc. **1558** (2013), 546–549.

[12] F. Sommen, V. Souček, *Monogenic differential forms*, Complex Variables Theory Appl., **19** (1992), 81–90.

Irene Sabadini
Politecnico di Milano
Dipartimento di Matematica
Via E. Bonardi, 9
I-20133 Milano, Italy
e-mail: `irene.sabadini@polimi.it`

Franciscus Sommen
Clifford Research Group
Faculty of Sciences
Ghent University
Galglaan 2
B-9000 Gent, Belgium
e-mail: `fs@cage.ugent.be`

Modern Trends in Hypercomplex Analysis
Trends in Mathematics, 265–310

Examples of Morphological Calculus

Franciscus Sommen

Abstract. In this paper we present an introduction to morphological calculus in which geometrical objects play the rule of generalized natural numbers.

Mathematics Subject Classification (2010). 30G35, 32A30.

Keywords. Lie groups; homogeneous spaces; fibre bundles; Poincaré polynomial; twistor space.

1. Introduction

Morphological calculus in an extension of the calculus of natural numbers $1, 2, 3$ etc. whereby all sorts of geometrical objects are seen as generalized natural numbers. To make a list, we have

- the natural numbers $1, 2, 3, \ldots$
- the real line $\mathbb{R}$
- the set of natural numbers $\mathbb{N}$
- Cartesian spaces $\mathbb{R}^2, \mathbb{R}^3, \ldots$
- projective spaces $\mathbb{RP}_n, \mathbb{CP}_n, \ldots$
- spheres $S^{n-1}, \mathbb{C}S^{n-1}$
- groups like $SO(n), U(n), GL(n, \mathbb{R}), \ldots$
- Graßmann manifold $G_{n,k}(\mathbb{R})$

and other groups and homogeneous spaces. In fact any kind of geometrical objects can be added to the list.

The rules for morphological calculus extend the rules for calculating with natural numbers. We have

1. The addition $t_1 + t_2 + \cdots + t_k$

The terms $t_1, \ldots, t_k$ are supposed to represent morphological objects and the addition represents any object that can be formed by making a disjoint union of the objects $t_1, \ldots, t_k$ and glueing them together when possible. This glueing process is itself not part of the calculus so there in no unique way to do it and one also needn't do it; one can simply put the objects $t_1, \ldots, t_k$ in a list, as the language of calculus suggests.

For example $1 + 2 + 3 + 4$ can be visualized as a triangle of 10 points: $1 + (1 + 1) + (1 + 1 + 1) + (1 + 1 + 1 + 1)$.

The terms $t_1, \ldots, t_k$ in an addition may simply be names for morphological objects but, they also could be expressions between brackets like in:

$$5 + (3 + 1) + 2 + (1 + 2 + 7).$$

The material between brackets is interpreted as a single morphological object.

2. The subtraction $t_1 - t_2$

This means that the object t_2 is deleted from the object t_1. For example $3 - 2$ means to delete 2 points from a set of 3 points or $\mathbb{R} - 1$ means to delete a point from a line. The subtraction represents a problem: we have to look for an object such that $t_1 - t_2 = c$ or such that t_1 can be written as $c + t_2$. There may not be a morphologically acceptable solution for this. For example

$$0 = 1 - 1$$

means to create a point 1 and then to wipe if off.

Also negative numbers like $-1, -2, \ldots$ are no objects of morphological calculus although they may be meaningful as actions: $-1 =$ to delete one point, $-2 =$ to delete two points, etc.

The subtraction is presented as a binary operation $t_1 - t_2$ here, but of course one may also consider an extended expression like $7 - 3 - 2 + 4 - 1$, as long as things add up to a morphological object.

3. The multiplication $v \cdot w$

For the natural numbers, the multiplication is a notation for repeated addition, so for example

- $1 \cdot a = a$
- $2 \cdot a = a + a$
- $3 \cdot a = a + a + a$

etc. In other words, the meaning of multiplication is in fact determined by the rule of distributivity

$$(t_1 + t_2 + \cdots + t_k) \cdot w = t_1 \cdot w + t_2 \cdot w + \cdots + t_k \cdot w.$$

In morphological calculus, the product $v \cdot w$ means that every point of the object v is replaced by a copy of w and then all those copies of w are possibly glued together in some way that is not specified by the language of calculus.

Typical examples are: the Cartesian product $v \times w$, a fibre bundle $E = M \cdot F$ with base space M and fibre F.

One can also consider long multiplication like $v_1 \cdot v_2 \cdots v_k$ that may correspond to iterated fibre bundles. Note that the fibre bundle interpretation is only an option; it isn't a must and it will not always be available.

4. The division v/w

Like the subtraction, also the division is seen as a problem: to find a morphological object c for which $v = c \cdot w$. Any good solution to this will be denoted as v/w and again there may not always be a solution. For example rational numbers like $1/2, 1/3, 4/7$ etc., are no morphological objects even though one may write $7/3 = 6/3 + 1/3 = 2 + 1/3$.

Hence the language of morphological calculus is similar to that of natural numbers. There are however some aspects of language of calculus that cause dilemmas and also need more explanation

1. *Names, definitions, substitutions*

Every morphological object has a name attached to it. For example $1, 2, 3, \ldots$ the natural members, $\mathbb{R}$ the real line and so on. Then every name is given a definition or several definitions of the form

$$\text{Name} = \text{Expression}$$

the first main examples being the definitions of the natural numbers

$$2 = 1 + 1,\ 3 = 1 + 1 + 1,\ 4 = 1 + 1 + 1 + 1$$

and so on.

Such definitions may come from geometry, but they are algebraic expression of some geometrical decomposition of an object, i.e., geometrical knowledge can only enter the calculus via algebraic relations of the form Name = Expression.

When a name N appears somewhere in an expression E, i.e., $E = E(N)$ and when one has a definition $N =$ Expr.; then one may perform the substitution $E(N) = E((\text{Expr.}))$, i.e., replacing the name N by the expression (Expr.) between brackets. Later on one may investigate how and when brackets may be removed. We do not use brackets in a redundant manner like, e.g., (7) is not used, (Name) is not used, ((Expr.)) is not used, Name = (Expr.) is not used.

Example (The Fibonacci trees). These morphological structures are defined by

$$f_1,\ f_2 = 1,\ f_n = f_{n-1} + f_{n-2}$$

leading to the solutions

$$\begin{aligned} f_2 &= 1 + 1 \\ f_3 &= (1 + 1) + 1 \\ f_4 &= ((1 + 1) + 1) + (1 + 1) \\ f_5 &= (((1 + 1) + 1) + (1 + 1)) + ((1 + 1) + 1), \end{aligned}$$

so what appears here are not just the Fibonacci numbers $2, 3, 5, 8$, but the tree-like structures that give rise to these numbers if one removes the brackets. This tree-like structure is a typical example of a morphological object.

2. *Commutativity, associativity*

In morphological calculus the addition $t_1 + \cdots + t_k$ is in the first place a listing of objects; it is not commutative. Also within an addition one may consider expressions between brackets and since brackets refer to morphological objects one can't just ignore them; the addition is not just associative. On the other hand, for the natural numbers the addition also refers to the total quantity or sum. For example the total quantity of $5 + (3 + 1) + 2$ may be evaluated as:

$$\begin{aligned} 5 + (3+1) + 2 &= (1+1+1+1+1) + ((1+1+1)+1) + (1+1) \\ &= (1+1+1+1+1) + (1+1+1+1) + (1+1) \\ &= 1+1+1+1+1+1+1+1+1+1+1 \\ &= 11, \end{aligned}$$

so it requires substitutions $5 \to (1+1+1+1+1)$ etc. and deleting the brackets. So the total quantity is evaluated within the language of calculus and not in some outside theory. It corresponds to a morphological process in which the morphological structure is constantly changed to the extent that in the final evaluation of the quantity, the identity of the numbers $5, 3, \ldots$ as well as their place in the context is lost. Commutativity, substitutions and putting and deleting brackets are guaranteed in so far that the total quantity is preserved, but they are also mutations. For more general morphological objects, such as the line $\mathbb{R}$ the notion of quantity is not defined and we will illustrate that, if it were defined it wouldn't correspond to the cardinality of a set.

Yet we calculate as if these objects would have a form of quantity and so, in particular, terms in an addition may be commuted, substituted and brackets may be put or deleted.

For the multiplication $v \cdot w$, commutativity $v \cdot w = w \cdot v$ is even less obvious especially if one thinks of a fibre bundle $E = M \cdot F$. But again these geometrical interpretations happen outside morphological calculus and the total quantity of $v \cdot w$ is the same as that of $w \cdot v$. Moreover, to be able to calculate one has to be able to commute factors in a product, even though this deforms the morphological structure. Also the law of distributivity

$$(t_1 + \cdots + t_k) \cdot w = t_1 \cdot w + \cdots + t_k \cdot w$$

is essential to give a meaning to the product while as the same time it is a deformation.

So, to conclude, the morphological universe consists of the totality of all meaningful algebraic expressions based on a set of names for morphological objects together with their definitions within calculus. The calculus rules, leading to the relations $A = B$ are the same as for the natural numbers and the relations $A = B$ are interpreted at the same time as morphological deformation and as preservation of quantity, whatever meaning this may have.

2. The real line

The real line $\mathbb{R}$ is in mathematics defined as the set of all real numbers, represented as points on that line. It is hence an infinite point set and its cardinality c is called the continuum; it is larger than the cardinality $\aleph_0$ of the natural numbers.

The real line decomposes as

$$\mathbb{R} = \mathbb{R}_- \cup \{0\} \cup \mathbb{R}_+$$

with

$$\mathbb{R}_- =]-\infty, 0[: \text{ the half-line of negative numbers.}$$

$$\mathbb{R}_+ =]0, +\infty[: \text{ the half-line of positive numbers.}$$

So $\mathbb{R}_+$ and $\mathbb{R}_-$ are open intervals that are closed off and glued together by the point $\{0\}$ to form the real line.

Morphologically we write this disjoint union as

$$\mathbb{R} = \mathbb{R}_- + 1 + \mathbb{R}_+$$

whereby "1" represents the middle point $\{0\}$.

Both $\mathbb{R}_+$ and $\mathbb{R}_-$ are half-lines having "the same shape", so we identify $\mathbb{R}_- = \mathbb{R}_+$, leading to the first definition

$$\mathbb{R} = \mathbb{R}_+ + 1 + \mathbb{R}_+,$$

which, after commuting terms, leads to

$$\mathbb{R} = 2\mathbb{R}_+ + 1.$$

Next one may argue that all open intervals $]a, b[$ have "the same shape", so they are all copies of $\mathbb{R}$ and, in particular, we may identify

$$\mathbb{R}_+ = \mathbb{R},$$

leading to the relations

$$\mathbb{R} = \mathbb{R} + 1 + \mathbb{R} = 2\mathbb{R} + 1.$$

This may be interpreted as the way to produce an open interval or curve $]a, c[$ by taking an open interval $]a, b[$, glue to it a point $\{b\}$ and then glue to the next open interval or curve $]b, c[$.

The question now is: what is the quantity of $\mathbb{R}$?

If it is the cardinality "c" then one should identify $\mathbb{R} + 1$ with $\mathbb{R}$ but $\mathbb{R} + 1$ would be a semi-interval like $]0, 1]$, open from one side and closed from the other, which is not the same as $]0, 1[$.

Next, the relation $\mathbb{R} = \mathbb{R} + 1 + \mathbb{R}$ indicates the fact that $\mathbb{R}$ contains at least one point and, by iteration

$$\mathbb{R} = \mathbb{R} + 1 + \mathbb{R} = (\mathbb{R} + 1 + \mathbb{R}) + 1 + (\mathbb{R} + 1 + \mathbb{R}) = \cdots$$

we obtain 3 points, 7 points, 15 points etc., any finite number of points. So the morphological version of $\mathbb{R}$ seems to house infinity many points.

Now let us consider the relation

$$\mathbb{R} = \mathbb{R} + 1 + \mathbb{R} = 2\mathbb{R} + 1$$

as an equation. Then by subtracting $\mathbb{R}$ from both sides we get

$$\mathbb{R} + 1 = 0$$

and by subtracting 1 we get

$$\mathbb{R} = 0 - 1 = -1,$$

so that the total quantity of $\mathbb{R}$ should be -1.

This clearly conflicts with the idea of $\mathbb{R}$ being a set of points; the morphological line is hence not merely a set of points but rather a brand new object that doesn't quantify as a pointset. Of course one could argue that also

$$\text{infinity} = 2\text{infinity} + 1,$$

but infinity is a too trivial and vague number to work with for it absorbs everything.

There is an interesting interpretation for $\mathbb{R} = -1$.

Every manifold or surface of finite dimension may be represented by a cell complex, which we may represent by a polynomial

$$a_o\mathbb{R}^n + a_1\mathbb{R}^{n-1} + \cdots + a_n, \;\; a_0 \in \mathbb{N}, \; a_1, \ldots, a_n \in \mathbb{N} \cup \{0\}.$$

By making the identification $\mathbb{R} = -1$ we obtain

$$e\{M\} = a_0(-1)^n + a_1(-1)^{n-1} + \cdots + a_n,$$

which is the Euler characteristic $e\{M\}$ of manifold M.

The Euler number $e\{M\}$ is a topological invariant and for a given manifold M it is independent of the cell decomposition of that manifold. To see this, note that for any two cell decompositions of M there exists a kind cell decomposition that refines both of them and so it suffices to consider the case $M = \mathbb{R}^n$. Moreover, every cell decomposition of $\mathbb{R}^n$ may be obtained from simple cell decompositions of the form $\mathbb{R}^j = 2\mathbb{R}^j + \mathbb{R}^{j-1}$, which proves the invariance of $e\{M\}$ morphologically.

The fact that morphological calculus respects the Euler characteristic is like a corner stone (it is the final invariant that is preserved!). But as it is now, morphological calculus is reduced to the calculus of the integers $\mathbb{Z}$ and a point 1 is identified with a closed interval $\mathbb{R} + 2$, a plane $\mathbb{R}^2$ is identified with a point 1.

Hence, the idea of a line as an infinite point set is completely lost and also the dimension of an object is not preserved. As a result we have the identification $\mathbb{R} + 1 = 0$ between a semi-interval (or circle) $\mathbb{R} + 1$ and the number zero and, in fact $\mathbb{R} = -1$ between a line $\mathbb{R}$ and the number -1, while numbers 0 and -1 are no point sets and hence no objects.

To overcome this collapse of the notion of dimension we are going to introduce the following assumption.

Axiom 2.1. Morphological calculations are only granted if all the algebraic expressions and operations make sense in terms of geometrical objects.

Hence, in particular, number zero 0 and negative numbers $-1, -2$, etc. are hereby excluded or at least pushed to the background. Moreover, a relation like

$$\mathbb{R} = \mathbb{R} + 1 + \mathbb{R} = 2\mathbb{R} + 1$$

does not automatically allow one to solve it like an equation; one could also simply interpret it by stating that one is allowed to replace $\mathbb{R}$ by $2\mathbb{R}+1$ or vice-versa within calculations and nothing more. Hence, it does not automatically imply, e.g., that $\mathbb{R} + 1 = 0$ or even $\mathbb{R} - 1 \equiv 2\mathbb{R}$, although this last relation $\mathbb{R} - 1 = 2\mathbb{R}$ makes morphological sense. This now leads to the following result.

Theorem 2.2 (Morphological Stability). *Under the assumption of the relation* $\mathbb{R} = 2\mathbb{R} + 1$, *every cell complex* $a_0\mathbb{R}^n + a_1\mathbb{R}^{n-1} + \cdots + a_0$ *is equivalent to either*

$$a\mathbb{R}^n, \quad a \in \mathbb{N} \quad \textit{or} \quad \mathbb{R}^n + b\mathbb{R}^{n-1}, \quad b \in \mathbb{N},$$

no further identifications being possible.

Proof. The statement holds trivially for $n = 0$. For $n = 1$, $a > 1$ and $b > 0$ we clearly have

$$a\mathbb{R} + b = (a-2)\mathbb{R} + 2\mathbb{R} + (b-1) + 1 = (a-1)\mathbb{R} + b - 1$$

so we are reduced to either $b = 0$ or $a = 1$.

Next, assuming the property for $n - 1$, $n > 1$, we may reduce any cell complex to

$$a\mathbb{R}^n + b\mathbb{R}^{n-1} + c\mathbb{R}^{n-2}.$$

If $c = 0$ we may reason as in the case $n = 1$ to arrive at the final form $a\mathbb{R}^n$ or $\mathbb{R}^n + b\mathbb{R}^{n-1}$. If $c > 0$ we may write $a\mathbb{R}^n + b\mathbb{R}^{n-1} + c\mathbb{R}^{n-2} = (a-1)\mathbb{R}^n + (2\mathbb{R}^n + \mathbb{R}^{n-1}) + b\mathbb{R}^{n-1} + c\mathbb{R}^{n-2} = (a+1)\mathbb{R}^n + (b+1)\mathbb{R}^{n-1} + c\mathbb{R}^{n-2}$ and repeat this idea until $b > 1$. Then one may reduce using $2\mathbb{R}+1 = \mathbb{R} : a\mathbb{R}^n + b\mathbb{R}^{n-1} + c\mathbb{R}^n = a\mathbb{R}^n + (b-1)\mathbb{R}^{n-1} + (c-1)\mathbb{R}^{n-2}$ and so on, until the final form is reached. □

Note that this theorem guarantees us that in the worst case, at least the

$$\text{dimension} = n$$

as well as the

$$\text{Euler characteristic} = (-1)^n a \quad \text{or} \quad (-1)^n + (-1)^{n-1} b$$

are being preserved during morphological calculations; it is a second approximation for any possible notion of morphological quantity (the first one being just the Euler characteristic).

But this calculus is still too poor and to be able to evaluate the quantity of $\mathbb{R}$ we have to ignore the distinction between a line $\mathbb{R}$ and a half-line $\mathbb{R}_+$. This is again a dilemma, similar the once mentioned in introduction concerning commutativity and use of brackets. There are two options:

1. The “canonical” option. Hereby we assume as definition for $\mathbb{R}$ the relation
$$\mathbb{R} = \mathbb{R}_+ + 1 + \mathbb{R}_+ = 2\mathbb{R}_+ + 1$$
and consider the identification $\mathbb{R}_+ = \mathbb{R}$ as a form of decay. So the relation $\mathbb{R} = 2\mathbb{R} + 1$ is suspended in what we regard as “the canonical style”.
This style of calculating is on the other hand flexible with respect to commutativity and the use of brackets. Its main purpose is (not exclusively): Morphological analysis: to analyse geometrical objects (surfaces, manifolds) by decomposing them into parts (or other ways) and to express this knowledge in calculus language in order to arrive at morphological definitions.
2. The “formal” option. Hereby we consider morphological calculus as a formed language in which the order of terms in an addition and the use of brackets is not ignored. For the morphological line we have two definitions:
$$\mathbb{R} = \mathbb{R} + 1 + \mathbb{R} \text{ or } \mathbb{R} = \mathbb{R} + (\mathbb{R} + 1).$$
Its main purpose is (not exclusively):
Morphological synthesis: to construct a geometrical interpretation for an algebraic expression in morphological calculus.

In this paper we mostly use the canonical style. Our main interest is to study manifolds and try to understand their morphological quantity, whatever that may mean. The formal style will be discussed briefly in the last section.

3. Cartesian space, spheres, projective spaces

The Cartesian plane is defined as the product $\mathbb{R}^2 = \mathbb{R} \cdot \mathbb{R}$, using the relation $\mathbb{R} = 2\mathbb{R}_+ + 1$ we thus arrive at
$$\mathbb{R}^2 = (2\mathbb{R}_+ + 1)^2 = 4\mathbb{R}_+^2 + 4\mathbb{R}_+ + 1,$$
decomposing the plane into 4 quadrants $\mathbb{R}_+^2$, 4 half-planes $\mathbb{R}_+$ and one point 1 (the origin).

Similarly the Cartesian n-space is defined as product
$$\mathbb{R}^n = \mathbb{R} \cdot \cdots \cdot R = \mathbb{R} \cdot \mathbb{R}^{n-1}$$
and we have its decomposition into “octants”:
$$\mathbb{R}^n = (2\mathbb{R}_+ + 1)^n = \sum_{j=0}^{n} \binom{n}{j} 2^j \mathbb{R}_+^j.$$
To define the sphere S^{n-1} we make the following analysis: for a vector $\underline{x} \in \mathbb{R}^n$ with $\underline{x} \neq 0$ we have the polar decomposition $\underline{x} = r\underline{\omega}$, $r = |\underline{x}| \in \mathbb{R}_+$, $\underline{\omega} = \dfrac{\underline{x}}{|\underline{x}|} \in S^{n-1}$.

In morphological calculus language we write
$$\mathbb{R}^n - 1 = S^{n-1}\mathbb{R}_+,$$

leading to the morphological definition of S^{n-1}:

$$S^{n-1} = \frac{\mathbb{R}^n - 1}{\mathbb{R}_+}.$$

Now from $\mathbb{R} = 2\mathbb{R}_+ + 1$ we obtain that

$$\mathbb{R}_+ = \frac{\mathbb{R} - 1}{2},$$

a line without a point indeed gives 2 half-lines.

Hence we obtain a quantity formula for S^{n-1}:

$$S^{n-1} = 2\frac{\mathbb{R}^n - 1}{\mathbb{R} - 1} = 2\mathbb{R}^{n-1} + 2\mathbb{R}^{n-2} + \cdots + 2\mathbb{R} + 2.$$

In particular, a circle is given by
$S^1 = 2\mathbb{R} + 2$: two semi-circles and two points and a 2-sphere is given by

$$S^2 = 2\mathbb{R}^2 + 2\mathbb{R} + 2 = 2\mathbb{R}^2 + S^1,$$

two hemi-spheres and a circle (equator). Also
$S^2 = S^1\mathbb{R} + 2$, a "cylinder $S^1\mathbb{R}$" and two poles "2".

Using here $\mathbb{R} = 2\mathbb{R}_+ + 1$ we obtain

$$S^2 = 2\left(2\mathbb{R}_+ + 1\right)^2 + 2(2\mathbb{R}_+ + 1) + 2 = 8\mathbb{R}_+^2 + 12\mathbb{R}_+ + 6,$$

which may be interpreted as an octahedron whereby

$\mathbb{R}_+^2$ translates as a triangle,

$\mathbb{R}_+$ translates as a quarter circle or short interval.

Other regular polyhedra are harder to obtain, yet they are obtainable by transformations of the form $\mathbb{R} = 2\mathbb{R}+1$, $\mathbb{R}^2 = 2\mathbb{R}^2+\mathbb{R}$, which as we know are questionable. In fact, every cell complex $a\mathbb{R}^2 + b\mathbb{R} + c$ that corresponds to an embedded connected 2-manifold in $\mathbb{R}^3$ has Euler characteristic $a - b + c = 2(1 - g)$, g being the genus or number of holes, a number which characterizes the manifold. Hence, for 2-manifolds the relation $\mathbb{R} = 2\mathbb{R} + 1$ is not such a destructive deformation.

However, this also means that, e.g., a dodecahedron will be identified with $12\mathbb{R}^2 + 30\mathbb{R} + 20$ and hence a solid pentagon is identified with a square $\mathbb{R}^2$, an identification which is only topologically true.

For general spheres we have the recursion formula

$S^{n-1} = 2\mathbb{R}^{n-1} + S^{n-2}$: two hemi-spheres and an equator,

as well as the "polar coordinate" formula

$S^{n-1} = S^{n-2}\mathbb{R} + 2$: a cylinder and 2 poles.

These formula are special cases of the following general method for introducing polar coordinates on S^{n-1}.

Let $\underline{\omega} \in S^{n-1}$ and consider the decomposition $\mathbb{R}^n = \mathbb{R}^p \times \mathbb{R}^q$, $p+q = n$. Then we may write

$$\underline{\omega} = \cos\theta\underline{\omega}_1 + \sin\theta\underline{\omega}_2,\ \ \theta = \left[0, \frac{\pi}{2}\right], \underline{\omega}_1 \in S^{p-1},\ \ \underline{\omega}_2 \in S^{q-1}.$$

There are three cases:

$$\theta = 0 : \underline{\omega} = \underline{\omega}_1 \in S^{p-1}, \quad \theta = \frac{\pi}{2} : \underline{\omega} = \underline{\omega}_2 \in S^{q-1},$$

$$\theta \in \left]0, \frac{\pi}{2}\right[: \underline{\omega} = \cos\theta \underline{\omega}_1 + \sin\theta \underline{\omega}_2 \sim (\underline{\omega}_1, \underline{\omega}_2) \in S^{p-1} \times S^{q-1}.$$

In morphological calculus this situation is expressed as follows:

$$\mathbb{R}^n - 1 = (\mathbb{R}^p - 1)(\mathbb{R}^q - 1) + (\mathbb{R}^p - 1) + (\mathbb{R}^q - 1)$$

or

$$S^{n-1}\mathbb{R}^+ = \left(S^{p-1}\mathbb{R}_+\right)\left(S^{q-1}\mathbb{R}_+\right) + \left(S^{p-1}\mathbb{R}_+\right) + \left(S^{q-1}\mathbb{R}_+\right)$$

leading to the addition formula for spheres:

$$S^{n-1} = S^{p-1} \cdot S^{q-1} \cdot \mathbb{R}_+ + S^{p-1} + S^{q-1}.$$

Notice that also here $\mathbb{R}_+$ is interpreted as the quarter circle (small interval) $\theta \in \left]0, \frac{\pi}{2}\right[$, while the full line $\mathbb{R}$ would rather correspond to a semi-circle $\theta \in]0, \pi[$.

The addition formula also leads to:

$$\begin{aligned} S^{n-1} &= S^{p-1}\left(S^{q-1}\mathbb{R}_+ + 1\right) + S^{q-1} \\ &= S^{p-1}\mathbb{R}^q + S^{q-1}, \end{aligned}$$

which generalizes the recursion formula and the "polar coordinate" formula mentioned earlier.

Of particular interest is the odd-dimensional sphere S^{2n-1} where we can take $p = q = n$.

This leads to the "Hopf factorization formula"

$$S^{2n-1} = S^{n-1}\mathbb{R}^n + S^{n-1} \text{ or } S^{2n-1} = (\mathbb{R}^n + 1)\, S^{n-1}.$$

In particular we have the Hopf fibrations

$$S^3 = S^2 S^1, \quad S^7 = S^4 S^3$$

that are well known and follow from complex resp. quaternionic projective geometry. They can be seen as interpretations of the Hopf factorization

$$S^3 = \left(\mathbb{R}^2 + 1\right) S^1, \quad S^7 = \left(\mathbb{R}^4 + 1\right) S^3,$$

whereby the spheres S^2 resp. S^4 are identified with $\left(\mathbb{R}^2 + 1\right)$ resp. $\left(\mathbb{R}^4 + 1\right)$. But of course the Hopf fibrations are by no means proved or even implied by morphological calculus.

In general, the sphere S^{n-1} can be mapped onto $\mathbb{R}^{n-1}$ by stereographic projection. Hereby one takes line from the south pole $\underline{w} = (0, \ldots, 0, -1)$ to general point $\underline{w}$, denoted by $L(\underline{w})$ and the stereographic projection $\mathrm{st}(\underline{w})$ is the intersection of $L(\underline{w})$ with plane $x_n = 1$ (the tangent plane to the north pole $(0, \ldots, 0, +1)$).

This leads to the identification between S^{n-1} and $\mathbb{R}^{n-1} \cup \{\infty\}$. In morphological calculus one might hence think of identification $S^{n-1} = \mathbb{R}^{n-1} + 1$. But that would lead to the unwanted identification

$$2\mathbb{R}^2 + 2\mathbb{R} + 2 = \mathbb{R}^2 + 1,$$

that would again correspond to $\mathbb{R} = 2\mathbb{R} + 1$ via:

$$\mathbb{R}^2 + 1 = (2\mathbb{R} + 1)\mathbb{R} + 1 = 2\mathbb{R}^2 + \mathbb{R} + 1 = 2\mathbb{R}^2 + (2\mathbb{R} + 1) + 1 = 2\mathbb{R}^2 + 2\mathbb{R} + 2.$$

In morphological calculus we introduce a kind of stereographic sphere or "Poincaré sphere" by

$$\mathbb{S}^n = \mathbb{R}^n + 1,$$

leading to the recursion formula

$$\mathbb{S}^n = (2\mathbb{R}_+ + 1)\mathbb{R}^{n-1} + 1 = 2\mathbb{R}^{n-1}\mathbb{R}_+ + \mathbb{S}^{n-1}$$

and leading to the total quantity (whatever that may mean)

$$\mathbb{S}^n = 2\mathbb{R}^{n-1}\mathbb{R}_+ + 2\mathbb{R}^{n-2}\mathbb{R}_+ + \cdots + 2\mathbb{R}_+ + 2.$$

Notice hence that the identification $\mathbb{R}_+ = \mathbb{R}$ would lead to $\mathbb{S}^n = S^n$ or $S^2 = \mathbb{R}^2 + 1$ (a point and a square is a sphere). It is true that the only 2-manifold interpretation for $\mathbb{R}^2 + 1$ is indeed a sphere. Also the Poincaré polynomial of the sphere S^n is given by $t^n + 1$, which corresponds to $\mathbb{R}^n + 1$.

Recall that the Poincaré polynomial of a manifold M is defined as $a_n t^n + \cdots + a_0$ with $a_j = \dim H_j$, H_j leading the jth homology space of M.

It turns out that the Poincaré polynomial often appears as the morphological quantity of an object, in particular for $\mathbb{R}^n$ itself and the sphere $\mathbb{S}^n$. But for the sphere S^n we obtain a "higher" morphological quantity: $2\mathbb{R}^n + \cdots + 2\mathbb{R} + 2$ that does not correspond to the Poincaré polynomial. The real projective space $\mathbb{RP}^n$ corresponds to the set of $1D$ subspaces of $\mathbb{R}^{n+1}$, also defined as the set of vectors $(x_1, \ldots, x_{n+1}) \sim (\lambda x_1, \ldots, \lambda x_{n+1})$, $\lambda \neq 0$.

In mathematics we write it as the quotient structure

$$\mathbb{RP}^n = \frac{\mathbb{R}^{n+1}\backslash\{0\}}{\mathbb{R}\backslash\{0\}}.$$

This leads to the morphological definition

$$\mathbb{RP}^n = \frac{\mathbb{R}^{n+1} - 1}{\mathbb{R} - 1}.$$

and to the formula for the quantity of $\mathbb{RP}^n$:

$$\mathbb{RP}^n = \mathbb{R}^n + \mathbb{R}^{n-1} + \cdots + \mathbb{R} + 1.$$

In this case the quantity polynomial corresponds to the Poincaré polynomial for $\mathbb{RP}^n : t^n + \cdots + t + 1$. It also leads to the recursion formula

$$\mathbb{RP}^n = \mathbb{R}^n + \mathbb{RP}^{n-1}$$

in which "$\mathbb{R}^n$" symbolizes the Affine subspace consisting of the points $(x_1, \ldots, x_n, 1)$ while "$\mathbb{RP}^{n-1}$" stands for the plane at infinity: $x_{n+1} = 0$.

Of course we also have that

$$\mathbb{RP}^n = \frac{S^n}{S^0} = \frac{S^n}{2}$$

whereby S^0 in the multiplicative group $S^0 = \{-1, 1\}$.

In particular the projective line is given by

$$\mathbb{RP}^1 = \mathbb{R} + 1$$

symbolizing $\mathbb{R} \cup \{\infty\}$ and it also represents the Poincaré circle

$$\mathbb{S}^1 = \mathbb{R} + 1 = S^1/2,$$

$S^1 = 2\mathbb{R} + 2$ being the standard circle.

The projective plane is given by

$$\mathbb{RP}^2 = \mathbb{R}^2 + \mathbb{R} + 1 = \mathbb{R}^2 + \mathbb{RP}^1 = (\mathbb{R} + 1)\mathbb{R} + 1,$$

whereby the object $(\mathbb{R} + 1)\mathbb{R}$ in this context corresponds to a Moebius band.

Just seen by itself, $(\mathbb{R} + 1)\mathbb{R}$ could correspond to several things, including any line bundle over the circle $\mathbb{R} + 1$, i.e., either a cylinder or a Moebius band. In geometry the Moebius band can be recognized by cutting it in half along the center circle; if it was a cylinder, then the cut object would give 2 cylinders and if it was a Moebius band then the cut object would be a single cylinder. Now, this cutting procedure can be translated into morphological calculus as the subtraction

$$(\mathbb{R} + 1)\mathbb{R} - (\mathbb{R} + 1) = (\mathbb{R} + 1)(\mathbb{R} - 1) = \mathbb{R}^2 - 1$$

and $\mathbb{R}^2 - 1$ symbolizes a plane minus a point but also a single cylinder

$$\mathbb{R}^2 - 1 = S^1\mathbb{R}_+ = (2\mathbb{R} + 2)\mathbb{R}_+,$$

here represented as a product of a circle $2\mathbb{R} + 2$ (which has two glueing points and twice the length of the original circle) with a half-line $\mathbb{R}_+$ (stretching from the cutting point $\{0\}$ to the boundary $\{\infty\}$).

So this simple calculation symbolizes quite well the whole cutting experiment and it illustrates us the object $(\mathbb{R} + 1)\mathbb{R}$ as being a Moebius band. In general, morphological objects are merely organized quantities that can have a number of meanings called morphological synthesis. This synthesis takes place outside the calculus but it can be guided by calculations that give the object an intrinsic meaning. In the Moebius experiment we also see that the circle $2\mathbb{R} + 2$ and the Poincaré circle $\mathbb{R} + 1$ clearly play different roles like also the line $\mathbb{R}$ and the half-line $\mathbb{R}_+$.

If we apply a similar experiment to the cylinders

$$(2\mathbb{R} + 2)\mathbb{R} - (2\mathbb{R} + 2) = (2\mathbb{R} + 2)(\mathbb{R} - 1) = 2S^1\mathbb{R}_+$$

we obtain two cylinders. Of course one always calculates in a certain way and that may force a certain interpretation; the language of calculus can be used as an illustration but not as a real proof. In fact the language of calculus also has to remain flexible enough but this flexibility is at the cost of the stability of the morphological synthesis. For example we have

$$\mathbb{R}^2 - 1 = S^1\mathbb{R}_+ = (2\mathbb{R} + 2)\mathbb{R}_+ = 2(\mathbb{R} + 1)\mathbb{R}_+ = 2\mathbb{S}^1\mathbb{R}_+$$

showing that distributivity results in the cutting and reglueing of one cylinder $S^1\mathbb{R}_+$ into two cylinders $\mathbb{S}^1\mathbb{R}_+$, half the size and with one single cutting edge $\mathbb{R}_+$.

For the general projective space we have a kind of "Moebius factorization"

$$\mathbb{RP}^n = \mathbb{RP}^{n-1} \cdot \mathbb{R} + 1$$

whereby the Moebius cutting experiment is represented as

$$\mathbb{RP}^{n-1} \cdot \mathbb{R} - \mathbb{RP}^{n-1} = \mathbb{RP}^{n-1}(\mathbb{R} - 1) = \mathbb{R}^n - 1 = S^{n-1}\mathbb{R}_+,$$

also a kind of cylinder.

We now turn to complex projective spaces.

The complex numbers $\mathbb{C}$ are morphologically given by

$$\mathbb{C} = \mathbb{R}^2$$

and this is all. Anything concerning $\sqrt{-1} = i$ exists outside the calculus. We also have that

$$\mathbb{C}^n = \left(\mathbb{R}^2\right)^n = \mathbb{R}^{2n}.$$

Complex projective space $\mathbb{CP}^n$ is defined as the set of equivalence classes of relation

$$(z_1, \ldots, z_{n+1}) \in \mathbb{C}^{n+1} \setminus \{0\} \sim (\lambda z_1, \ldots, \lambda z_{n+1}), \quad \lambda \in \mathbb{C} \setminus \{0\},$$

i.e., the quotient structure

$$\mathbb{CP}^n = \mathbb{C}^{n+1} \setminus \{0\}/\mathbb{C} \setminus \{0\}.$$

Hence, in morphological calculus we have the definition

$$\mathbb{CP}^n = \frac{\mathbb{C}^{n+1} - 1}{\mathbb{C} - 1}$$

which immediately leads to the quantity

$$\mathbb{CP}^n = \mathbb{C}^n + \mathbb{C}^{n+1} + \cdots + 1 = \mathbb{R}^{2n} + \mathbb{R}^{2n-2} + \cdots + 1$$

that also corresponds to the Poincaré polynomial. The Euler number of $\mathbb{CP}^n$ equals n.

We also have that in real terms:

$$\mathbb{CP}^{n-1} = \frac{\mathbb{R}^{2n} - 1}{\mathbb{R}^2 - 1} = \frac{S^{2n-1}}{S^1},$$

leading to the $\mathbb{CP}^n$-factorization of S^{2n-1}

$$S^{2n-1} = \mathbb{CP}^{n-1} \cdot S^1,$$

which is the fibration obtained from the group structure $(z_1, \ldots, z_n) \in S^{2n-1} \to \left(e^{i\theta} z_1, \ldots, e^{i\theta} z_n\right)$.

In particular we have that

$$\mathbb{CP}^1 = \mathbb{C} + 1 = \mathbb{R}^2 + 1 = \mathbb{S}^2$$

and the above fibration leads to the first Hopf fibration

$$S^3 = \mathbb{S}^2 \cdot S^1.$$

Like in the real case one has the recursion formula

$$\mathbb{CP}^n = \mathbb{C}^n + \mathbb{CP}^{n-1}$$

and also the Moebius factorization

$$\mathbb{CP}^n = \mathbb{CP}^{n-1} \cdot \mathbb{C} + 1.$$

Hereby the complex line bundle $\mathbb{CP}^{n-1} \cdot \mathbb{C}$ reduces for $n = 2$ to

$$\mathbb{CP}^2 = \mathbb{S}^2 \cdot \mathbb{C} = (\mathbb{C} + 1)\mathbb{C}$$

and it is a non-trivial plane bundle over the 2-sphere.

In fact also here we have "Moebius cutting experiment"

$$\mathbb{CP}^1 \cdot \mathbb{C} - \mathbb{CP}^1 = (\mathbb{C} + 1)(\mathbb{C} - 1)) = \mathbb{C}^2 - 1 = \mathbb{R}^4 - 1 = S^3\mathbb{R}_+.$$

showing that fibration $\mathbb{S}^2(\mathbb{C} - 1)$ is non-trivial: $S^3\mathbb{R}_+$.

This remains true in general:

$$\mathbb{CP}^{n-1} \cdot \mathbb{C} - \mathbb{CP}^{n-1} = \mathbb{CP}^{n-1}(\mathbb{C} - 1) = \mathbb{C}^n - 1 = \mathbb{R}^{2n} - 1 = S^{2n-1} \cdot \mathbb{R}_+.$$

The above may be repeated for the quaternions; we present the morphological headlines:

We have

$$\mathbb{H} = \mathbb{R}^4, \quad \mathbb{H}^n = (\mathbb{R}^4)^n = \mathbb{R}^{4n},$$

$$\begin{aligned} \mathbb{HP}^n &= \frac{\mathbb{H}^{n+1} - 1}{\mathbb{H} - 1} = \mathbb{H}^n + \mathbb{H}^{n-1} + \cdots + 1 \\ &= \mathbb{R}^{4n} + \mathbb{R}^{4n-4} + \cdots + 1 = \mathbb{H}^n + \mathbb{HP}^{n-1}. \end{aligned}$$

Also

$$\mathbb{HP}^{n-1} = \frac{\mathbb{R}^{4n} - 1}{\mathbb{R}^4 - 1} = \frac{S^{4n-1}}{S^3}$$

leading to the $\mathbb{HP}^n$-factorization (fibration)

$$S^{4n-1} = \mathbb{HP}^{n-1} \cdot S^3,$$

which in particular for $n = 2$ leads to the second Hopf fibration

$$S^7 = \mathbb{HP}^1 \cdot S^3 = (\mathbb{H} + 1)S^3 = (\mathbb{R}^4 + 1)S^3 = \mathbb{S}^4 S^3.$$

The Moebius factorization is given by

$$\mathbb{HP}^n - 1 = \mathbb{HP}^{n-1} \cdot \mathbb{H}$$

while we also have the Moebius cutting experiment:

$$\mathbb{HP}^{n-1} \cdot \mathbb{H} - \mathbb{HP}^{n-1} = \mathbb{HP}^{n-1} \cdot (\mathbb{H} - 1) = \mathbb{H}^n - 1 = \mathbb{R}^{4n} - 1 = S^{4n-1} \cdot \mathbb{R}_+.$$

But not every interesting quotient in calculus leads to a morphological synthesis that produces a nice manifold. Yet these quotients are also interesting because they say a lot about the meaning of morphological calculus and we call them "phantom geometrical objects".

Example ($\mathbb{RP}_h^{2n}$).

$$\mathbb{RP}_h^{2n} = \frac{\mathbb{R}^{2n+1} + 1}{\mathbb{R} + 1} = \mathbb{R}^{2n} - \mathbb{R}^{2n-1} + \cdots - \mathbb{R} + 1,$$

which we call the phantom (real) projective space of dimension $2n$.

The simplest case is

$$\mathbb{RP}_h^2 = \mathbb{R}^2 - \mathbb{R} + 1,$$

with Euler characteristic 3. This would be one too high for a connected 2-manifold and $\mathbb{R}^2 - \mathbb{R} + 1$ corresponds to: take plane $\mathbb{R}^2$, delete line $\mathbb{R}$ and add point 1; it makes sense as a weird object but not as a 2-manifold.

In fact one could say

$$\mathbb{R}^2 - \mathbb{R} + 1 = (2\mathbb{R}_+ + 1)\mathbb{R} - \mathbb{R} + 1 = 2\mathbb{R}_+\mathbb{R} + 1,$$

two half-planes (or half-discs or triangles) glued together by a single point (a butterfly).

Note that we also have that

$$\mathbb{RP}_h^2 = \frac{\mathbb{S}^3}{\mathbb{S}^1}.$$

If we would now use $\mathbb{R} = 2\mathbb{R} + 1$ we could make the identification $\mathbb{S}^3 = S^3$, $\mathbb{S}^1 = S^1$ and arrive at

$$\frac{\mathbb{S}^3}{\mathbb{S}^1} = \frac{S^3}{S^1} = \mathbb{S}^2 = \mathbb{R}^2 + 1 \text{ (Hopf fibration)}$$

and therefore

$$\mathbb{R}^2 - \mathbb{R} + 1 = \mathbb{R}^2 + 1.$$

This is total nonsense because this identification is even wrong on the level of Euler numbers: $3 = 2$.

The reason why such bad identification happens is because the Euler numbers of $\mathbb{S}^3, S^3, \mathbb{S}^1, S^1$ are all equal to zero, so, on the level of Euler numbers:

$$\frac{S^3}{S^1} = \frac{0}{0} \;\&\; \frac{\mathbb{S}^3}{\mathbb{S}^1} = \frac{0}{0},$$

so one would not even be allowed to consider the quotients S^3/S^1, $\mathbb{S}^3/\mathbb{S}^1$. But that would also exclude $\mathbb{CP}^n$ from the picture as well as the Hopf fibration, an unpermitable exclusion. This is a sound reason why the relations $\mathbb{R} = 2\mathbb{R} + 1$ or $\mathbb{R}_+ = \mathbb{R}$ or $\mathbb{S}^n = S^n$ must be forbidden: they simply spoil the calculus.

The general phantom projective space

$$\mathbb{RP}_h^{2n} = \mathbb{R}^{2n} - \mathbb{R}^{2n-1} + \cdots - \mathbb{R} + 1$$

surely makes sense as a geometrical object, but the corresponding quantity $\mathbb{R}^{2n} - \mathbb{R}^{2n-1} + \cdots - \mathbb{R} + 1$ still has negative numbers as coefficients, so it is not yet fully evaluated. This can be done by replacing $\mathbb{R} = 2\mathbb{R}_+ + 1$ at suitable places, giving rise to

$$\mathbb{R}^{2n} - \mathbb{R}^{2n-1} + \cdots - \mathbb{R} + 1 = 2\mathbb{R}_+\mathbb{R}^{2n-1} + 2\mathbb{R}_+\mathbb{R}^{2n-3} + \cdots + 2\mathbb{R}_+\mathbb{R} + 1,$$

which also provides a synthesis for $\mathbb{RP}_h^{2n}$. Comparing $\mathbb{R}^{2n} - \mathbb{R}^{2n-1} + \cdots - \mathbb{R} + 1$ with the Poincaré polynomial also suggests that some of the homology spaces of $\mathbb{RP}_h^{2n}$ would have negative dimension. But we also have that phenomena with the object

$$\mathbb{R}^2 - 1 = (\mathbb{R} - 1)(\mathbb{R} + 1) = 2\mathbb{R}_+\mathbb{R} + 2\mathbb{R}_+.$$

Quantity simply doesn't always have a positive evaluation as an addition of powers $\mathbb{R}^s$. This leads to

Definition 3.1. A morphological object is called integrable if it has an evaluation of the form

$$\mathbb{F} = a_0\mathbb{R}^n + a_1\mathbb{R}^{n-1} + \cdots + a_n,\ a_0 \in \mathbb{N},\ a_1, \ldots, a_n \in \mathbb{N} \cup \{0\};$$

this polynomial is then called the "total quantity" or "integral". The object $\mathbb{F}$ is called semi-integrable if it has an evaluation as an addition of terms of the form $\mathbb{R}_+^j\mathbb{R}^k$. Such an expression is not unique unless we require the power "j" of $\mathbb{R}_+$ to be minimal, in which case the obtained expression is also called "total quantity" or "integral".

Note that not every object is semi-integrable; for example $F = \mathbb{R} - 2$ is an object and hence it has certain hidden quantity, but it cannot be evaluated as an addition in terms of $\mathbb{R}$ and $\mathbb{R}_+$. One option would be to introduce new type of line, e.g.,

$$\mathbb{R}_+ = 2\mathbb{R}_{++} + 1,$$

but that would not lead to be more interesting calculus.

Notice that the phantom projective plane can also be interpreted as the result of the cutting experiment

$$\begin{aligned}\mathbb{RP}_h^{2n} &= \mathbb{R}^{2n} - \mathbb{R}^{2n-1} + \cdots - \mathbb{R} + 1\\ &= \left(\mathbb{R}^{2n} + \mathbb{R}^{2n-2} + \cdots + 1\right) - \left(\mathbb{R}^{2n-2} + \cdots + 1\right)\mathbb{R}\\ &= \mathbb{CP}^n - \mathbb{CP}^{n-1} \cdot \mathbb{R}.\end{aligned}$$

We also have the phantom Moebius strip

$$\mathbb{RP}_h^{2n} - 1 = (\mathbb{R} - 1)\mathbb{CP}^{n-1} \cdot \mathbb{R}$$

and this time we have a Moebius "glueing experiment"

$$\begin{aligned}&(\mathbb{R} - 1)\mathbb{CP}^{n-1} \cdot \mathbb{R} + (\mathbb{R} - 1)\mathbb{CP}^{n-1}\\ &\quad = \mathbb{CP}^{n-1}(\mathbb{R}^2 - 1) = \mathbb{R}^{2n} - 1 = S^{2n-1} \cdot \mathbb{R}_+,\end{aligned}$$

the same cylinder as we had earlier on.

Of course one may also consider complex and quaternionic phantom projective spaces:

$$\begin{aligned}\mathbb{CP}_h^{2n} &= \frac{\mathbb{C}^{2n+1} + 1}{\mathbb{C} + 1} = \mathbb{C}^{2n} - \mathbb{C}^{2n-1} + \cdots - \mathbb{C} + 1 = \cdots,\\ \mathbb{HP}_h^{2n} &= \frac{\mathbb{H}^{2n+1} + 1}{\mathbb{H} + 1} = \mathbb{H}^{2n} - \mathbb{H}^{2n-1} + \cdots - \mathbb{H} + 1 = \cdots.\end{aligned}$$

The fact that the corresponding synthesis for phantom projective spaces does not add up to a manifold implies that these quotients do not correspond to a group action (or else the quotients would be homogeneous spaces). Indeed, the denominators in the definition of the projective spaces are the multiplicative groups $\mathbb{R} - 1, \mathbb{C} - 1, \mathbb{H} - 1$ while for the phantom spaces we have the spheres $\mathbb{S}^1 = \mathbb{R} + 1$,

$\mathbb{S}^2 = \mathbb{C}+1, \mathbb{S}^3 = \mathbb{H}+1$ which are non-groups leading to non-group actions. In fact group actions can not be recognized within morphological calculus itself, only by the outside interpretations. The consideration of phantom geometry also leads to the next definition.

Definition 3.2. A morphological object is said to be of "integer type" if it has an evaluation of the form

$$\mathbb{F} = a_0\mathbb{R}^n + a_1\mathbb{R}^{n-1} + \cdots + a_n,\ a_0 \in \mathbb{N},\ a_1, \ldots, a_n \in \mathbb{Z}.$$

A semi-integrable object that is not of integer type is to said to be of "half-integer type". Other objects are "just another type".

Notice that $\mathbb{R}-2$ is of integer type but not semi-integrable while the building blocks $\mathbb{R}_+^j\mathbb{R}^k$, $j > 0$ are semi integrable but not integer type: they are half-integer type.

The cylinder $(2\mathbb{R}+2)\mathbb{R}_+ = \mathbb{R}^2 - 1$ is clearly of integer type but only semi-integrable while the small cylinder $(\mathbb{R}+1)\mathbb{R}_+$ is only of half-integer type. This example confirms that it is a good idea to keep two circles $S^1 = 2\mathbb{R}+2$, $\mathbb{S}^1 = \mathbb{R}+1$ in use rather than deciding that $2\mathbb{R}+2 = 2(\mathbb{R}+1)$ is always a pair of circles. Note that the object $\mathbb{R}_+ - 1$ is just another type while $1 - \mathbb{R}$ isn't even an object. So we have a kind of hierarchy that is quantity based.

Example (Phantom fibrations). We already discussed the Hopf factorization

$$S^{2n-1} = (\mathbb{R}^n + 1)S^{n-1}$$

which only for $n = 2$ and $n = 4$ leads to a true fibration: the Hopf fibrations

$$S^3 = \mathbb{S}^2 S^1, \quad S^7 = \mathbb{S}^4 S^3.$$

These fibrations in fact correspond to projective geometry and the factors S^1 and S^3 are group actions.

In the other cases like, e.g.,

$$S^5 = \mathbb{S}^3 S^2$$

we don't have this. However also the product

$$\mathbb{S}^3 S^2 = (\mathbb{R}^3 + 1)S^2 = \mathbb{R}^3 S^2 + S^2$$

does lead to a synthesis of S^5 and it is like a fibration still, but an irregular fibration that would not locally correspond to a Cartesian product, whence the name "phantom fibration".

For the spheres S^{2^n-1} we also have repeated factorizations

$$\begin{aligned} S^7 &= \left(\mathbb{R}^4+1\right) S^3 = \left(\mathbb{R}^4+1\right)\left(\mathbb{R}^2+1\right)(\mathbb{R}+1)\,2, \\ S^{15} &= \left(\mathbb{R}^8+1\right)\left(\mathbb{R}^4+1\right)\left(\mathbb{R}^2+1\right)(\mathbb{R}+1)\,2, \end{aligned}$$

and so on. If we apply non-associativity we get

$$\begin{aligned} S^7 &= \left(\left(\mathbb{R}^4+1\right)\left(\mathbb{R}^2+1\right)\right) S^1 = \mathbb{CP}^3 S^1, \\ S^7 &= \left(\mathbb{R}^4+1\right)\left(\left(\mathbb{R}^2+1\right)(\mathbb{R}+1)\,2\right) = \left(\mathbb{R}^4+1\right) S^3 = \mathbb{S}^4 S^3, \end{aligned}$$

two fibrations of S^7 that follow from complex and quaternion geometry and that are unrelated.

There are also more Hopf factorizations, the simplest one being

$$S^8 = \left(\mathbb{R}^6 + \mathbb{R}^3 + 1\right) S^2.$$

In general they follow from products of the form

$$\left(\mathbb{R}^{s\cdot k} + \mathbb{R}^{(s-1)\cdot k} + \cdots + \mathbb{R}^k + 1\right)\left(2\mathbb{R}^{k-1} + 2\mathbb{R}^{k-2} + \cdots + 2\right)$$

leading to

$$S^{(s+1)k-1} = \left(\mathbb{R}^{s\cdot k} + \cdots + \mathbb{R}^k + 1\right) S^{k-1}$$

and they play a crucial role in the "Graßmann division problem".

Needless to say that there are repeated factorizations of this type.

Also the addition formula for spheres may be generalized.

For $p + q + r = m$ we have

$$\begin{aligned}\mathbb{R}^m - 1 = {} & (\mathbb{R}^p - 1)(\mathbb{R}^q - 1)(\mathbb{R}^r - 1) + (\mathbb{R}^p - 1)(\mathbb{R}^q - 1)\\ & + (\mathbb{R}^p - 1)(\mathbb{R}^r - 1) + (\mathbb{R}^q - 1)(\mathbb{R}^r - 1) + (\mathbb{R}^p - 1)\\ & + (\mathbb{R}^q - 1) + (\mathbb{R}^r - 1),\end{aligned}$$

from which we obtain:

$$\begin{aligned}S^{m-1} = {} & S^{p-1}S^{q-1}S^{r-1}\mathbb{R}_+^2 + S^{p-1}S^{q-1}\mathbb{R}_+\\ & + S^{p-1}S^{r-1}\mathbb{R}_+ + S^{q-1}S^{r-1}\mathbb{R}_+ + S^{p-1} + S^{q-1} + S^{r-1}.\end{aligned}$$

Needless to say also that our list of interesting manifolds and geometries is far from complete.

Let us take the Klein bottle as an example, we have the following morphological analysis. A Klein bottle can be obtained from a Moebius band by properly glueing a circle to the edge, thus closing it up into a compact 2-manifold. As we know, a Moebius band may be obtained by removing a point from $\mathbb{RP}^2 : \mathbb{RP}^2 - 1$. Then one blow up the hole to a small disc and one glues a circle $S^1 = 2\mathbb{R} + 2$ to that, giving 2-manifold with boundary. Finally one identifies every point on this S^1 with its anti-podal point: $S^1/\mathbb{Z}_2$ which leads to a continuation across the boundary and to the Klein bottle. In morphological language we have:

$$\begin{aligned}\left(\mathbb{RP}^2 - 1\right) + S^1/\mathbb{Z}_2 &= \left(\mathbb{RP}^2 - 1\right) + (\mathbb{R} + 1)\\ &= \left(\left(\mathbb{R}^2 + \mathbb{R} + 1\right) - 1\right) + (\mathbb{R} + 1)\\ &= \left(\mathbb{R}^2 + \mathbb{R}\right) + (\mathbb{R} + 1)\\ &= (\mathbb{R} + 1)\,\mathbb{R} + (\mathbb{R} + 1) = (\mathbb{R} + 1)(\mathbb{R} + 1),\end{aligned}$$

so we end up with a circle $\mathbb{S}^1$-bundle over $\mathbb{S}^1$. But $\mathbb{S}^1 \cdot \mathbb{S}^1$ may also simply represent a torus: there is no way one can tell from the quantity $(\mathbb{R} + 1)^2$ alone whether this represents a torus or a Klein bottle. Only in the initial formula $\left(\mathbb{RP}^2 - 1\right) + (\mathbb{R} + 1)$ one can specify a Klein bottle but as one starts calculating, this specification is lost.

Higher-dimensional Klein bottles may be introduced as the "blow up" experiment:

$$(\mathbb{RP}^n - 1) + S^{n-1}/\mathbb{Z}_2 = (\mathbb{RP}^n - 1) + \mathbb{RP}^{n-1}$$
$$= \mathbb{RP}^{n-1} \cdot \mathbb{R} + \mathbb{RP}^{n-1} = \mathbb{RP}^{n-1} \cdot \mathbb{S}^1,$$

an $\mathbb{S}^1$-bundle over $\mathbb{RP}^{n-1}$.

Similarly, complex and quaternionic Klein bottles may be introduced as (exercise) the "blow up experiment":

$$(\mathbb{CP}^n - 1) + S^{2n-1}/S^1 = \mathbb{CP}^{n-1} \cdot (\mathbb{C} + 1),$$
$$(\mathbb{HP}^n - 1) + S^{4n-1}/S^3 = \mathbb{HP}^{n-1} \cdot (\mathbb{H} + 1).$$

To summarize this section, we notice that there is no one to one correspondence between morphological calculus and geometry. This may be seen as a drawback but it is also a stronghold because it means that there exists another perspective that reveals a hidden aspect of geometry: the quantity of an object.

4. Groups and homogeneous spaces

Groups enter morphological calculus via a proper morphological analysis; the group structure will be lost and the organized quantity remains.

We begin with the groups

$$O(n),\ SO(n),\ GL(n,\mathbb{R}),\ GL(n,\mathbb{R}),\ SL(n,\mathbb{R}).$$

The orthogonal group $O(n)$ is the group of all orthogonal matrices (a_{ij}). If we represent such a matrix as a row $(\underline{a}_1, \dots, \underline{a}_n)$ of column vectors it simply means that $\underline{a}_1, \dots, \underline{a}_n$ are orthogonal unit vectors. This means that one can start off by choosing

$$\underline{a}_1 \in S^{n-1}$$

followed by choosing

$$\underline{a}_2 \in S^{n-1} \cap \{\lambda \underline{a}_1,\ \lambda \in \mathbb{R}\}^{\perp} = S^{n-2}$$

and then

$$\underline{a}_3 \in S^{n-1} \cap \{\lambda_1 \underline{a}_1 + \lambda_2 \underline{a}_2,\ \lambda_j \in \mathbb{R}\}^{\perp} = S^{n-3}$$

and so on, until for $\underline{a}_n$ there are just 2 choices

$$\underline{a}_n \in S^{n-1} \cap \operatorname{span}\{\underline{a}_1, \dots, \underline{a}_{n-1}\}^{\perp} = S^0.$$

This immediately leads to the morphological definition

$$O(n) = S^{n-1} \cdot S^{n-2} \cdots S^0,$$

as well as to the recursion formula

$$O(n) = S^{n-1} \cdot O(n-1),\ \ O(0) = 1.$$

For the group $SO(n)$ everything remains the same except that for the last vector $\underline{a}_n$ there is just one choice, determined by $\det(a_{ij}) = 1$ condition. We thus have the definition

$$SO(n) = S^{n-1}S^{n-2}\cdots S^1 = O(n)/\mathbb{Z}_2.$$

Clearly $O(n), SO(n)$ are integrable and the integral is obtained by substituting $S^{j-1} = 2\mathbb{R}^{j-1} + \cdots + 2$ and working out the product.

The general linear group $GL(n,\mathbb{R})$ is obtained similarly by writing the matrix (a_{ij}) as $(\underline{a}_1, \ldots, \underline{a}_n)$ whereby

$$\begin{aligned}
&\underline{a}_1 \in \mathbb{R}^n \setminus \{0\},\\
&\underline{a}_2 \in \mathbb{R}^n \setminus \operatorname{span}\{\underline{a}_1\},\\
&\vdots\\
&\underline{a}_n \in \mathbb{R}^n \setminus \operatorname{span}\{\underline{a}_1, \ldots, \underline{a}_{n-1}\}
\end{aligned}$$

which leads to the morphological definition

$$GL(n,\mathbb{R}) = (\mathbb{R}^n - 1)(\mathbb{R}^n - \mathbb{R})\cdots(\mathbb{R}^n - \mathbb{R}^{n-1}).$$

We readily obtain the quotient formula

$$\begin{aligned}
\frac{GL(n,\mathbb{R})}{O(n)} &= \frac{(\mathbb{R}^n - 1)(\mathbb{R}^n - \mathbb{R})\cdots(\mathbb{R}^n - \mathbb{R}^{n-1})}{S^{n-1}\cdot S^{n-2}\cdots S^0}\\
&= \left(\frac{\mathbb{R}^n - 1}{S^{n-1}}\right)\left(\frac{\mathbb{R}^n - \mathbb{R}}{S^{n-2}}\right)\cdots\frac{(\mathbb{R}^n - \mathbb{R}^{n-1})}{S^0}\\
&= \mathbb{R}_+ \cdot (\mathbb{R}\cdot\mathbb{R}_+)\cdots(\mathbb{R}^{n-1}\cdot\mathbb{R}_+),
\end{aligned}$$

which symbolizes the GRAMM–SCHMIDT orthogonalization procedure. Here we applied commutativity of the product but that doesn't matter too much; in fact one can also write

$$GL(n,\mathbb{R}) = (S^{n-1}\mathbb{R}_+)(\mathbb{R}S^{n-2}\mathbb{R}_+)\cdots(\mathbb{R}^{n-1}S^0\mathbb{R}_+).$$

For the group $SL(n,\mathbb{R})$ we have the extra condition $\det(a_{ij}) = 1$, which readily leads to

$$SL(n,\mathbb{R}) = \frac{GL(n,\mathbb{R})}{\mathbb{R} - 1},$$

and so also

$$\frac{SL(n,\mathbb{R})}{SO(n)} = \mathbb{R}_+ \cdot (\mathbb{R}\cdot\mathbb{R}_+)\cdots(\mathbb{R}^{n-2}\cdot\mathbb{R}_+)\,\mathbb{R}^{n-1}.$$

Now let us look some homogeneous spaces.

The Stiefel manifold $V_{n,k}(\mathbb{R})$ is by definition the manifold of orthonormal k-frames $(\underline{v}_1, \ldots, \underline{v}_k)$ in $\mathbb{R}^n$. We hence have that for $k < n$:

$$V_{n,k}(\mathbb{R}) = \frac{SO(n)}{SO(n-k)} = S^{n-1}\cdots S^{n-k} = \frac{O(n)}{O(n-k)}.$$

The Stiefel manifold $V_{n,k}(\mathbb{R})$ is the manifold of k-frames $(\underline{v}_1, \dots, \underline{v}_k)$ that are linearly independent and hence span a k-plane. We have for $k < n$:

$$VL_{n,k}(\mathbb{R}) = \frac{GL(n,\mathbb{R})}{\mathbb{R}^{n-k} \cdot GL(n-k,\mathbb{R})} = (\mathbb{R}^n - 1)(\mathbb{R}^n - \mathbb{R}) \cdots \left(\mathbb{R}^n - \mathbb{R}^{k-1}\right).$$

The Graßmann manifold $G_{n,k}(\mathbb{R})$ is the manifold of k-dimensional subspaces of $\mathbb{R}^n$. Now, each k-dimensional subspace has an orthogonal frame and that can be chosen in $O(k)$ in different ways. This leads to the combinatorial formula:

$$G_{n,k}(\mathbb{R}) = \frac{VL_{n,k}(\mathbb{R})}{O(k)} = \frac{O(n)}{O(k) \cdot O(n-k)} = \frac{S^{n-1} \cdot S^{n-2} \dots S^{n-k}}{S^{k-1} \dots S^0}.$$

The Graßmann manifold may also be constructed starting from the general linear group:

$$G_{n,k}(\mathbb{R}) = \frac{V_{n,k}(\mathbb{R})}{GL(k,\mathbb{R})} = \frac{(\mathbb{R}^n - 1)(\mathbb{R}^n - \mathbb{R}) \cdots \left(\mathbb{R}^n - \mathbb{R}^{k-1}\right)}{(\mathbb{R}^k - 1) \cdots (\mathbb{R}^k - \mathbb{R}^{k-1})}$$

and the equivalence of both definitions readily follows from the Gramm–Schmidt factorization.

By $\widetilde{G_{n,k}}(\mathbb{R})$ we denote the manifold of all ORIENTED k-dimensional subspaces of $\mathbb{R}^n$, i.e.,

$$\widetilde{G_{n,k}}(\mathbb{R}) = \frac{V_{n,k}(\mathbb{R})}{SO(k)} = \frac{SO(n)}{SO(k) \cdot SO(n-k)} = \frac{S^{n-1} \cdot S^{n-2} \dots S^{n-k}}{S^{k-1} \dots S^1}.$$

Now, for the Stiefel manifolds everything is clear, but for the Graßmann manifolds we have one major problem.

Problem 4.1 (Graßmann division problem). Can one work out the polynomial division $\frac{S^{n-1} \cdot S^{n-2} \dots S^{n-k}}{S^{k-1} \dots S^0}$, and does it result in an integral (polynomial in "$\mathbb{R}$" with natural number coefficients).

To solve the problem we will work with the quotient $VL_{n,k}(\mathbb{R})/GL(k,\mathbb{R})$ that is equivalent and easier to work with. For the case of simplicity take $k = 3$. Every 3D-subspace V of $\mathbb{R}^n$ is spanned by 3 linearly independent vectors $\underline{v}_1, \underline{v}_2, \underline{v}_3$ that may be chosen in $GL(3,\mathbb{R})$ different ways. For each V there is a unique triple $(\underline{v}_1, \underline{v}_2, \underline{v}_3)$ that may be written as a matrix of the form

$$\begin{bmatrix} \underline{v}_1 \\ \underline{v}_2 \\ \underline{v}_3 \end{bmatrix} = \begin{bmatrix} c_{11} & \cdots & c_{1j_3} & 0 & c_{1j_3+2} & \cdots & c_{1j_2} & 0 & c_{1j_2+2} & \cdots & c_{1j_1} & 1 & 0 & \cdots & 0 \\ c_{21} & \cdots & c_{2j_3} & 0 & c_{2j_3+2} & \cdots & c_{2j_2} & 1 & 0 & \cdots & 0 & 0 & 0 & \cdots & 0 \\ c_{31} & \cdots & c_{3j_3} & 1 & 0 & \cdots & 0 & 0 & 0 & \cdots & 0 & 0 & 0 & \cdots & 0 \end{bmatrix}$$

and any other frame in V may be obtained by a unique $GL(3,\mathbb{R})$-action from this, so in fact the division is carried out by looking to matrices of the above special form. As the coefficients c_{ij} vary the matrices of the above form constitute a cell of $G_{n,3}(\mathbb{R})$ that is a copy of a certain $\mathbb{R}^j$ and it is called a Schubert cell. We thus have proved the following results.

Theorem 4.2 (Schubert cells). *The object* $G_{n,k}(\mathbb{R}) = \mathbb{R}^d + c_1\mathbb{R}^{d-1} + \cdots + c_d$ *whereby* $c : j \in \mathbb{N}$ *is the number of Schubert cells of dimension* $d - j$.

Apart from this there are typical morphological questions such as:

Q1: To decompose $G_{n,k}(\mathbb{R}) = O_1 \cdots O_s$ as a (e.g., maximal) product of morphological objects O_j of integer type (that are irreducible, e.g.).

Q2: To look for $G_{n,k}(\mathbb{R})$-factorization $O_1 \cdots O_t$ in terms of objects O_j that are integrable.

Let us consider a few examples of such Graßmann factorizations.

Of course we readily have $G_{n,k}(\mathbb{R}) = \mathbb{RP}^{n-1}$ and the Hopf factorizations provide further ways of factorizing this.

Next for $k = 2$ we have:

$$G_{2n,2}(\mathbb{R}) = \frac{S^{2n-1} \cdot S^{2n-2}}{S^1 \cdot S^0} = \mathbb{CP}^{n-1} \cdot \mathbb{RP}^{2n-2}$$
$$G_{2n+1,2}(\mathbb{R}) = \frac{S^{2n} \cdot S^{2n-1}}{S^0 \cdot S^1} = \mathbb{RP}^{2n} \cdot \mathbb{CP}^{n-1},$$

showing a clear 2-periodicity.

For $k = 3$ the first interesting case is

$$G_{6,3}(\mathbb{R}) = \frac{S^5 \cdot S^4 \cdot S^3}{S^2 \cdot S^1 \cdot S^0},$$

which, using the Hopf factorizations

$$S^5 = \left(\mathbb{R}^3 + 1\right) S^2 = \mathbb{S}^3 S^2, \quad S^3 = \left(\mathbb{R}^2 + 1\right) S^1 = \mathbb{S}^2 S^1$$

may be evaluated as

$$G_{6,3}(\mathbb{R}) = \left(\mathbb{R}^3 + 1\right) \cdot \mathbb{RP}^4 \cdot \left(\mathbb{R}^2 + 1\right) = \mathbb{RP}^4 \cdot \mathbb{S}^3 \cdot \mathbb{S}^2.$$

Note here that it is forbidden to divide $\mathbb{S}^2 / S^2 = 1$.

More interesting still is the next case

$$G_{7,3}(\mathbb{R}) = \frac{S^6 \cdot S^5 \cdot S^4}{S^2 \cdot S^1 \cdot S^0},$$

which, using the Hopf factorization $S^5 = (\mathbb{R}^3 + 1)S^2$ yields.

$$G_{7,3}(\mathbb{R}) = \frac{\mathbb{RP}^6 \cdot \left(\mathbb{R}^3 + 1\right) \mathbb{RP}^4}{(\mathbb{R} + 1)}.$$

Now $\mathbb{RP}^4$ and $\mathbb{RP}^6$ cannot be divided by $(\mathbb{R} + 1)$; in fact these objects are irreducible in morphological sense. So, the division that works here is:

$$\mathbb{RP}^2_h = \frac{\mathbb{R}^3 + 1}{\mathbb{R} + 1} = \mathbb{R}^2 - \mathbb{R} + 1,$$

the phantom projective plane, leading to the following maximal factorization

$$G_{7,3}(\mathbb{R}) = \mathbb{RP}^6 \cdot \mathbb{RP}^4 \cdot \mathbb{RP}^2_h$$

in terms of irreducible objects of integer type.

But now the factors are no longer integrable, which also shows that the integrability of Graßmann manifolds is in fact not so trivial. But we have:

$$\frac{(\mathbb{R}^3+1)}{\mathbb{R}+1}\mathbb{RP}^4 = \frac{(\mathbb{R}^3+1)}{\mathbb{R}+1}\left(\mathbb{R}^2\frac{(\mathbb{R}^3-1)}{\mathbb{R}-1}+(\mathbb{R}+1)\right)$$

$$= \mathbb{R}^2\frac{(\mathbb{R}^6-1)}{\mathbb{R}^2-1}+(\mathbb{R}^3+1) = \mathbb{CP}^2\cdot\mathbb{R}^2+\mathbb{S}^3,$$

so that in fact we have integrable factorization

$$G_{7,3} = \mathbb{RP}^6\cdot(\mathbb{CP}^2\cdot\mathbb{R}^2+\mathbb{S}^3).$$

The next case is again simpler:

$$G_{8,3}(\mathbb{R}) = \frac{S^7\cdot S^6\cdot S^5}{S^2\cdot S^1\cdot S^0} = (\mathbb{CP}^3\cdot\mathbb{RP}^6)(\mathbb{R}^3+1).$$

For the next case

$$G_{9,3}(\mathbb{R}) = \frac{S^8\cdot S^7\cdot S^6}{S^2\cdot S^1\cdot S^0},$$

we have to use the next Hopf factorization

$$S^8 = (\mathbb{R}^6+\mathbb{R}^3+1)\,S^2,$$

which gives us:

$$G_{9,3}(\mathbb{R}) = (\mathbb{R}^6+\mathbb{R}^3+1)\,\mathbb{CP}^3\cdot\mathbb{RP}^6.$$

The next cases are:

$$G_{10,3}(\mathbb{R}) = \frac{S^9\cdot S^8\cdot S^7}{S^2\cdot S^1\cdot S^0} = \mathbb{CP}^4\cdot(\mathbb{R}^6+\mathbb{R}^3+1)\cdot\mathbb{RP}^7,$$

the first appearance of an odd-dimensional $\mathbb{RP}^n$, and

$$G_{11,3}(\mathbb{R}) = \frac{S^{10}\cdot S^9\cdot S^8}{S^2\cdot S^1\cdot S^0} = \mathbb{RP}^{10}\cdot\mathbb{CP}^4\cdot(\mathbb{R}^6+\mathbb{R}^3+1).$$

In the next case we again have 2 odd spheres and the Hopf factorization

$$S^{11} = (\mathbb{R}^6+1)(\mathbb{R}^3+1)\,S^2 = \mathbb{S}^6\cdot\mathbb{S}^3\cdot S^2,$$

giving rise to

$$G_{12,3}(\mathbb{R}) = \frac{S^{11}\cdot S^{10}\cdot S^9}{S^2\cdot S^1\cdot S^0} = \mathbb{RP}^{10}\cdot\mathbb{CP}^4\cdot\mathbb{S}^6\cdot\mathbb{S}^3.$$

and finally in the next case we again have two irreducible spheres S^{12}, S^{10}, leading to

$$G_{13,3}(\mathbb{R}) = \frac{S^{12}\cdot S^{11}\cdot S^{10}}{S^2\cdot S^1\cdot S^0} = \mathbb{RP}^{12}\cdot\mathbb{RP}^{10}\cdot\mathbb{S}^6\frac{(\mathbb{R}^3+1)}{\mathbb{R}+1}.$$

where once again, the phantom projective plane appears

$$\frac{\mathbb{R}^3+1}{\mathbb{R}+1} = \mathbb{R}^2-\mathbb{R}+1 = \mathbb{RP}^2_h.$$

There is clearly a 6-periodicity in the factorization of Graßmann manifolds for $k = 3$. The formulas obtained here lead to a classification but they do not correspond to the fibre bundles of any kind. Besides, we used repeatedly the fact that quantity is commutative. Another interesting homogeneous space is the flag manifold $F_{n;k,\ell}(\mathbb{R})$, $k<\ell<n$ whereby W is subspace of $\mathbb{R}^n$ of dimension 1 and $V\subset W$ is a subspace of dimension k. This clearly leads to the fibration

$$\begin{aligned}F_{n;k,\ell}(\mathbb{R}) &= G_{n,\ell}(\mathbb{R})\cdot G_{\ell,k}(\mathbb{R})\\ &= \frac{O(n)}{O(n-\ell)O(\ell)}\cdot\frac{O(\ell)}{O(\ell-k)\cdot O(k)} = \frac{O(n)}{O(k)O(\ell-k)\cdot O(n-\ell)}.\end{aligned}$$

The flag manifold $F_{n;k,\ell}(\mathbb{R})$ may also be seen as manifold (V,V') with $V\subset\mathbb{R}^n$ a subspace of dimension k and $V\perp V'$ of dimension $\ell-k$. The link with the previous definition simply follows from $W=V\oplus V'$ and we have the fibration

$$\begin{aligned}F_{n;k,\ell}(\mathbb{R}) &= G_{n,k}(\mathbb{R})\cdot G_{n-k,\ell-k}(\mathbb{R})\\ &= \frac{O(n)}{O(k)O(n-k)}\cdot\frac{O(n-k)}{O(\ell-k)\cdot O(n-\ell)} = \frac{O(n)}{O(k)O(\ell-k)\cdot O(n-\ell)}.\end{aligned}$$

More in general for $0<k_1<\cdots<k_s<n$ we may define the flag manifold $F_{n;k_1,\ldots,k_s}(\mathbb{R})$ as the manifold of flags $(V_1,\ldots,V_s)$ with $V_1\subset V_2\subset\cdots\subset V_s\subset\mathbb{R}^n$ subspaces of dimension $\dim V_j=k_j$. We clearly have the iterated fibration

$$\begin{aligned}F_{n;k_1,\ldots,k_s}(\mathbb{R}) &= G_{n,k_s}(\mathbb{R})\cdot G_{k_s,k_{s-1}}(\mathbb{R})\cdots G_{k_2,k_1}(\mathbb{R})\\ &= \frac{O(n)}{O(n-k_s)O(k_s)}\cdot\frac{O(k_s)}{O(k_s-k_{s-1})\cdot O(k_{s-1})}\cdots\frac{O(k_2)}{O(k_2-k_1)\cdot O(k_1)}\\ &= \frac{O(n)}{O(n-k_s)O(k_s-k_{s-1})\cdots O(k_2-k_1)O(k_1)}.\end{aligned}$$

Using orthogonal subspaces, we have:

$$F_{n;k_1,\ldots,k_s}(\mathbb{R}) = G_{n,k_1}(\mathbb{R})\cdot G_{n-k_1,k_2-k_1}(\mathbb{R})\cdots G_{n-k_{s-1},k_s-k_{s-1}}(\mathbb{R}).$$

Orthogonal groups may also be defined for the spaces $\mathbb{R}^{p,q}$ with pseudo-Euclidean inner product

$$\langle x,y\rangle = x_1y_1+\cdots+x_py_q-x_{p+1}y_{p+1}-\cdots-x_{p+q}y_{p+q}.$$

The corresponding groups are $O(p,q)$ and $SO(p,q)$. The group $SO(p,q)$, e.g., is determined as the manifold of frames of signature (p,q) :

$$\left(\underline{v}_1,\ldots,\underline{v}_p;\underline{v}_{p+1},\ldots,\underline{v}_{p+q}\right)$$

whereby $\underline{v}_1\in S^{p-1}\cdot\mathbb{R}^q$ is the first spacelike vector $\underline{v}_2\perp\underline{v}_1\in S^{p-2}\cdot\mathbb{R}^q$ up to $\underline{v}_p\perp\operatorname{span}\left(\underline{v}_1,\ldots,\underline{v}_{p-1}\right)\in S^0\cdot\mathbb{R}^q$ and the remaining vectors $\left(\underline{v}_{p+1},\ldots,\underline{v}_{p+q}\right)$ form a right oriented time-like q-frame, i.e., $\underline{v}_{p+1}\in S^{q-1}$, $\underline{v}_{p+2}\in S^{q-2}$ and $\underline{v}_{p+q}$ is fixed by the fact that the determinant of the whole frame equals $+1$.

In total, the morphological bill adds up to:

$$\begin{aligned}SO(p,q) &= \left(S^{p-1}\cdot\mathbb{R}^q\right)\cdots\left(S^0\cdot\mathbb{R}^q\right)S^{q-1}\cdots S^1\\ &= O(p)\cdot SO(q)\cdot\mathbb{R}^{p\cdot q},\end{aligned}$$

and it is a two component group still.

All of the above may be generalized to the complex Hermitian case. Let us start with $\mathbb{C}^n$ provided with the Hermitian inner product:

$$(\underline{z}, \underline{w}) = z_1\overline{w}_1 + \cdots + z_n\overline{w}_n.$$

Then by $U(n)$ we denote the unitary group of matrices learning the Hermitian form invariant; its matrices may be written as Hermitian orthonormal frames $\underline{v}_1, \ldots, \underline{v}_n$ whereby $|\underline{v}_j| = 1$, $(\underline{v}_j, \underline{v}_k) = 0$ for $j \neq k$.
This leads to the following morphological analysis:

$\underline{v}_1 \in S^{2n-1}$ is the unit vector in $\mathbb{C}^n = \mathbb{R}^{2n}$,

$\underline{v}_2 \perp \underline{v}_1$ in the Hermitian sense, i.e., $\underline{v}_2 \in \underline{v}_1^{\perp} \cap S^{2n-1} = S^{2n-3}$

up to

$\underline{v}_n \perp \underline{v}_1, \ldots, \underline{v}_{n-1}$, i.e., $\underline{v}_n \in S^1$

and, therefore,

$U(n) = S^{2n-1} \cdot S^{2n-3} \cdots S^1$.

In the above, please note that $\langle \underline{v}, \underline{w} \rangle = \mathrm{Re}(\underline{v}, \underline{w})$ is the orthogonal inner product in $\mathbb{R}^{2n}$ and so

$$(\underline{v}, \underline{w}) = 0 \text{ iff } \langle \underline{v}, \underline{w} \rangle = 0 \text{ and } \langle i\underline{v}, \underline{w} \rangle = 0.$$

Clearly $U(n)$is a subgroup of $SO(2n)$ and for the quotient we have:

$$\frac{SO(2n)}{U(n)} = S^{2n-2}\dot{S}^{2n-4} \cdots S^2,$$

which actually is a manifold, namely the manifold of all complex structures on $\mathbb{R}^{2n}$ (Exercise).

The special unitary group $SU(n)$ is the subgroup of matrices in $U(n)$ with determinant $= 1$, i.e.,

$$SU(n) = S^{2n-1} \cdots S^3,$$

and in particular $SU(2) = S^3$.

The definition of the complex general and special linear groups is obvious; they are denoted by $GL(n, \mathbb{C})$, $SL(n, \mathbb{C})$. Like for the orthogonal groups also for the complex group $U(n)$ we have the associated homogeneous spaces, in particular Graßmann manifolds

$$G_{n,k}(\mathbb{C}) = \frac{U(n)}{U(k) \cdot U(n-k)} = \frac{S^{2n-1} \cdots S^{2n-2k+1}}{S^{2k-1} \cdot S^{2k-3} \cdots S^1},$$

so for example

$$G_{4,2}(\mathbb{C}) = \frac{S^7 \cdot S^5}{S^3 \cdot S^1} = (\mathbb{R}^4 + 1)\mathbb{CP}^2 = \mathbb{S}^4\mathbb{CP}^2 = \mathbb{HP}^1 \cdot \mathbb{CP}^2.$$

$$G_{5,2}(\mathbb{C}) = \frac{S^9 \cdot S^7}{S^3 \cdot S^1} = \mathbb{CP}^4 \cdot (\mathbb{R}^4 + 1) = \mathbb{CP}^4 \cdot \mathbb{HP}^1.$$

$$G_{6,2}(\mathbb{C}) = \frac{S^{11} \cdot S^9}{S^3 \cdot S^1} = \mathbb{HP}^2 \cdot \mathbb{CP}^4.$$

and so we have again a clear 2-periodicity.

We leave the discussion of $G_{n,3}(\mathbb{C})$ as an exercise.

Unitary groups may also be constructed over spaces $\mathbb{C}^{p,q}$ with pseudo-Hermitian form

$$(\underline{z}, \underline{w}) = z_1\overline{w}_1 + \cdots + z_p\overline{w}_p - z_{p+1}\overline{w}_{p+1} - \cdots - z_{p+q}\overline{w}_{p+q}$$

and the corresponding invariance groups are denoted by $U(p,q)$ and $SU(p,q)$ in case det $= 1$.

The corresponding frames now have to be chosen on the pseudo-Hermitian unit sphere:

$$|z_1|^2 + \cdots + |z_p|^2 - |z_{p+1}|^2 - \cdots - |z_{p+q}|^2 = 1$$

which leads to the morphological formula

$$U(p,q) = (S^{2p-1} \cdot \mathbb{C}^q) \cdot (S^{2p-3} \cdot \mathbb{C}^q) \cdots (S^1 \cdot \mathbb{C}^q) \cdot S^{2q-1} \cdot S^{2q-3} \cdots S^1.$$

Of course we also have the complexified versions $O(n,\mathbb{C})$ and $SO(n,\mathbb{C})$ of $O(n)$ and $SO(n)$; it is another story which we will leave out for the moment.

To finish the list of matrix groups leading to morphological analysis, we mention the compact symplectic groups $Sp(n)$; they follow from the quaternionic Hermitian form

$$(\underline{q}, \underline{r}) = q_1\overline{r}_1 + \cdots + q_n\overline{r}_n,$$

whereby $q_j = q_{j_0} + iq_{j_1} + jq_{j_2} + kq_{j_3}$ is a quaternion and $\overline{q}_j = q_{j_0} - iq_{j_1} - jq_{j_2} - kq_{j_3}$ its quaternion conjugate.

$Sp(n)$ is by definition the group of quaternion $n \times n$ matrices leaving this form invariant and its matrix elements may be regarded as quaternionic frames $q_1, \ldots, q_n$ whereby $q_r \in \mathbb{H}^n$ with $|q_1| = 1$, i.e., $q_1 \in S^{4n-1}$, $q_2 \in \mathbb{H}^n$ with $|q_2| = 1$ and $(q_1, q_2) = 0$, i.e., $q_2 \in S^{4n-5}$, and so on. This leads to the morphological bill:

$$Sp(n) = S^{4n-1} \cdot S^{4n-5} \cdot S^3,$$

in particular $Sp(1) = S^3$ and $Sp(2) = S^7 \cdot S^3$.

Also here may be investigated quaternionic Graßmannians.

The groups $Sp(n)$ should not be confused with the non-compact groups $Sp(2n,\mathbb{R})$ of matrices $A \in GL(2n,\mathbb{R})$ leaving the maximal 2-form invariant.

For $Sp(2n,\mathbb{R})$ we did not find a morphological evaluation yet.

To finish this section we discuss the Spin groups Spin(m).

We start by considering the real 2^m-dimensional Clifford algebra $\mathbb{R}_m$ with generators $e_1, \ldots, e_m$ and relations $e_j\, e_k + e_k\, e_j = -2\delta_{jk}$.

The space of bivectors

$$\mathbb{R}_{m,2} = \left\{ \sum_{i,j} b_{ij} e_i\, e_j \,:\, b_{ij} \in \mathbb{R} \right\}$$

forms a Lie algebra for the commutation product and the corresponding group is the Spin group:

$$\text{Spin}(m) = \exp(\mathbb{R}_{m,2}).$$

Its elements may be written into the form $s = \underline{w}_1 \cdots \underline{w}_{2s}$, $\underline{w}_j = \sum w_{jk} e_k \in \mathbb{R}^m$ with $\underline{w}_j^2 = -1$, i.e., $\underline{w}_j \in S^{m-1}$.

We have the following Spin(m) representation

$$h : \text{Spin}(m) \to SO(m)$$

whereby

$$h : s \to h(s) : \underline{x} \to s\underline{x}\overline{s}$$

whereby for $a \in \mathbb{R}_m$, $\overline{a}$ is the conjugation with properties $\overline{ab} = \overline{b}\,\overline{a}$ & $\overline{e_j} = -e_j$.

In this way Spin(m) is a 2-fold covering group of $SO(m)$, i.e.,

$$SO(m) = \text{Spin}(m)/_{\mathbb{Z}_2}$$

and also Spin(m) is simply connected.

This might suggest the morphological evaluation

$$\text{Spin}(m) = SO(m) \cdot \mathbb{Z}_2 = S^{m-1} \cdots S^1 \cdot S^0 = O(m),$$

which, through not entirely wrong in the sense of quantity, is somewhat uninteresting.

But there is a more interesting evaluation of Spin(m).

Let us start with

$$\begin{aligned} \text{Spin}(3) &= \{q_0 + q_1 e_{23} + q_2 e_{31} + q_3 e_{12} \ : \ q\overline{q} = 1\} \\ &= S^3 = \mathbb{S}^2 \cdot S^1 = SU(2) = Sp(1) \end{aligned}$$

with differs rather substantially from

$$O(3) = S^2 \cdot S^1 \cdot S^0.$$

So in fact, the rotation group $SO(3)$ has two different representations in morphological calculus:

one as the matrix group

$$SO(3) = S^2 \cdot S^1$$

and one in terms of the Spin group (quaternion S^3):

$$\begin{aligned} \mathbb{S}O(3) &= \text{Spin}(3)/_{\mathbb{Z}_2} = S^3/_2 \\ &= \mathbb{S}^2 S^1/_2 = \mathbb{S}^2\mathbb{S}^1 = (\mathbb{R}^2 + 1)(\mathbb{R} + 1) = \mathbb{RP}^3. \end{aligned}$$

In general we got

$$\mathbb{S}O(m) = \text{Spin}(m)/_2,$$

which is another morphological version of the rotation group.

For $m = 4$ we consider the pseudoscalar e_{1234} with $e_{1234}^2 = +1$ and e_{1234} is central in the even subalgebra

$$\mathbb{R}_4^+ = \text{Alg}\{e_{jk} \ : \ j < k\} \cong \mathbb{R}_3 \cong \mathbb{H} \oplus \mathbb{H}.$$

Putting

$$E_{\pm} = \frac{1}{2}\left(1 \pm e_{1234}\right)$$

we have

$$E_+ + E_- = 1,\ \ E_\pm^2 = E_\pm,\ \ E_+E_- = 0,$$

so every $a \in \mathbb{R}_4^+$ may be written uniquely as:

$$a = a_+E_+ + a_-E_-,\ \ a_\pm \in \mathbb{H} = \text{span}\{1, e_{23}, e_{31}, e_{12}\}$$

and in particular

$$s \in \text{Spin}(4) : s_+E_+ + s_-E_-,\ s_\pm \in S^3.$$

So we have the morphological analysis

$$\begin{aligned}\text{Spin}(4) &= \text{Spin}(3) \times \text{Spin}(3)\\ &= S^3 \cdot S^3 = S^3 \cdot \mathbb{S}^2 \cdot S^1.\end{aligned}$$

For $m = 5$, we use the fact that

$$\text{Spin}(5) = \{s \in \mathbb{R}_5^+ \,:\, s\overline{s} = 1\},$$

together with the isomorphisms

$$\mathbb{R}_5^+ = \mathbb{R}_4 \cong \mathbb{H}(2),$$

i.e., the set of 2×2 quaternionic matrices

$$a = \begin{pmatrix} a_{11} & a_{12} \\ a_{21} & a_{22} \end{pmatrix},\ a_{ij} \in \mathbb{H}$$

and under this isomorphism we also have

$$\overline{a} = \begin{pmatrix} \overline{a}_{11} & \overline{a}_{21} \\ \overline{a}_{12} & \overline{a}_{22} \end{pmatrix}.$$

This shows that in fact

$$\text{Spin}(5) = \text{Sp}(2) = S^7 \cdot S^3 = \mathbb{S}^4 \cdot S^3 \cdot S^3 = \mathbb{S}^4 \cdot S^3 \cdot \mathbb{S}^2 \cdot S^1.$$

For $m = 6$, the pseudoscalar e_{123456} satisfies $e_{123456}^2 = -1$ and it is central in even subalgebra $\mathbb{R}_6^+ = \mathbb{R}_5$, so it may identified with complex number i, leading to

$$\mathbb{R}_6^+ \cong \mathbb{C} \otimes \mathbb{R}_5^+ \cong \mathbb{C} \otimes \mathbb{H}(2) \cong \mathbb{C}(4),$$

and under this map $\mathbb{R}_6^+ \to \mathbb{C}(4)$, the conjugate $\overline{a}$ of $a \in \mathbb{R}_6^+$ corresponds to the Hermitian conjugate $(a)^+$ of matrix $(a) \in \mathbb{C}(4)$.

Hence the group $G = \{a \,:\, a\overline{a} = 1,\ a \in \mathbb{R}_6^+\}$ corresponds to $U(4)$. But for $m > 5$, the group G no longer corresponds to $\text{Spin}(m)$ and for $m = 6$

$$\begin{aligned}G &= \exp\left\{\sum b_{ij}e_{ij} + e_{123456}\right\}\\ &= \exp\left\{\sum b_{ij}e_{ij}\right\} \times \exp\{e_{123456}\} = \text{Spin}(6) \times U(1)\end{aligned}$$

which shows that really

$$\text{Spin}(6) = SU(4) = S^7 \cdot S^5 \cdot S^3 = S^5 \cdot \mathbb{S}^4 \cdot S^3 \cdot \mathbb{S}^2 \cdot S^1.$$

For $m = 7$ on the situation is much more complicated. Could it be that

$$\text{Spin}(7) = \mathbb{S}^6 \cdot S^5 \cdot \mathbb{S}^4 \cdot S^3 \cdot \mathbb{S}^2 \cdot S^1?$$

5. Nullcones and things

The nullcone NC^{n-1} of complex dimension $n-1$ in the locus of points

$$(z_1, \dots, z_n) \in \mathbb{C}^n$$

that satisfy $z_1^2 + \cdots + z_n^2 = 0$.

The complex $(n-1)$-sphere $\mathbb{C}S^{n-1}$ consists of the solutions $(z_1, \dots, z_n)$ of the equation $z_1^2 + \cdots + z_n^2 = 1$.

It is a non-compact manifold that admits a canonical compactification $\overline{\mathbb{C}S}^{n-1} \subset \mathbb{CP}^n$ given by the equation in homogeneous coordinates $z_1, \dots, z_{n+1}$:

$$z_1^2 + \cdots + z_n^2 = z_{n+1}^2$$

that is equivalent to $z_1^2 + \cdots + z_n^2 + z_{n+1}^2 = 0$ if we replace $z_{n+1} \to \mathbf{i} z_{n+1}$. The submanifold $\mathbb{C}S^{n-1}$ corresponds to the intersection with the region $z_{n+1} \neq 0$ while the "points at infinity" corresponds to the intersection with plane $z_{n+1} = 0$, leading to:

$$\overline{\mathbb{C}S}^{n-2} \; : \; z_1^2 + \cdots + z_n^2 = 0.$$

Hence we have the disjoint union

$$\overline{\mathbb{C}S}^{n-1} = \mathbb{C}S^{n-1} \cup \overline{\mathbb{C}S}^{n-2}.$$

We are going to perform the morphological calculus of those objects in two different ways, leading to two different formulas for the quantity (once again). The first method could be called the real geometry approach.

Let us write $z = \underline{x} + \mathbf{i}\underline{y}$, $\underline{x} = (x_1, \dots, x_n)$, $\underline{y} = (y_1, \dots, y_n) \in \mathbb{R}^n$; then the equation for NC^{n-1} may be rewritten as

$$|\underline{x}|^2 = |\underline{y}|^2 \;\; \& \;\; \langle \underline{x}, \underline{y} \rangle = 0$$

with $|\underline{x}|^2 = x_1^2 + \cdots + x_n^2$, $\langle \underline{x}, \underline{y} \rangle = x_1 y_1 + \cdots + x_n y_n$.

First solution is the point $\underline{z} = 0$ with quantity 1.

For $\underline{z} \neq 0$ we may write $\underline{x} = \rho \underline{\omega}$, $\underline{y} = \rho \underline{\nu}$, $\rho \in \mathbb{R}_+$ and $\underline{\omega}, \underline{\nu} \in S^{n-1}$ such that $\underline{\omega} \perp \underline{\nu}$, i.e., $(\underline{\omega}, \underline{\nu}) \in V_{n,2}(\mathbb{R})$. Hence we have

$$NC^{n-1} = 1 + V_{n,2}(\mathbb{R}) \cdot \mathbb{R}_+ = 1 + S^{n-1} \cdot S^{n-2} \cdot \mathbb{R}_+.$$

The complex sphere $\mathbb{C}S^{n-1}$ written in real coordinates would lead to:

$$|\underline{x}|^2 = 1 + |\underline{y}|^2, \;\; \langle \underline{x}, \underline{y} \rangle = 0.$$

First we have the case $|\underline{y}| = 0$, $|\underline{x}| = 1$ leading to the quantity S^{n-1}. Next for $|\underline{y}| \in \mathbb{R}_+$ we again may put $\underline{x} = r\underline{\omega}$, $y = \rho \underline{\nu}$ whereby $r^2 = 1 + \rho^2$, $\rho \in \mathbb{R}_+$ and $\underline{\omega}, \underline{\nu} \in S^{n-1}$ with $\underline{\omega} \perp \underline{\nu}$. This leads to the morphological bill

$$\begin{aligned} \mathbb{C}S^{n-1} &= S^{n-1} + V_{n,2}(\mathbb{R}) \cdot \mathbb{R}_+ \\ &= S^{n-1} + S^{n-1} \cdot S^{n-2} \cdot \mathbb{R}_+ = S^{n-1} \cdot \left(1 + S^{n-2} \cdot \mathbb{R}_+\right) \\ &= S^{n-1} \cdot \left(1 + \left(\mathbb{R}^{n-1} - 1\right)\right) = S^{n-1} \cdot \mathbb{R}^{n-1}, \end{aligned}$$

which represents the tangent bundle to S^{n-1}. Once again remark that $S^{n-1}\cdot\mathbb{R}^{n-1}$ might represent any $(n-1)$-dimensional vector bundles over S^{n-1} or more general stuff, so it only represents the quantity of the tangent bundle.

For $\overline{\mathbb{C}S}^{n-1}$ we have two approaches.

First it is the set of points $(z_1,\ldots,z_{n+1})\in\mathbb{CP}^n$ solving the equation $z_1^2+\cdots+z_{n+1}^2=0$, which means that the homogeneous coordinates $(z_1,\ldots,z_{n+1})\neq 0$ belong to $NC^n\setminus\{0\}$ and they are determined up to a homogeneity factor $\lambda\in\mathbb{C}\setminus\{0\}$. This leads to

$$\overline{\mathbb{C}S}^{n-1}=\frac{NC^n-1}{\mathbb{C}-1}=\frac{V_{n+1,2}(\mathbb{R})\cdot\mathbb{R}_+}{S^1\cdot\mathbb{R}_+}=\frac{V_{n+1,2}(\mathbb{R})}{S^1}$$
$$=\widetilde{G}_{n+1,2}(\mathbb{R})=\frac{S^n\cdot S^{n-1}}{S^1}.$$

Secondly we also have that

$$\overline{\mathbb{C}S}^{n-1}=\mathbb{C}S^{n-1}+\overline{\mathbb{C}S}^{n-2}$$
$$S^{n-1}\cdot\mathbb{R}^{n-1}+S^{n-2}\cdot\mathbb{R}^{n-2}+\cdots+S^1\cdot\mathbb{R}+2$$

giving the total quantity, while also

$$\mathbb{C}S^{n-1}+\overline{\mathbb{C}S}^{n-2}=S^{n-1}\cdot\left(\mathbb{R}^{n-1}+\frac{S^{n-2}}{S^1}\right)$$
$$S^{n-1}\cdot\frac{((2\mathbb{R}+2)\mathbb{R}^{n-1}+S^{n-2})}{S^1}=S^{n-1}\cdot\frac{S^n}{S^1},$$

as expected.

Note also that there is a 2-periodicity expressed by

$$\overline{\mathbb{C}S}^{2n-1}=\frac{S^{2n}\cdot S^{2n-1}}{S^1}=S^{2n}\cdot\mathbb{CP}^{n-1},$$
$$\overline{\mathbb{C}S}^{2n}=\frac{S^{2n+1}\cdot S^{2n}}{S^1}=\mathbb{CP}^n\cdot S^{2n}.$$

Note that we also have the identity

$$NC^{n-1}=1+\overline{\mathbb{C}S}^{n-2}\cdot(\mathbb{C}-1)$$

that often turns out useful in calculations.

We now use a purely complex method to compute the complexified sphere; we use a different notation $\overline{\mathbb{CS}}^n$.

For $\overline{\mathbb{CS}}^0$ we have the equation

$$z_1^2+z_2^2=0\;\Leftrightarrow\;uv=0,\;\;u=z_1+\mathbf{i}z_2,\;v=z_1-\mathbf{i}z_2.$$

Up to a factor $\lambda\neq 0$ there are solutions (u,v) namely $(1,0)$ and $(0,1)$, leading to the quantity $\overline{\mathbb{CS}}^0=2$.

For $\overline{\mathbb{CS}}^1$ we have the equation

$$uv=z_3^2$$

including for $z_3^2 = 0$, $uv = 0$, i.e., $\overline{\mathbb{CS}}^0$ and for $z_3 \neq 0$ we normalise $z_3 = 1$, so we have the equation for $\overline{\mathbb{CS}}^1$: $uv = 1$, i.e., $u \in \mathbb{C} \setminus \{0\}$, $v = 1/u$. This leads to

$$\mathbb{CS}^1 = \mathbb{C} - 1 \ \ \& \ \ \overline{\mathbb{CS}}^1 = \mathbb{CS}^1 + \mathbb{CS}^0 = (\mathbb{C} - 1) + 2 = \mathbb{C} + 1$$

so in fact $\overline{\mathbb{CS}}^1 = \mathbb{CP}^1 = \mathbb{S}^2$.

For $\overline{\mathbb{CS}}^2$ we again have the splitting

$$\overline{\mathbb{CS}}^2 = \mathbb{CS}^2 + \overline{\mathbb{CS}}^1$$

whereby $\overline{\mathbb{CS}}^2$ is given by the equation

$$uv = z_4^2 - z_3^2 = 1 - z_3^2,$$

with normalization $z_4 = 1$. There are two cases of this: $z_3^2 \neq 1$, giving $z_3 \in \mathbb{C} \setminus \{1, -1\}$ and $z_3 \in \{+1, -1\}$. So, morphologically, we have a factor

$$z_3 \in \mathbb{C} - 2 \quad \text{or} \quad z_3 \in 2.$$

In the case $z_3 \in \mathbb{C} - 2$ we have the equation

$$uv = \text{cte} \neq 0$$

to solve, which gives us $u \in \mathbb{C} - 1$, $v = \text{cte}/u$, leading to the quantity:

$$(\mathbb{C} - 1)(\mathbb{C} - 2).$$

For $z_3 \in 2$ we have equation $uv = 0$ to be solved, which gives us $(u, v) = (0, 0)$ or $v = 0$ and $u \in \mathbb{C} - 1$ or $u = 0$ and $u \in \mathbb{C} - 1$. So the total quantity is

$$1 + 2(\mathbb{C} - 1),$$

with an extra factor 2, which gives the total

$$\begin{aligned} \mathbb{CS}^2 &= (\mathbb{C} - 1)(\mathbb{C} - 2) + 4(\mathbb{C} - 1) + 2 \\ &= (\mathbb{C} - 1)(\mathbb{C} + 2) + 2 = \mathbb{C}^2 + \mathbb{C} = (\mathbb{C} + 1)\mathbb{C}. \end{aligned}$$

Hence, we arrive at

$$\overline{\mathbb{C}S}^2 = (\mathbb{C} + 1)\mathbb{C} + (\mathbb{C} + 1) = (\mathbb{C} + 1)^2 = \mathbb{S}^2\mathbb{CP}^1 = \frac{S^3\mathbb{S}^2}{S^1}.$$

For $\mathbb{CS}^3$ we have the equation

$$u_1v_1 = 1 - u_2v_2$$

leading to the cases $u_2v_2 = 1$ and $u_2v_2 \neq 1$ for which we have the morphological factors $\mathbb{C} - 1$ and $\mathbb{C}^2 - \mathbb{C} + 1$ (the phantom complex projective plane). In case $1 - u_2v_2 = c \neq 0$ the remaining equation $u_1v_1 = c$ yields the factor $\mathbb{C} - 1$ while for $1 = u_2v_2$ we have $u_1v_1 = 0$, i.e., $1 + 2(\mathbb{C} - 1)$. In total this gives

$$\begin{aligned} \mathbb{CS}^3 &= (\mathbb{C} - 1)(\mathbb{C}^2 - \mathbb{C} + 1) + (1 + 2(\mathbb{C} - 1))(\mathbb{C} - 1) \\ &= (\mathbb{C} - 1)(\mathbb{C}^2 + \mathbb{C}) = (\mathbb{C}^2 - 1)\mathbb{C}, \end{aligned}$$

which is also clear from the fact that $u_1v_1 + u_2v_2 = 1$ is basically the equation $ad - bc = 1$ for

$$SL(2, \mathbb{C}) = \left(\mathbb{C}^2 - 1\right) \mathbb{C}.$$

This leads to

$$\begin{aligned}\overline{\mathbb{CS}}^3 &= \mathbb{CS}^3 + \overline{\mathbb{CS}}^2 = \left(\mathbb{C}^2 - 1\right) \mathbb{C} + (\mathbb{C} + 1)^2 \\ &= (\mathbb{C} + 1)\left((\mathbb{C} - 1)\mathbb{C} + \mathbb{C} + 1\right) = (\mathbb{C} + 1)(\mathbb{C}^2 + 1) \\ &= \frac{\mathbb{S}^4 S^3}{S^1} = \mathbb{S}^4 \cdot \mathbb{CP}^1 = \mathbb{CP}^3.\end{aligned}$$

For $\mathbb{CS}^4$ we have the equation

$$u_1v_1 + u_2v_2 = 1 - z_5^2$$

leading to the factors $\mathbb{C} - 2$ for $z_5^2 \neq 1$ and 2 for $z_5^2 = 1$.

For $1 - z_5^2 = c \neq 0$ the remaining equation gives the factor $SL(2, \mathbb{C}) = (\mathbb{C}^2 - 1)\mathbb{C}$ while for $c = 0$ we have the equation $u_1v_1 + u_2v_2 = 0$, which is the nullcone

$$N\mathbb{C}^3 = 1 + \overline{\mathbb{CS}}^2 \cdot (\mathbb{C} - 1) = 1 + (\mathbb{C}^2 - 1)(\mathbb{C} + 1).$$

In total we get

$$\begin{aligned}\mathbb{CS}^4 &= (\mathbb{C}^2 - 1)\mathbb{C}(\mathbb{C} - 2) + 2(\mathbb{C}^2 - 1)(\mathbb{C} + 1) + 2 \\ &= (\mathbb{C}^2 - 1)(\mathbb{C}^2 + 2) + 2 = \mathbb{C}^4 + \mathbb{C}^2 = \mathbb{C}^2(\mathbb{C}^2 + 1)\end{aligned}$$

so that

$$\begin{aligned}\overline{\mathbb{CS}}^4 &= \mathbb{CS}^4 + \overline{\mathbb{CS}}^3 = \left(\mathbb{C}^2 + 1\right)\left(\mathbb{C}^2 + \mathbb{C} + 1\right) \\ &= \mathbb{CP}^2 \cdot \mathbb{S}^4 = \frac{S^5 \mathbb{S}^4}{S^1}.\end{aligned}$$

It seems that in general we will have

$$\begin{aligned}\overline{\mathbb{CS}}^{2n} &= \frac{S^{2n+1}}{S^1} \mathbb{S}^{2n} = \mathbb{CP}^n \cdot \mathbb{S}^{2n}, \\ \overline{\mathbb{CS}}^{2n-1} &= \mathbb{S}^{2n} \frac{S^{2n-1}}{S^1} = \mathbb{S}^{2n} \cdot \mathbb{CP}^{n-1} = \mathbb{CP}^{2n-1}.\end{aligned}$$

To prove this recursively we begin with $\mathbb{CS}^{2n}$ given by the equation

$$u_1v_1 + \cdots + u_nv_n = 1 - z_{2n+1}^2.$$

For the right-hand side we have the factor $\mathbb{C} - 2$ for $z_{2n+1}^2 \neq 1$ and the factor 2 for $z_{2n+1}^2 = 1$. The equation $c = 1 - z_{2n+1}^2 \neq 0$ gives the factor $u_1v_1 + \cdots + u_nv_n = c$, which is in fact $\mathbb{CS}^{2n-1}$ while for $c = 0$ we have the equation $u_1v_1 + \cdots + u_nv_n = 0$, which is

$$N\mathbb{C}^{2n-1} = 1 + \overline{\mathbb{CS}}^{2n-2} \cdot (\mathbb{C} - 1).$$

So in total we have

$$\begin{aligned}\mathbb{CS}^{2n} &= \mathbb{CS}^{2n-1}\cdot(\mathbb{C}-2)+2+2\overline{\mathbb{CS}}^{2n-2}\cdot(\mathbb{C}-1)\\ &= \overline{\mathbb{CS}}^{2n-1}\cdot(\mathbb{C}-2)-\overline{\mathbb{CS}}^{2n-2}\cdot(\mathbb{C}-2)+2\overline{\mathbb{CS}}^{2n-2}\cdot(\mathbb{C}-1)+2\\ &= \overline{\mathbb{CS}}^{2n-1}\cdot(\mathbb{C}-2)+\overline{\mathbb{CS}}^{2n-2}\cdot\mathbb{C}+2\end{aligned}$$

and, therefore,

$$\overline{\mathbb{CS}}^{2n} = \mathbb{CS}^{2n}+\overline{\mathbb{CS}}^{2n-1} = \overline{\mathbb{CS}}^{2n-1}\cdot(\mathbb{C}-1)+\overline{\mathbb{CS}}^{2n-2}\cdot\mathbb{C}+2.$$

Using the induction hypothesis

$$\overline{\mathbb{CS}}^{2n-1} = \mathbb{CP}^{2n-1} \quad\text{and}\quad \overline{\mathbb{C}S}^{2n-2} = \mathbb{CP}^{n-1}\cdot\mathbb{S}^{2n-2},$$

this gives rise to

$$\begin{aligned}\overline{\mathbb{CS}}^{2n} &= \mathbb{C}^{2n}-1+(\mathbb{C}^{n-1}+1)(\mathbb{C}^{n}+\mathbb{C}^{n-1}+\cdots+\mathbb{C})+2\\ &= (\mathbb{C}^{2n}+\mathbb{C}^{2n-1}+\cdots+\mathbb{C}^{n})+(\mathbb{C}^{n}+\mathbb{C}^{n-1}+\cdots+\mathbb{C}+1)\\ &= (\mathbb{C}^{n}+1)\mathbb{CP}^{n} = \mathbb{CP}^{n}\cdot\mathbb{S}^{2n}.\end{aligned}$$

For the other case $\overline{\mathbb{CS}}^{2n+1}$ we note that $\mathbb{CS}^{2n+1}$ is given by the equation $u_1v_1+\cdots+u_nv_n = 1-u_{n+1}v_{n+1}$ giving the factor (phantom projective plane) $\mathbb{C}^2-\mathbb{C}+1$ for $c = 1-u_{n+1}v_{n+1}\neq 0$ and $\mathbb{C}-1$ for $u_{n+1}v_{n+1}=1$. Again for $c\neq 0$ we have the equation $u_1v_1+\cdots+u_nv_n = c\neq 0$, leading to the factor $\mathbb{CS}^{2n-1}$ and for $c=0$ we get factor $N\mathbb{C}^{2n-1}$ as before. This leads to

$$\begin{aligned}\mathbb{CS}^{2n+1} &= \mathbb{CS}^{2n-1}\cdot(\mathbb{C}^2-\mathbb{C}+1)+(1+\overline{\mathbb{CS}}^{2n-2}\cdot(\mathbb{C}-1))\cdot(\mathbb{C}-1)\\ &= \overline{\mathbb{CS}}^{2n-1}\cdot(\mathbb{C}^2-\mathbb{C}+1)-\overline{\mathbb{CS}}^{2n-2}\cdot\mathbb{C}+\mathbb{C}-1,\end{aligned}$$

which, using the formulae for $\overline{\mathbb{CS}}^{2n-1}$ and $\overline{\mathbb{C}S}^{2n-2}$ yields

$$\begin{aligned}\mathbb{CS}^{2n+1} &= \mathbb{C}^{n}\cdot(\mathbb{C}^{n+1}-1)\\ &= S^{2n+1}\cdot\mathbb{R}^{2n}\cdot\mathbb{R}_{+}\end{aligned}$$

and so we finally get

$$\begin{aligned}\overline{\mathbb{CS}}^{2n+1} &= \mathbb{CS}^{2n+1}+\overline{\mathbb{CS}}^{2n}\\ &= \mathbb{C}^{n}\cdot(\mathbb{C}^{n+1}-1)+(\mathbb{C}^{n}+1)(\mathbb{C}^{n}+\cdots+1)\\ &= \mathbb{C}^{2n+1}+\mathbb{C}^{2n}+\cdots+\mathbb{C}^{n}-\mathbb{C}^{n}+\mathbb{C}^{n}+\cdots+1 = \mathbb{CP}^{2n+1}\\ &= \mathbb{S}^{2n+2}\cdot\mathbb{CP}^{n} = \mathbb{S}^{2n+2}\cdot\frac{S^{2n+1}}{S^{1}}.\end{aligned}$$

These calculations show a certain consistency in which the Poincaré sphere $\mathbb{S}^{2n}$ and complex projective spaces $\mathbb{CP}^{n}$ play a central role. Also the phantom complex projective plane $\mathbb{C}^2-\mathbb{C}+1$ reappears here as the set of points $(u,v)\in\mathbb{C}^2$ which lie outside the hyperbola $uv=1$; it is the new geometric interpretation for strange phantom plane that arises from the morphological analysis.

Finally also the "bipolar plane" $\mathbb{C}-2$ arises naturally within the discussion. Of course one could always consider $\mathbb{C}-n$, but in morphological calculus we are not interested in generality, only in canonical objects.

Our next investigation concerns "Null Graßmannians".

By $NG_{n,k}(\mathbb{C})$ we denote the manifold of all k-dimensional subspaces of the nullcone NC^{n-1} in $\mathbb{C}^n$. Hence in particular $NG_{n,1}(\mathbb{C}) = \overline{\mathbb{C}S}^{n-2}$. Let us make the morphological analysis; once again there are two ways.

Let $V \subset NC^{n-1}$ be a k-dimensional complex subspace spanned by k-vectors $\underline{\tau}_1, \ldots, \underline{\tau}_k$. Which are of course linearly independent and satisfy:

$$\underline{\tau}_j^2 = (\underline{t}_j + \mathbf{i}\underline{s}_j)^2 = 0, \text{ i.e., } \underline{t}_j \perp \underline{s}_j \,\&\, |\underline{t}_j| = |\underline{s}_j|$$
$$\langle \underline{\tau}_j, \underline{\tau}_k \rangle = \langle \underline{t}_j, \underline{t}_k \rangle - \langle \underline{s}_j, \underline{s}_k \rangle + \mathbf{i}(\langle \underline{t}_j, \underline{s}_k \rangle + \langle \underline{t}_k, \underline{s}_j \rangle) = 0.$$

Next consider on $\mathbb{C}^n$ the Hermitian inner product $(\underline{z}, \underline{w}) = \sum_{j=1}^n z_j \overline{w}_j$; then we can normalize vector $\underline{\tau}_1$, i.e., $(\underline{\tau}_1, \overline{\underline{\tau}}_1) = \langle \underline{t}_1, \underline{t}_1 \rangle + \langle \underline{s}_1, \underline{s}_1 \rangle = 2$, which together with $\underline{t}_1 \perp \underline{s}_1$ $|\underline{t}_1| = |\underline{s}_1|$ means that the pair $(\underline{t}_1, \underline{s}_1) \in V_{n,2}(\mathbb{R})$ is the manifold of orthonormal 2-frames.

Next one may choose $\underline{\tau}_2$ such that $(\underline{\tau}_2, \underline{\tau}_1) = 0 \,\&\, |\underline{\tau}_2|^2 = 2$ with $\underline{\tau}_2 = \underline{t}_2 + \mathbf{i}\underline{s}_2$. This automatically implies that

$$\langle \overline{\underline{\tau}}_2, \underline{\tau}_1 \rangle = \langle \underline{\tau}_2, \underline{\tau}_1 \rangle = 0, \text{ i.e., } \langle \underline{t}_2, \underline{\tau}_1 \rangle = \langle \underline{s}_2, \underline{\tau}_1 \rangle = 0$$

so that the pair $(\underline{t}_2, \underline{s}_2)$ is an orthonormal 2-frame that is also orthogonal to $\mathrm{span}_{\mathbb{R}}\{\underline{t}_1, \underline{s}_1\}$, i.e., $(\underline{t}_1, \underline{s}_1, \underline{t}_2, \underline{s}_2) \in V_{n,4}(\mathbb{R})$.

Continuing the reasoning, we may choose $\underline{t}_j, \underline{s}_j$ in such a way that

$$(\underline{t}_1, \underline{s}_1, \underline{t}_2, \underline{s}_2, \ldots, \underline{t}_k, \underline{s}_k) \in V_{n,2k}(\mathbb{R}) = \frac{SO(n)}{SO(n-2k)};$$

a necessary condition for this is $n \geq 2k$.

Now let $(\underline{\tau}'_1, \ldots, \underline{\tau}'_k)$ be another k-tuple for which

$$\mathrm{span}\{\underline{\tau}'_1, \ldots, \underline{\tau}'_k\} = V \,\&\, |\underline{\tau}'_j|^2 = 2,\ (\underline{\tau}'_j, \underline{\tau}'_k) = 0,\ j \neq k;$$

then there exists the unique matrix $A \in U(k)$ such that $\underline{\tau}'_j = \sum_{\ell=1}^k A_{j\ell} \underline{\tau}_\ell$. Hence we obtain the identity in terms of homogeneous spaces and in morphological sense

$$NG_{n,k}(\mathbb{C}) = \frac{SO(n)}{U(k) \times SO(n-2k)} = \frac{S^{n-1} \cdot S^{n-2} \ldots S^{n-2k}}{S^{2k-1} \cdot S^{2k-3} \ldots S^1}.$$

So, in the case $n = 2m$ is even, we have that

$$NG_{n,k}(\mathbb{C}) = \frac{S^{2m-1} \ldots S^{2m-2k}}{S^{2k-1} \ldots S^1} = G_{m,k}(\mathbb{C}) \cdot S^{2m-2} \ldots S^{2m-2k}$$

while for $n = 2m+1$, odd, we have

$$NG_{n,k}(\mathbb{C}) = \frac{S^{2m} \cdot S^{2m-1} \ldots S^{2m-2k+1}}{S^{2k-1} \ldots S^1} = G_{m,k}(\mathbb{C}) \cdot S^{2m} \ldots S^{2m-2k+2}.$$

This also implies that $NG_{m,k}(\mathbb{C})$ is integrable.

Another way of calculating the quantity makes use of the complex compact spheres $\overline{\mathbb{CS}}^{n-2}$ that were obtained in terms of complex analysis. Using the notation $N\mathbb{G}_{n,k}(\mathbb{C})$ for the corresponding null Graßmannians we have:

$$\begin{aligned} N\mathbb{G}_{n,1}(\mathbb{C}) &= \overline{\mathbb{CS}}^{n-2}, \\ N\mathbb{G}_{n,2}(\mathbb{C}) &= \{(\underline{\tau}_1, \underline{\tau}_2) \in \overline{\mathbb{CS}}^{n-2} \cdot S^1 \times \overline{\mathbb{CS}}^{n-4} \cdot S^1\} \operatorname{mod} U(2) \\ &= \frac{\overline{\mathbb{CS}}^{n-2} \cdot S^1 \cdot \overline{\mathbb{CS}}^{n-4} \cdot S^1}{S^3 \cdot S^1} = \frac{\overline{\mathbb{C}S}^{n-2} \cdot \overline{\mathbb{C}S}^{n-4}}{\mathbb{CP}^1}, \end{aligned}$$

and in general

$$N\mathbb{G}_{n,k}(\mathbb{C}) = \frac{\overline{\mathbb{C}S}^{n-2} \cdots \overline{\mathbb{C}S}^{n-2k}}{\mathbb{CP}^{k-1} \cdots \mathbb{CP}^1}.$$

Hence, in case $n = 2m$ we obtain

$$\begin{aligned} N\mathbb{G}_{2m,k}(\mathbb{C}) &= \frac{\overline{\mathbb{C}S}^{2m-2} \cdots \overline{\mathbb{C}S}^{2m-2k}}{\mathbb{CP}^{k-1} \cdots \mathbb{CP}^1} \\ &= \frac{\mathbb{CP}^{m-1} \cdot \mathbb{S}^{2m-2} \cdots \mathbb{CP}^{m-k} \cdot \mathbb{S}^{2m-2k}}{\mathbb{CP}^{k-1} \cdots \mathbb{CP}^1} \\ &= \frac{\mathbb{CP}^{m-1} \cdots \mathbb{CP}^{m-k}}{\mathbb{CP}^{k-1} \cdots \mathbb{CP}^1} \mathbb{S}^{2m-2} \cdots \mathbb{S}^{2m-2k} \\ &= G_{m,k}(\mathbb{C}) \cdot \mathbb{S}^{2m-2} \cdots \mathbb{S}^{2m-2k} \end{aligned}$$

and similarly for $n = 2m + 1$ we get

$$N\mathbb{G}_{2m+1,k}(\mathbb{C}) = G_{m,k}(\mathbb{C}) \cdot \mathbb{S}^{2m} \cdots \mathbb{S}^{2m-2k+2},$$

and so these objects are also integrable. The calculus of nullcones and things can also be done in real variables. Let $\mathbb{R}^{p,q}$ be the space $\mathbb{R}^{p,q} = \mathbb{R}^{p+q}$ with quadratic form

$$|\underline{x}|^2 - |\underline{y}|^2 = \sum_{j=1}^{p} x_j^2 - \sum_{j=1}^{q} y_j^2, \quad (\underline{x}, \underline{y}) \in \mathbb{R}^{p,q}.$$

Then the nullcone $NC^{p,q}$ is the set of solutions $(\underline{x}, \underline{y})$ of equation $|\underline{x}|^2 = |\underline{y}|^2$; it contains of course $(0, 0)$ and for $|\underline{x}| \in \mathbb{R}_+$ we have $(\underline{x}, \underline{y}) = \rho(\underline{\omega}, \underline{\nu})$ with $\rho > 0$ and $(\underline{x}, \underline{y}) \in S^{p-1} \times S^{q-1}$. Hence, we have relation

$$NC^{p,q} = S^{p-1} \cdot S^{q-1} \cdot \mathbb{R}_+ + 1.$$

By $S^{p-1,q-1}$ we denote the set of $1D$ subspaces of $NC^{p,q}$; it may be represented by the equivalence classes $(\underline{\omega}, \underline{\nu}) \sim (-\underline{\omega}, -\underline{\nu})$, $(\underline{\omega}, \underline{\nu}) \in S^{p-1} \times S^{q-1}$. In morphological notation we have:

$$S^{p-1,q-1} = \frac{NC^{p,q} - 1}{\mathbb{R} - 1} = \frac{S^{p-1} \cdot S^{q-1} \cdot \mathbb{R}_+}{\mathbb{R} - 1} = \frac{S^{p-1} \cdot S^{q-1}}{2} = S^{p-1} \cdot \mathbb{RP}^{q-1}.$$

For example for $q = 2$ we may put $\underline{\nu} = (\cos\theta, \sin\theta)$ and $S^{p-1,1}$ may be identified with the equivalent pairs $(\underline{\omega}, \cos\theta, \sin\theta) \sim (-\underline{\omega}, -\cos\theta, -\sin\theta)$, which is equivalent with the Lie sphere

$$S^{p-1,1} \cong LS^p = \{e^{\mathbf{i}\theta}\underline{\omega} : \underline{\omega} \in S^{p-1},\ \theta \in [0,\pi[\}.$$

But there is also another calculation of this manifold that leads to another quantity $\mathbb{S}^{p-1,q-1}$ and it corresponds to the "conformal compactification" $\overline{\mathbb{R}}^{p-1,q-1}$ of $\mathbb{R}^{p-1,q-1}$. To find this, let $(\underline{x}, \underline{y}) = (\underline{x}', x_p; \underline{y}', y_q)$ with $(\underline{x}', \underline{y}') \in \mathbb{R}^{p-1,q-1}$. Then first we may intersect the nullcone with the plane $x_p - y_q = 1$, i.e., we put

$$x_p = \frac{1}{2}(1-\rho), \quad y_q = -\frac{1}{2}(1+\rho).$$

The equation $|\underline{x}|^2 = |\underline{y}|^2$ for the manifold $\mathbb{S}^{p-1,q-1}$ gives us $\rho = |\underline{x}'|^2 - |\underline{y}'|^2$ so that $(\underline{x}', \underline{y}') \in \mathbb{R}^{p-1,q-1}$ freely and then (x_p, y_q) are fixed. So this part of $\mathbb{S}^{p-1,q-1}$ is equivalent to $\mathbb{R}^{p+q-2}$. The remaining part of $\mathbb{S}^{p-1,q-1}$ is represented by the nonzero vectors $\lambda(\underline{x}, \underline{y})$, $\lambda \in \mathbb{R} \setminus \{0\}$ for which $x_p = y_q$; there are two cases:

- If $x_p = y_q \neq 0$ we may normalize $x_p = y_q = 1$ and we have $(\underline{x}, \underline{y}) = (\underline{x}', 1, \underline{y}', 1)$ together with the equation $|\underline{x}'|^2 - |\underline{y}'|^2 = 0$. So this part of $\mathbb{S}^{p-1,q-1}$ is equivalent to the modified nullcone

$$N\mathbb{C}^{p-1,q-1} = 2\mathbb{S}^{p-2,q-2} \cdot \mathbb{R}_+ + 1.$$

- If $x_p = y_q = 0$ we have $\lambda(\underline{x}', 0, \underline{y}', 0)$ with $\lambda \in \mathbb{R} \setminus \{0\}$ and $|\underline{x}'| \neq 0$ and $|\underline{x}'| = |\underline{y}'|$, which is the definition of $\mathbb{S}^{p-2,q-2}$.

So the total morphological calculation becomes

$$\begin{aligned}\mathbb{S}^{p-1,q-1} &= \mathbb{R}^{p+q-2} + \mathbb{S}^{p-2,q-2}(2\mathbb{R}_+ + 1) + 1, \\ &= \mathbb{R}^{p+q-2} + \mathbb{S}^{p-2,q-2} \cdot \mathbb{R} + 1,\end{aligned}$$

or, in terms of compactification of $\mathbb{R}^{p,q}$:

$$\overline{\mathbb{R}}^{p,q} = \mathbb{R}^{p+q} + \overline{\mathbb{R}}^{p-1,q-1} \cdot \mathbb{R} + 1.$$

Case 1: for $q = 0$ we simply obtain

$$\overline{\mathbb{R}}^{p,0} = \overline{\mathbb{R}}^{p} = \mathbb{R}^p + 1 = \mathbb{S}^p.$$

Case 2: compactified Minkowski space-time

$$\begin{aligned}\overline{\mathbb{R}}^{p,1} &= \mathbb{R}^{p+1} + \overline{\mathbb{R}}^{p-1,0} \cdot \mathbb{R} + 1, \\ &= \mathbb{R}^{p+1} + (\mathbb{R}^{p-1} + 1) \cdot \mathbb{R} + 1 = (\mathbb{R}^p + 1)(\mathbb{R} + 1) \\ &= \mathbb{S}^p \cdot \mathbb{RP}^1.\end{aligned}$$

More in general we obtain for $p \geq q$

$$\begin{aligned}\overline{\mathbb{R}}^{p,2} &= \mathbb{R}^{p+2} + \overline{\mathbb{R}}^{p-1,1} \cdot \mathbb{R} + 1, \\ &= \mathbb{R}^{p+2} + (\mathbb{R}^p + \mathbb{R}^{p-1} + \mathbb{R} + 1) \cdot \mathbb{R} + 1 = (\mathbb{R}^p + 1)(\mathbb{R} + 1) \\ &= \mathbb{R}^{p+2} + \mathbb{R}^{p+1} + \mathbb{R}^p + \mathbb{R}^2 + \mathbb{R} + 1 \\ &= (\mathbb{R}^p + 1)(\mathbb{R}^2 + \mathbb{R} + 1) = \mathbb{S}^p \cdot \mathbb{RP}^2,\end{aligned}$$

and, continuing in this way we obtain for $p \geq q$:

$$\overline{\mathbb{R}}^{p,q} = (\mathbb{R}^p + 1)(\mathbb{R}^q + \cdots + \mathbb{R}^2 + \mathbb{R} + 1) = \mathbb{S}^p \cdot \mathbb{RP}^q,$$

as expected from the similar (but different) formula $S^{p,q} = S^p \cdot \mathbb{RP}^q$.

So once again we have two different quantities that are obtained in two different canonical ways from what is mathematically considered to be one manifold.

Note that in particular (and this is weird)

$$\begin{aligned}\overline{\mathbb{R}}^{m,m} &= (\mathbb{R}^m + 1)(\mathbb{R}^m + \cdots + \mathbb{R} + 1) \\ &= \mathbb{R}^{2m} + \cdots + \mathbb{R}^{m+1} + 2\mathbb{R}^m + \mathbb{R}^{m-1} + \cdots + 1 \\ &= (\mathbb{R}^m + 1)^2 + \mathbb{R} \cdot (\mathbb{R}^m + 1)(\mathbb{R}^{m-2} + \cdots + 1).\end{aligned}$$

For the classical Minkowski space-time we get

$$\overline{\mathbb{R}}^{3,1} = (\mathbb{R}^3 + 1)(\mathbb{R} + 1),$$

and this is indeed projective line bundle over 3-sphere.

Compactified complexified Minkowski space-time is given by

$$\begin{aligned}\overline{\mathbb{CS}}^4 &= (\mathbb{C}^2 + 1)(\mathbb{C}^2 + \mathbb{C} + 1) = \mathbb{C}^4 + \mathbb{C}^3 + 2\mathbb{C}^2 + \mathbb{C} + 1 \\ &= \mathbb{C}^4 + \mathbb{C} \cdot (\mathbb{C} + 1)^2 + 1 = \mathbb{C}^4 + N\mathbb{C}^3,\end{aligned}$$

with $N\mathbb{C}^3 = \mathbb{C} \cdot \overline{\mathbb{CS}}^2 + 1$, so it is not just replacing "$\mathbb{R}$" by "$\mathbb{C}$" in $\overline{\mathbb{R}}^{3,1}$.

We must still calculate the null Graßmannians $NG_{p,q;k}$; they are defined as manifold of k-dimensional subspaces of the nullcone $N\mathbb{C}^{p,q} = S^{p-1} \cdot S^{q-1} \cdot \mathbb{R}^+ + 1$.

Let V be such a k-dimensional plane; then V is spanned by the basis of the form:

$$e_1 + \epsilon_1, e_2 + \epsilon_2, \ldots, e_k + \epsilon_k;\ e_1, \ldots, e_k \in S^{p-1};\ \epsilon_1, \ldots, \epsilon_k \in S^{q-1};$$

orthonormal frames, so these bases belong to:

$$e_1 + \epsilon_1 \in S^{p-1} \cdot S^{q-1} = 2\frac{N\mathbb{C}^{p,q} - 1}{\mathbb{R} - 1} = 2S^{p-1,q-1},$$

up to

$$e_k + \epsilon_k \in S^{p-k} \cdot S^{q-k} = 2S^{p-k,q-k},$$

and within V the total quantity of such bases is given by $O(k) = S^{k-1} \cdot S^{k-2} \cdots S^0$. We thus have the morphological representation (with $p \geq q$, $q \geq k$)

$$\begin{aligned} NG_{p,q;k} &= \frac{\left(2S^{p-1,q-1}\right) \cdots \left(2S^{p-k,q-k}\right)}{S^{k-1} \cdot S^{k-2} \cdots S^0} \\ &= \frac{S^{p-1,q-1} \cdots S^{p-k,q-k}}{\mathbb{RP}^{k-1} \cdots \mathbb{RP}^1} \\ &= \frac{S^{p-1} \cdot S^{p-2} \cdots S^{p-k}}{S^{k-1} \cdots S^0} S^{q-1} \cdots S^{q-k} \\ &= G_{p,k}(\mathbb{R}) \cdot S^{q-1} \cdots S^{q-k}. \end{aligned}$$

Again there is another way of computing this whereby in the above, $S^{p,q}$ is replaced by $\mathbb{S}^{p,q} = \overline{\mathbb{R}}^{p,q}$, leading up to the stereographic null Graßmannian:

$$\begin{aligned} N\mathbb{G}_{p,q;k} &= \frac{\left(2\mathbb{S}^{p-1,q-1}\right) \cdots \left(2\mathbb{S}^{p-k,q-k}\right)}{S^{k-1} \cdots S^0} \\ &= \frac{\overline{\mathbb{R}}^{p-1,q-1} \cdots \overline{\mathbb{R}}^{p-k,q-k}}{\mathbb{RP}^{k-1} \cdots \mathbb{RP}^1} \\ &= \frac{\mathbb{S}^{p-1} \cdots \mathbb{S}^{p-k} \cdot \mathbb{RP}^{q-1} \cdots \mathbb{RP}^{q-k}}{\mathbb{RP}^{k-1} \cdots \mathbb{RP}^1} = G_{q,k}(\mathbb{R}) \cdot \mathbb{S}^{p-1} \cdots \mathbb{S}^{p-k}. \end{aligned}$$

All these manifolds give hence rise to integrable morphological objects.

For the Minkowski space-time we have the manifold of null-lines (light rays):

$$\begin{aligned} N\mathbb{G}_{4,2;2} &= \frac{\overline{\mathbb{R}}^{3,1} \cdot \overline{\mathbb{R}}^{2,0}}{\mathbb{RP}^1} = \frac{(\mathbb{R}^3+1)(\mathbb{R}+1)(\mathbb{R}^2+1)}{\mathbb{R}+1} \\ &= (\mathbb{R}^3+1)(\mathbb{R}^2+1) = \mathbb{S}^3 \cdot \mathbb{S}^2, \end{aligned}$$

i.e., the real twistor space.

The last case we consider here is that of the space $\mathbb{C}^{p,q} = \mathbb{C}^{p+q}$ provided with the pseudo-Hermitian form

$$((\underline{z}, \underline{u}), (\underline{z}', \underline{u}')) = (\underline{z}, \underline{z}') - (\underline{u}, \underline{u}') = \sum_{j=1}^{p} z_j \overline{z}'_j - \sum_{j=1}^{q} u_j \overline{u}'_j.$$

The nullcone $((\underline{z}, \underline{u}), (\underline{z}, \underline{u})) = 0$ is denoted by $NC^{p,q}(\mathbb{C})$ and it has real codimension one, so its real dimension equals $2p+2q-1$. The equation is $|\underline{z}|^2 = |\underline{u}|^2$ so:

$$NC^{p,q}(\mathbb{C}) = S^{2p-1} \cdot S^{2q-1} \cdot \mathbb{R}_+ + 1.$$

By $T^{p,q}$ we denote the manifold of one-dimensional complex subspaces of $NC^{p,q}(\mathbb{C})$; it is a real submanifold of $\mathbb{CP}^{p+q-1}$ of real codimension one that hence subdivides $\mathbb{CP}^{p+q-1}$ in 3 parts and $T^{2,2} \subset \mathbb{CP}^3$ corresponds to "real twistor space" (the manifold of light-lines in the Minkowski space). From the definition we have

$$T^{p,q} = \frac{NC^{p,q}(\mathbb{C}) - 1}{\mathbb{C} - 1} = \frac{S^{2p-1} \cdot S^{2q-1} \cdot \mathbb{R}_+}{S^1 \cdot \mathbb{R}_+} = S^{2p-1} \cdot \mathbb{CP}^{q-1},$$

so in particular $T^{2,2} = S^3 \cdot \mathbb{CP}^1 = S^3 \cdot \mathbb{S}^2$.

There is another approach leading to the twister space $\mathbb{T}^{p,q}$ with a different quantity.

To that end we write $(\underline{z},\underline{u}) = (\underline{z}', z_p, \underline{u}', u_p)$ and consider the intersection $NC^{p,q}(\mathbb{C}) \cap \{z_p - u_q = 1\}$, which allows us to write

$$z_p = \frac{1}{2}(1+\rho+\mathbf{i}\alpha), \quad u_q = \frac{1}{2}(-1+\rho+\mathbf{i}\alpha).$$

The equation for the point $(\underline{z},\underline{u})$ now becomes

$$|\underline{z}'|^2 - |\underline{u}'|^2 + \frac{1}{4}\left((1+\rho)^2+\alpha^2\right) - \frac{1}{4}\left((1-\rho)^2+\alpha^2\right) = 0$$

or

$$\rho = |\underline{z}'|^2 - |\underline{u}'|^2 \;\&\; \alpha \in \mathbb{R}.$$

Hence the 1D complex subspaces of $NC^{p,q}(\mathbb{C})$ that intersect the plane $z_p - u_q = 1$ are representable by vectors of the form $\left(\underline{z}', \frac{1}{2}(1+\rho+\mathbf{i}\alpha), \underline{u}', \frac{1}{2}(-1+\rho+\mathbf{i}\alpha)\right)$ with $(\underline{z}',\underline{u}') \in \mathbb{C}^{p-1,q-1}$ and $\alpha \in \mathbb{R}$. So this part of $\mathbb{T}^{p,q}$ has quantity $\mathbb{C}^{p+q-2}\cdot\mathbb{R}$. The other points of $\mathbb{T}^{p,q}$ have the form $(\underline{z}',\lambda,\underline{u}',\lambda)$ so there are two cases:

- $\lambda \neq 0$, in this case we normalize $\lambda = 1$ and we have the equation $|\underline{z}'|^2 - |\underline{u}'|^2 = 0$, giving a version of nullcone:

$$N\mathbb{C}^{p-1,q-1}(\mathbb{C}) = \mathbb{T}^{p-1,q-1}\cdot(\mathbb{C}-1)+1\,.$$

- In the case $\lambda = 0$ we have the point $(\underline{z}',0,\underline{u}',0)$ with equation $|\underline{z}'| = |\underline{u}'|$ and determined up to a constant $c \in \mathbb{C}\setminus\{0\}$, i.e., we get $\mathbb{T}^{p-1,q-1}$.

  This leads to the recursion formula for $\mathbb{T}^{p,q}$ with $p \geq q$:

$$\mathbb{T}^{p,q} = \mathbb{C}^{p+q-2}\cdot\mathbb{R} + \mathbb{T}^{p-1,q-1}\cdot\mathbb{C}+1\,.$$

So in particular we get

$$\begin{aligned}
\mathbb{T}^{p,1} &= \mathbb{C}^{p-1}\cdot\mathbb{R}+1 = \mathbb{R}^{2p-1}+1 = \mathbb{S}^{2p-1},\\
\mathbb{T}^{p,2} &= \mathbb{C}^{p}\cdot\mathbb{R} + \left(\mathbb{C}^{p-2}\cdot\mathbb{R}+1\right)\mathbb{C}+1\\
&= \left(\mathbb{C}^{p}+\mathbb{C}^{p-1}\right)\mathbb{R}+\mathbb{C}+1 = \left(\mathbb{C}^{p-1}\cdot\mathbb{R}+1\right)(\mathbb{C}+1)\\
&= \mathbb{S}^{2p-1}\cdot\mathbb{CP}^1,\\
\mathbb{T}^{p,3} &= \mathbb{C}^{p+1}\cdot\mathbb{R} + \left(\mathbb{C}^{p-1}\cdot\mathbb{R}+\mathbb{C}^{p-2}\cdot\mathbb{R}+\mathbb{C}+1\right)\mathbb{C}+1\\
&= \left(\mathbb{C}^{p-1}\cdot\mathbb{R}+1\right)\left(\mathbb{C}^2+\mathbb{C}+1\right) = \mathbb{S}^{2p-1}\cdot\mathbb{CP}^2
\end{aligned}$$

and so, continuing in this way, we obtain for $p \geq q$:

$$\mathbb{T}^{p,q} = \left(\mathbb{C}^{p-1}\cdot\mathbb{R}+1\right)\left(\mathbb{C}^{q-1}+\cdots+\mathbb{C}+1\right) = \mathbb{S}^{2p-1}\cdot\mathbb{CP}^{q-1}.$$

In particular we re-obtain the expected formula

$$\mathbb{T}^{2,2} = (\mathbb{C}\cdot\mathbb{R}+1)(\mathbb{C}+1) = \mathbb{S}^3\cdot\mathbb{CP}^1 = \mathbb{S}^3\cdot\mathbb{S}^2.$$

The manifold $\mathbb{CP}^3$ is itself called the complex twistor space; it decomposes into real twistor space $\mathbb{T}^{2,2}$ together with two equals parts corresponding to $|\underline{z}| < |\underline{u}|$ and $|\underline{z}| > |\underline{u}|$. In morphological language we have the cutting experiment

$$\begin{aligned}\mathbb{CP}^3 - \mathbb{T}^{2,2} &= \mathbb{C}^3 + \mathbb{C}^2 + \mathbb{C} + 1 - (\mathbb{CR}+1)(\mathbb{C}+1)\\ &= (\mathbb{C}^2 - \mathbb{C}\cdot\mathbb{R})(\mathbb{C}+1) = \mathbb{C}\cdot(\mathbb{C}-\mathbb{R})(\mathbb{C}+1)\\ &= 2\mathbb{C}\cdot\mathbb{C}_+\cdot(\mathbb{C}+1),\end{aligned}$$

which indeed gives 2 copies of $\mathbb{C}\cdot\mathbb{C}_+\cdot(\mathbb{C}+1)$ whereby we put $\mathbb{C}_+ = \frac{\mathbb{C}-\mathbb{R}}{2} = \mathbb{R}\cdot\frac{\mathbb{R}-1}{2} = \mathbb{R}\cdot\mathbb{R}_+$.

We can also calculate the null Graßmannian $NG^{p,q;k}(\mathbb{C})$ of k-dimensional complex subspaces of $NC^{p,q}(\mathbb{C})$. Let V be a complex k-subspace; then the frames $(\underline{t}_1;\underline{s}_1),\ldots,(\underline{t}_k;\underline{s}_k)$ may be chosen such that $(\underline{t}_j;\underline{t}_k)-(\underline{s}_j;\underline{s}_k)=0$, of course, but we may also choose $(\underline{t}_j;\underline{t}_j)$ to be the Hermitian orthonormal frame, i.e., $(\underline{t}_j;\underline{t}_k)+(\underline{s}_j;\underline{s}_k)=0$ and $|\underline{t}_j|=|\underline{s}_j|=1$.

So in fact we can choose

$$\begin{aligned}(\underline{t}_1;\underline{s}_1) &\in S^{2p-1}\times S^{2q-1},\\ (\underline{t}_2;\underline{s}_2) &\in S^{2p-3}\times S^{2q-3},\end{aligned}$$

and so on. Moreover these frames per plane V can be chosen in $U(k)$-different ways, leading up to the morphological formula for $p\geq q\geq k$

$$\begin{aligned}NG_{p,q;k}(\mathbb{C}) &= \frac{S^{2p-1}\ldots S^{2p-2k}\cdot S^{2q-1}\ldots S^{2q-2k}}{S^{2k-1}\ldots S^3\cdot S^1}\\ &= S^{2p-1}\ldots S^{2p-k}\cdot\frac{\mathbb{CP}^{q-1}\ldots\mathbb{CP}^{q-k}}{\mathbb{CP}^{k-1}\ldots\mathbb{CP}^1}.\end{aligned}$$

A similar calculation can be made using the stereographic spheres, leading to:

$$\begin{aligned}N\mathbb{G}_{p,q;k}(\mathbb{C}) &= \frac{\left(\mathbb{T}^{p,q}\cdot\frac{(\mathbb{C}-1)}{\mathbb{R}_+}\right)\cdot\left(\mathbb{T}^{p-1,q-1}\cdot\frac{(\mathbb{C}-1)}{\mathbb{R}_+}\right)\cdots\left(\mathbb{T}^{p-k+1,q-k+1}\cdot\frac{(\mathbb{C}-1)}{\mathbb{R}_+}\right)}{U(k)}\\ &= \mathbb{S}^{2p-1}\ldots\mathbb{S}^{2p-2k}\cdot\frac{\mathbb{CP}^{q-1}\ldots\mathbb{CP}^{q-k}}{\mathbb{CP}^{k-1}\ldots\mathbb{CP}^1},\end{aligned}$$

and in particular for $p=q=2$, $k=2$ we obtain:

$$N\mathbb{G}_{2,2;2}(\mathbb{C}) = \frac{\mathbb{S}^3\cdot\mathbb{S}^1\cdot\mathbb{CP}^1}{\mathbb{CP}^1} = (\mathbb{R}^3+1)(\mathbb{R}+1),$$

which corresponds to the real compactified Minkowski space.

The compactified complex Minkowski space corresponds to:

$$G_{4,2} = \frac{S^7\cdot S^5}{S^3\cdot S^1} = \mathbb{S}^4\cdot\mathbb{CP}^2 = (\mathbb{C}^2+1)(\mathbb{C}^2+\mathbb{C}+1) = \overline{\mathbb{CS}}^4,$$

as can be shown using bivectors and Klein quadric.

6. Conclusions and remarks

(i) Completeness

Morphological calculus is best compared with a museum. It consists of a lots of special names, algebraic expressions and calculations that stand for geometrical objects and operations on these objects.

In this paper we presented morphological calculus for the most important classical manifolds. Like any museum, also our collection is incomplete. For example a full morphological treatment for the spin groups Spin(m) and Spin(p,q) is still to be done and there is a vast collection of special manifolds or objects to be added to the catalogue.

In building up our museum we give preference to the most interesting special manifolds (canonical manifolds) as well as to the "simplest ways of introducing them". So in fact the calculus is entirely based on examples of objects and experiments; there is no idea of "a general manifold" and no theory behind the scene.

(ii) Correctness

Morphological calculus is correct in the sense that it takes space within the language of calculus that is a correct language based on clear rules. This leads to the notion of quantity, which is in fact what a manifold becomes once it is introduced within the calculus language. This is practically done by assigning a name to an object along with an algebraic relation that expresses the definition of the object in calculus. The notion of quantity is somewhat comparable to the notions of cardinality and of volume that are used to express the contents or size of an object. But there is no mathematical definition for it; it is an imaginary substance that resides entirely within the calculus.

The main problem is not the calculus itself but the way of translating objects of geometry into calculus expressions (morphological analysis); it usually happens that one and the same object can be translated into morphological language in many ways and that may cause confusion.

To give an example, the compactified Minkowski space is given by

$$\overline{\mathbb{R}}^{3,4} = \mathbb{R}^4 + \mathbb{R}(\mathbb{R}^2 + 1) + 1 = (\mathbb{R}^3 + 1)(\mathbb{R} + 1)$$

whereby $\mathbb{R}^4$ is the usual Minkowski space and

$$\mathbb{R}(\mathbb{R}^2 + 1) + 1 = (2\mathbb{R}_+\mathbb{S}^2 + 1) + \mathbb{S}^2$$

is a compactified light cone at infinity whereby we made use of stereographic sphere $\mathbb{S}^2$. What would happen if we replace $\mathbb{S}^2$ by the usual sphere S^2? Well, we would get in total:

$$\mathbb{R}^4 + \mathbb{R}(2\mathbb{R}^2 + 2\mathbb{R} + 2) + 1 = \mathbb{R}^4 + 2\mathbb{R}^3 + 2\mathbb{R}^2 + 2\mathbb{R} + 1$$
$$= (\mathbb{R}^3 + \mathbb{R}^2 + \mathbb{R} + 1)(\mathbb{R} + 1) = \mathbb{RP}^3 \cdot \frac{S^1}{2} = \frac{S^3}{2}\frac{S^1}{2} = \mathbb{S}^2 \cdot \mathbb{S}^1 \cdot \mathbb{S}^1.$$

This is no longer Minkowski space-time, yet there exists a meaningful interpretation for this object, namely the manifold of pairs $(e^{i\theta}\underline{\omega}, -e^{i\theta}\underline{\omega})$ in $\mathbb{C}^4$, with

$e^{i\theta}\underline{\omega} \in LS^3$, the Lie sphere. For this manifold the above calculation makes sense. So the problem is not only to know what calculation to make to describe an object correctly but also how to correctly interpret a calculation (morphological synthesis). It often happens that different objects turn out to share the same quantity.

There is no way to avoid these problems; one simply has to experiment until one finds the best fitting calculations or interpretations. This may be seen as a drawback, but we see it as a stronghold that illustrates the richness of the morphological language.

(iii) Consistency

Morphological calculus may be compared to making the bill of a meal in a restaurant; usually the bill adds up correctly but sometimes the sum of the ingredients of the meal is more expensive than the meal.

Here is an example in morphological calculus: Consider the space $\mathbb{R}^2_n$ of bivectors in a Clifford algebra: $b = \sum_{i<j} b_{ij} e_i e_j$.

Then $\mathbb{R}^2_n$ is a real vector space of dimension $\binom{n}{2}$:

$$\mathbb{R}^2_n = \mathbb{R}^{\binom{n}{2}},$$

but on the other hand, $b \in \mathbb{R}^2_n \setminus \{0\}$ may be written as:

$$b = r_1 I_1 + \cdots + r_s I_s, \quad 2s \leq n,$$

whereby $r_1 \geq r_2 \geq \cdots \geq r_s > 0$ is unique and $I_j = \underline{\omega}_j \wedge \underline{\nu}_j$, $|\underline{\omega}_j| = |\underline{\nu}_j| = 1$, $|\underline{\omega}_j| \perp |\underline{\nu}_j|$ is a 2-blade such that $I_j I_k = I_k I_j$, i.e., $(\underline{\omega}_1, \underline{\nu}_1, \ldots, \underline{\omega}_j, \underline{\nu}_s) \in V_{n,2s}(\mathbb{R})$. This leads to a partition of $\mathbb{R}^2_n$ into orbits of the orthogonal group $O(r_1, \ldots, r_s)$, which one may calculate morphologically and add up properly.

For $n = 3$, $b = r\underline{\omega} \wedge \underline{\nu} \in \mathbb{R}_+ \times \widetilde{G_{3,2}}(\mathbb{R})$, leading to

$$\mathbb{R}^2_3 - 1 = \widetilde{G_{3,2}}(\mathbb{R}) \cdot \mathbb{R}_+ = S^2 \cdot \mathbb{R}_+ = \mathbb{R}^3 - 1,$$

which adds up correctly. But already for $n = 4$ there is a problem. Every $b \in \mathbb{R}^2_4 \setminus \{0\}$ may be written as

$$b = r_1 \underline{\omega}_1 \wedge \underline{\nu}_1 + r_2 \underline{\omega}_2 \wedge \underline{\nu}_2, \quad r_1 \geq r_2 \geq 0$$

and there are three cases:

1. For $r_1 > r_2 > 0$ the blades $\underline{\omega}_1 \wedge \underline{\nu}_1$ and $\underline{\omega}_2 \wedge \underline{\nu}_2$ are uniquely determined in terms of b, so we have in fact:

$$r_2 > 0 \in \mathbb{R}_+, \; r_1 > r_2 \in \mathbb{R}_+,$$

$$\underline{\omega}_1 \wedge \underline{\nu}_1 \in \widetilde{G_{4,2}}(\mathbb{R}) = \frac{S^3 \cdot S^2}{S^1} = \mathbb{S}^2 \cdot S^2$$

$[\underline{\omega}_2 \wedge \underline{\nu}_2, \underline{\omega}_1 \wedge \underline{\nu}_1] = 0$ leaves 2 possibilities: $\underline{\omega}_2 \wedge \underline{\nu}_2 = \pm \underline{\omega}_1 \wedge \underline{\nu}_1 \cdot e_{1234}$.

So, in morphological terms we get:

$$2\mathbb{S}^2 \cdot S^2 \cdot \mathbb{R}_+ \cdot \mathbb{R}_+ = 2\mathbb{R}_+ \cdot \mathbb{S}^2 \cdot (\mathbb{R}^3 - 1).$$

2. For $r_1 > 0, r_2 = 0$ we get $b = r\underline{\omega}\wedge\underline{\nu}$ with $r \in \mathbb{R}_+$ and $\underline{\omega}\wedge\underline{\nu} \in \widetilde{G_{4,2}}(\mathbb{R}) = \mathbb{S}^2 \cdot S^2$, so in total $\mathbb{S}^2 \cdot S^2 \cdot \mathbb{R}_+ = \mathbb{S}^2 \cdot (\mathbb{R}^3 - 1)$.
3. In the case $r_1 = r_2 = r > 0$ we get

$$b = r(\underline{\omega}_1 \wedge \underline{\nu}_1 + \underline{\omega}_2 \wedge \underline{\nu}_2),$$

whereby either $\underline{\omega}_2 \wedge \underline{\nu}_2 = \pm e_{1234}\underline{\omega}_1 \wedge \underline{\nu}_1$, so $b = r\underline{\omega} \wedge \underline{\nu}(1 \pm e_{1234})$. Hereby $\underline{\omega}\wedge\underline{\nu}$ may be chosen to belong to $\widetilde{G_{3,2}}(\mathbb{R}) = S^2$ because in fact every bivector $b \in \mathbb{R}_4^2$ may be decomposed uniquely into self-dual and anti-self-dual parts:

$$b = \frac{1}{2}(1 + e_{1234})b_+ + \frac{1}{2}(1 - e_{1234})b_-, \quad b_\pm \in \mathbb{R}_3^2.$$

So in the above case, $\underline{\omega} \wedge \underline{\nu} \in \widetilde{G_{3,2}}(\mathbb{R})$ is unique, so that the morphological contribution is given by

$$2\mathbb{R}_+ \cdot S^2 = 2(\mathbb{R}^3 - 1).$$

Hence, adding up (1) + (2) + (3), we get a total morphological sum of

$$(2\mathbb{R}_+ + 1)\mathbb{S}^2 \cdot (\mathbb{R}^3 - 1) + 2(\mathbb{R}^3 - 1)$$
$$(\mathbb{R} \cdot (\mathbb{R}^2 + 1) + 2) \cdot (\mathbb{R}^3 - 1) \neq (\mathbb{R}^3 + 1)(\mathbb{R}^3 - 1) = \mathbb{R}^6 - 1.$$

The gap in the calculation lies in the difference between $\mathbb{R} \cdot (\mathbb{R}^2 + 1) + 2$ and $\mathbb{S}^3 = \mathbb{R}^3 + 1$. If in the above we would replace $\mathbb{R}^2 + 1 = \mathbb{S}^2$ by $2\mathbb{R}^2 + 2\mathbb{R} + 2 = S^2$ we would get a factor $\mathbb{R} \cdot S^2 + 2 = S^3$ and replacing then S^2 by $\mathbb{S}^2$ would make the bill add up correctly.

So in fact $\mathbb{R} \cdot (\mathbb{R}^2 + 1) + 2$ may be interpreted as an oversized version of the Poincaré sphere $\mathbb{S}^3 = \mathbb{R}^3 + 1$.

Also for the bivector space $\mathbb{R}_5^2$ we have three cases:

1. in case $r_1 > r_2 > 0$ we obtain the quantity:

$$\mathbb{R}_+^2 \cdot \frac{S^4 \cdot S^3}{S^1} \cdot \frac{S^2 \cdot S^1}{S^1} = (\mathbb{R}^5 - 1) \cdot \mathbb{S}^2 \cdot (\mathbb{R}^3 - 1)$$

2. in case $r_1 > r_2 = 0$ we obtain:

$$\mathbb{R}_+ \cdot \frac{S^4 \cdot S^3}{S^1} = (\mathbb{R}^5 - 1) \cdot \mathbb{S}^2$$

3. in case $r_1 = r_2 = r > 0$ we get bivectors of the form $r(\underline{\omega}_1 \wedge \underline{\nu}_1 + \underline{\omega}_2 \wedge \underline{\nu}_2)$ in $\mathbb{R}^5$; the number of choices for span$\{\underline{\omega}_1, \underline{\nu}_1, \underline{\omega}_2, \underline{\nu}_2\}$ equals $\widetilde{G_{5,1}}(\mathbb{R}) = \frac{S^4}{2}$ while for each choice we have the quantity $2\mathbb{R}_+ S^2$ as before, leading to a total of

$$\frac{S^4}{2}(2\mathbb{R}_+ S^2) = (\mathbb{R}^5 - 1) \cdot S^2.$$

So, the total bill for $\mathbb{R}_5^2$ reads

$$(\mathbb{R}^5 - 1)((1 + \mathbb{R}^2)\mathbb{R}^3 + 2\mathbb{R}^2 + 2\mathbb{R} + 2)$$
$$= (\mathbb{R}^5 - 1)(\mathbb{R}^5 + (\mathbb{R} + 1)(\mathbb{R}^2 + \mathbb{R} + 1) + 1),$$

while we would need the second factor to be equal to $\mathbb{R}^5 + 1$ to make the bill add up correctly.

From $n \geq 6$ on the calculation of $\mathbb{R}_n^2$ is much more complicated so we won't do it here, but in any case we won't get just $\mathbb{R}^{\binom{n}{2}}$. This may be seen as an inconsistency which is likely to repeat itself in cases of partitions of geometrical objects. We have no solution as even explanation of this, but it is clear that one can study this phenomena within the language of morphological calculus, which in itself is consistent.

(iv) Calculus styles

A calculus style is obtained by making certain restrictions on the use of the calculus language and by a certain kind of application or focus.

In the canonical style we decided to replace the relation $\mathbb{R} = 2\mathbb{R}+1$ by its more rigorous form $\mathbb{R} = 2\mathbb{R}_+ + 1$ in order to avoid too many unwanted identifications.

This leads to the possibility to apply the rules of calculus on a free basis (commutativity, brackets, etc.) whereby our focus is the calculation of quantity for a large collection of manifolds and this calculation arises from a morphological analysis of the geometrical objects (and constructions) we are interested in.

In the formal style we start off from a given quantity, a polynomial $a_0\mathbb{R}^n + \cdots + a_n$ with $a_0 > 0$, $a_1, \ldots, a_n \in \mathbb{N}$ say, and we consider the collection of all the algebraic expressions that evaluate to this quantity. Since we already start with a polynomial with positive integer coefficients, we won't consider any subtractions or divisions here, just addition and multiplication. Also we won't be using $\mathbb{R}_+$ here and the relation $\mathbb{R} = 2\mathbb{R}_+ + 1$ will be replaced by a non-commutative and non-associative version of $\mathbb{R} = 2\mathbb{R} + 1$:

$$\mathbb{R} = \mathbb{R} + 1 + \mathbb{R}, \quad \mathbb{R} = \mathbb{R} + (\mathbb{R} + 1),$$

the use of which leads to a change in the quantity. Also other calculations involving commuting terms or factors or placing or removing brackets are seen as morphisms on the collection of morphological objects. So for each quantity we have basically a category.

Parallel to this, for each polynomial $a_0\mathbb{R}^n + \cdots + a_n$ we also have the set of all geometrical objects that can be formed by glueing together (or not) a_0 copies of $\mathbb{R}^n$, a_1 copies of $\mathbb{R}^{n-1}, \ldots, a_n$ points. The focus now is to study possible correlations between the category of algebraic expressions and geometrical objects (graphs) for a given quantity; this is morphological synthesis.

For example for $\mathbb{R} + 1$ we have two expressions

$$\mathbb{R} + 1, \quad 1 + \mathbb{R}$$

and two geometrical objects (apart from trivial disjoint union) semi-interval $[0, 1[$ or circle $\mathbb{S}^1$ and one possible correlation is to identify $\mathbb{R} + 1$ with $[0, 1[$ and $1 + \mathbb{R}$ with the circle $\mathbb{S}^1$.

The more general case $a\mathbb{R} + b$ leads to a kind of calligraphy that we'll study in forthcoming work. For this reason we will speak of calligraphic calculus.

7. Outlook

It is not easy to provide complete references to the topic of morphological calculus but certain examples of it as well as related topics are certainly available throughout the mathematical literature. First of all there is our paper [6] in which we gave an introduction to morphological calculus which was subdivided into an axiomatic approach, a canonical part and a formal part based on the formal language of calculus. In this paper we focused mainly on the canonical part by giving many more new examples of interesting calculations. Morphological calculus can be seen as a formal language and the task of constructing good geometrical interpretations of calculations, called morphological synthesis, can be seen as part of a research field called the theory of Lindenmayer systems (L-systems) for which there is a vast literature. We only refer to [5]. Also in the book [4] by Roger Penrose the language of calculus has been discussed, in particular the meaning of commutativity of the multiplication has been critically investigated. But the present paper is mostly concerned with examples concerning spheres, real and complex projective spaces, special Lie groups and homogeneous spaces including Stiefel manifolds and Graßmannians, various types of complex spheres and real and complex nullcones. All of this belongs to the theory of special manifolds (see, e.g., [7]).

In particular we also discussed real and complex compactified Minkowski spaces as well as twistor spaces which have many applications in mathematical physics and for which we refer to the pioneering work [3] of R. Penrose and W. Rindler. Morphological calculus is of course also related to various topics in algebraic topology in particular Betti numbers, homology and cohomology, Poincaré polynomials, Euler characteristics and much more that is to be found all over the literature (use Wikipedia and see also [7]). Finally, many of our calculations also make use of bivector spaces, Clifford algebras and Spin groups for which we refer to the books [1] and [2].

Acknowledgement

The author wishes to thank Dr. Narciso Gomes (University of Cape Verde – Uni-CV) for his help in the critical reading and the painstaking task of typewriting this manuscript. We also wish to thank the referee for his valuable suggestions during the preparation of the manuscript.

References

[1] R. Delanghe, F. Sommen, V. Soucek, Clifford algebra and spinor valued functions, Mathematics and Its Applications 53, (Kluwer Acad. Publ., Dordrecht 1992).

[2] P. Lounesto, Clifford algebras and spinors (second edition), London Math. Soc. Lecture Note Series 286, (Cambridge University Press, Cambridge 2001).

[3] R. Penrose, W. Rindler, Spinors and space-time, Volume 2, Spinor and twistor methods in space-time geometry, Cambridge Monographs on Mathematical Physics, (Cambridge University Press, Cambridge, 1986).

[4] R. Penrose, The emperors new mind: concerning computers, minds, and the laws of physics, (Oxford University Press, Oxford, 1999).

[5] G. Rozenberg, A. Salomaa, The mathematical theory of L-systems, (Academic Press, New York, 1980).

[6] F. Sommen, A morphological calculus for geometrical objects, J. Nat. Geom. 15 (1-2), 1999, 1–64.

[7] F.W. Warner, Foundations of differentiable manifolds and Lie groups, (Springer, New York, 1983).

Franciscus Sommen
Clifford Research Group
Department of Mathematical Analysis
Ghent University
Galglaan 2
B-9000 Gent, Belgium
e-mail: fs@cage.ugent.be

MIX
Papier aus verantwortungsvollen Quellen
Paper from responsible sources
FSC® C105338

If you have any concerns about our products,
you can contact us on
ProductSafety@springernature.com

In case Publisher is established outside the EU,
the EU authorized representative is:
Springer Nature Customer Service Center GmbH
Europaplatz 3, 69115 Heidelberg, Germany

Printed by Libri Plureos GmbH
in Hamburg, Germany